C for the Microprocessor Engineer

S. J. Cahill

Prentice Hall
New York   London   Toronto   Sydney   Tokyo   Singapore

First published 1994 by
Prentice Hall International (UK) Limited
Campus 400, Maylands Avenue
Hemel Hempstead
Hertfordshire, HP2 7EZ
A division of
Simon & Schuster International Group

© Prentice Hall International (UK) Limited, 1994

All rights reserved. No part of this publication may be reproduced, stored in a retrieval system, or transmitted, in any form, or by any means, electronic, mechanical, photocopying, recording or otherwise, without prior permission, in writing, from the publisher.
For permission within the United States of America contact Prentice Hall Inc., Englewood Cliffs, NJ 07632

Printed and bound in Great Britain by
Redwood Books, Trowbridge, Wiltshire

Library of Congress Cataloging-in-Publication Data

Cahill, S. J., 1948–
   C for the microprocessor engineer / S.J. Cahill.
     p.   cm.
   Includes bibliographical references and index.
   ISBN 0-13-115825-2
   1. C (Computer program language)   I. Title.
QA76.73.C15C35   1994
005.26'2—dc20
                                          93–38106
                                            CIP

British Library Cataloguing in Publication Data

A catalogue record for this book is available from the British Library

ISBN 0-13-115825-2

1 2 3 4 5   98 97 96 95 94

Contents

Preface ix

Part 1 Target Processors 1

1 The 6809 Microprocessor: Its Hardware 2
  1.1 Architecture 3
  1.2 Outside the 6809 6
  1.3 Making the Connection 10

2 The 6809 Microprocessor: Its Software 19
  2.1 Its Instruction Set 19
  2.2 Address Modes 34
  2.3 Example Programs 41

3 The 68000/8 Microprocessor: Its Hardware 57
  3.1 Inside the 68000/8 58
  3.2 Outside the 68000/8 65
  3.3 Making the Connection 72

4 The 68000/8 Microprocessor: Its Software 88
  4.1 Its Instruction Set 88
  4.2 Address Modes 108
  4.3 Example Programs 116

5 Subroutines, Procedures and Functions 124
  5.1 The Call-Return Mechanism 125
  5.2 Passing Parameters 131

6 Interrupts plus Traps equals Exceptions 143
  6.1 Hardware Initiated Interrupts 145
  6.2 Interrupts in Software 164

Part 2 C — 169

7 Source to Executable Code — 170
  7.1 The Assembly Process — 172
  7.2 Linking and Loading — 180
  7.3 The High-Level Process — 192

8 Naked C — 202
  8.1 A Tutorial Introduction — 203
  8.2 Variables and Constants — 206
  8.3 Operators, Expressions and Statements — 216
  8.4 Program Flow Control — 227

9 More Naked C — 240
  9.1 Functions — 240
  9.2 Arrays and Pointers — 249
  9.3 Structures — 265
  9.4 Headers and Libraries — 275

10 ROMable C — 283
  10.1 Mixing Assembly Code and Starting Up — 283
  10.2 Exception Handling — 292
  10.3 Initializing Variables — 297
  10.4 Portability — 303

Part 3 Project in C — 314

11 Preliminaries — 315
  11.1 Specification — 317
  11.2 System Design — 320

12 The Analog World — 328
  12.1 Signals — 328
  12.2 Digital to Analog Conversion — 334
  12.3 Analog to Digital Conversion — 341

13 The Target Microcomputer — 350
  13.1 6809 – Target Hardware — 350
  13.2 68008 – Target Hardware — 355

14 Software in C — 360
  14.1 Data Structure and Program — 360
  14.2 6809 – Target Code — 364
  14.3 68008 – Target Code — 376

15 Looking For Trouble	**390**
15.1 Simulation	391
15.2 Resident Diagnostics	404
15.3 In-Circuit Emulation	416
16 C'est la Fin	**423**
16.1 Results	423
16.2 More Ideas	427
A Acronyms and Abbreviations	**430**
Index	**431**

Preface

The Intel corporation introduced the first microprocessor in 1971 as a replacement for complex random logic-based systems. It shared many of its characteristics with the software controlled computer architecture of the previous two decades, only somewhat simplified and highly integrated on a few integrated circuits. Although 'proper' computers had been used to implement digital control, the high cost and size associated with this approach limited such strategies in the main to large industrial, medical and research projects.

One of the more visible manifestations of the microprocessor is the Personal Computer. To many, the term microprocessor is synonymous with computer and the terms are often used interchangeably. Thus the salesperson may well say that the microwave oven is controlled by a computer; and no doubt the customer is impressed. Indeed the vast majority of microprocessors are sold as the controller of intelligent instrumentation, for example, heart monitors, CAT scanners, vessel navigation, telecommunications, electric motor control and the smart egg timer all rely on one or more 'embedded' microprocessors as an integral part of their circuitry.

It is difficult to define precisely an embedded microprocessor, but such a system will generally exhibit several characteristics. A circuit dedicated to one task and which is driven by external physical events is likely to be in this category, as opposed to a general-purpose computer. Most embedded systems operate in real time, with millisecond actions, and are characterized by a rich mix of input and output peripherals. Furthermore they are lean, in that they carry little excess baggage — such as resident operating systems or human interface.

Of course mainframes through to the microprocessor all share the same computer-like structure, which is based on a relatively standard hardware architecture and is configured to its task via a series of instructions — its software. This contrasts with the random logic approach, whereby the personality of a circuit is in the physical pattern of interconnections between the various functional units (equivalent to instructions), although this distinction is blurred somewhat with the use of programmable logic devices.

Given that they share a common hardware core structure, it is not surprising that software for embedded microprocessors shares many of its characteristics with those for its bigger relatives. The original language used to program mainframe computers was machine code, rapidly followed by its slightly more user-friendly sibling, assembly-level language. Less than a decade after the introduction of commercial electronic computers, in the 1950s, the resources of a mainframe were such that the efficiency of the one-to-one relationship between machine structure and assembly-level instructions was no longer paramount. The use of most programs to implement business packages or large-scale scientific simulations meant that high-level languages, whose instructions are couched more on problem algorithms, were more productive. Such programs are very large, and development/maintenance cycles are measured in decades, covering many generations and types of hardware. Varying hardware configurations are 'covered up' by using a shell of software, known as an operating system.

This mega software approach was the situation in 1971, when the Intel 4004 microprocessor appeared on the market. The development of software for these flea-sized computers in many ways was a rerun of mainframe software some two decades previously. For many years, assembly-level (and even machine code, affectionately known as 'bit banging') software was the language of choice. This was not just because efficiency was *de rigueur* for the relatively simple processor, but because an embedded processor by definition requires to interface tightly to hardware whereby it interacts with the physical world. Real-time speed plus the efficiency criterion are the bedfellows of bit banging. Yet another problem was the need for a powerful computing engine to run the highly complex task of translating high-level software down to machine code; this is an order of magnitude greater than assembly translation. Initially this translation had to be undertaken separately on main or miniframe computers, and the resulting output taken to the embedded system for testing. Cross-software development was therefore expensive and hideously complicated to debug.

The dramatic increase in power of the Personal Computer (PC) and other workstations, coupled with an equally striking fall in their cost, has meant that the translation of high-level language to machine code (compilation) is no longer an issue. As in the case of mainframes, some four decades ago, the vast majority of native software applications for these microcomputers are coded in a high-level language of some sort.

The high-level advantage of hardware independence (portability), productivity and maintainability are equally applicable to embedded products. However, most high-level languages are too far away from their hardware host to produce the compact and fast code necessary for most embedded hosts. During the early 1970s the language C was developed primarily to code operating systems in an efficient portable manner. By definition, an operating system interfaces to hardware, such as serial ports and disk controllers, and thus C has many attributes that satisfy the needs of embedded targets. C is often regarded as a high-level assembly language, but is powerful enough to be one of the most popular native languages for PCs and workstations. It is reported that at least 50% of embedded code is now written in C and that the average embedded program comprises around 30,000 lines of code.[1] This is not surprising, as with programs of this complexity, the non-recurring engineering expenses (the one-off up-front design and testing costs) are very large, and the advantage of a technique increasing the productivity of software design will often outweigh the less efficient code thereby generated. Besides, C lends itself well to mixing with modules written at assembly level, which can be used to refine critical routines.

As befits one of the major languages there is a plethora of excellent books on C. In what way does this text differ? Well C is essentially the same language whether it runs in a rich well resourced native environment under an operating system or in the lean naked cross environment of an embedded world. Nevertheless, the latter presents the software engineer with a set of problems not usually encountered on a PC. For example, the code will normally be located in read-only memory, will require start-up assembly-level routines and will frequently be interrupt driven. Despite the popularity of C for embedded code, at the current time there seems to be little information on this topic outside the compiler manufacturers' manuals. This is the bailiwick of this book.

Throughout the book the tight integration between hardware and software, which is a feature of embedded systems, is highlighted. In the main, two processors are used for illustrative purposes; the 8-bit 6809 and 16/32-bit 68000 families. This dualism emphasizes the portability aspects of the use of a high-level language, but the processors are closely enough related to reduce the exposition to a reasonable length.

The hardware and machine-level software structures of these two processors are the subject of Part 1. Sufficient detail is presented so that this material can be used as a stand-alone introduction to microprocessor-based design for readers with no background in this area, or as a conversion course from other processors. It also lays the foundation for the rest of the text.

Part 2 covers the assembly and compilation process and the C language as it relates to naked targets. This material can be used as a mini-course in C, as no a priori knowledge of this language is assumed, although some exposure to a high-level language would be useful.

---

[1] Ganssle, J.G.; *The Art of Programming Embedded Systems*, Academic Press, 1992.

Part 3 is used to bring all these concepts together with a mini project. The interrupt-driven time-compressed memory illustrates basic design philosophy, analog input/output, testing and debugging techniques and hardware design as well as C coding. It also enables us to compare assembly-level against high-level implementations.

Cross C-compilers and simulators are expensive computer-aided engineering packages, typically costing $3000 – $5000. If you wish to experiment with the techniques outlined in this text, RTS have kindly offered a free evaluation kit to go along with this text. The features of the 68000-family version are (other processor kits offer similar capabilities):

- COSMIC ANSII-C compiler for the 68000/68020/68302 processors, together with support for the 68881 co-processor.

- ANSII libraries supporting integer and floating-point operations.

- COSMIC CXDB high-low-level cross simulator.

- Assembler, linker, hexer and other programming utilities.

- Documentation.

- The 6809 and 680x0 kits include the .s and .c files from this text in a separate subdirectory. All kits include demonstration source code.

The evaluation software includes most of the features of the full package, with a few restrictions; the main ones being:

- Five include files.

- Three functions.

- 150 lines of C source code.

- Machine code file limited to 1024 bytes.

Distribution details are contained on the order form (see postcard enclosed).

S.J. Cahill   Jordanstown, 21 December, 1993

Part 1

Target Processors

A major advantage of the use of a high-level language is its independence of the hardware its generated code will eventually run on; that is, its portability. However, one of the main strands of this book is the interaction of software with its hardware environment, and thus it is essential to use real products in both domains. For clarity, rather than describing a multitude of devices, most of the examples are based on just two microprocessors. Two, rather than one, not to loose sight of the portability aspects of high-level code.

In this part I describe the Motorola 6809 and 68000/8 microprocessors, the chosen devices. This gives us a hardware target spectrum ranging from 8 through 32-bit architecture. As both microprocessors share a common ancestor, the complexity is reduced compared with a non-related selection. Where necessary, other processors are used as examples, but in general the principles are similar irrespective of target. If the hardware detail seems excessive to a reader with a software background, much may be ignored if building the miniproject circuitry of Part 3 is to be omitted.

CHAPTER 1

The 6809 Microprocessor: Its Hardware

The microprocessor revolution began in 1971 with the introduction of the Intel 4004 device. This featured a 4-bit data bus, direct addressing of 512 bytes of memory and 128 peripheral ports. It was clocked at 108 kHz and was implemented with a transistor count of 2300. Within a year, the 8-bit 200 kHz 8008 appeared, addressing 16 kbyte of memory and needing a 3500 transistor implementation. The improved 8080 replacement appeared in 1974, followed a few months later by the Motorola 6800 MPU [1]. Both processors could directly address 64 kbytes of memory through a 16-bit address bus and could be clocked at up to 2 MHz. These two families, together with descendants and inspired close relatives, have remained the industry standards ever since.

The Motorola 6800 MPU [2] was perceived to be the easier of the two to use by virtue of its single 5 V supply requirement and a clean internal structure. The 8085 MPU is the current state of the art Intel 8-bit device. First produced in 1976, it has an on-board clock generator and requires only a single power supply, but has a virtually identical instruction set to the 8080 device. Soon after Zilog produced its Z80 MPU which was upwardly compatible with Intel's offering, then the market leader, with a much extended instruction set and additional internal registers [3].

The Motorola 6802/8 MPUs (1977) also have internal clock generators, with the former featuring 128 bytes of on-board RAM. This integration of support memory and peripheral interface leads to the single-chip microcomputer unit (MCU) or micro-controller, exemplified by the 6801, 6805 and 8051 MCU families [4]. The 6809 MPU introduced in 1979 [5, 6, 7] was seen as Motorola's answer to Zilog's Z80 and these both represent the most powerful 8-bit devices currently available. By this date the focus was moving to 16- and 32-bit MPUs, and it is unlikely that there will be further significant developments in general-purpose 8-bit devices. Nevertheless, these latter generation 8-bit MPUs are powerful enough to act as the controller for the majority of embedded control applications, and their architecture is sophisticated enough to efficiently support the requirements of high-level languages; more of which in later chapters. Furthermore, many MCU families have a core and language derived from their allied 8-bit MPU cousins.

1.1 Architecture

The internal structure of a general purpose microprocessor can be partitioned into three functional areas:

1. The mill.

2. Register array.

3. Control circuitry.

Figure 1.1 shows a simplified schematic of the 6809 MPU viewed from this perspective.

THE MILL

A rather old fashioned term used by Babbage [8] for his mechanical computer of the last century to identify the arithmetic and logic processor which 'ground' the numbers. In our example the 6809 has an 8-bit **arithmetic logic unit (ALU)** implementing Addition, Subtraction, Multiplication, AND, OR, Exclusive-OR, NOT and Shift operations. Associated with the ALU is the **Code Condition (or Status) register (CCR)**. Five of the eight CCR bits indicate the status of the result of ALU processes. They are: **C** indicating a Carry or borrow, **V** for 2's complement oVerflow, **Z** for a Zero result, **N** for Negative (or bit 7 = 1) and **H** for the Half carry between bits 3 and 4. These flags are set as a result of executing an instruction, and are normally used either for testing and acting on the status of a process, or for multiple-byte operations. The remaining three bits are associated with interrupt handling. The **I** bit is used to lock out or mask the IRQ interrupt, and the **F bit** carries out the same function for the FIRQ interrupt. During an interrupt service routine the **E flag** may be consulted to see if the Entire register state has been saved (IRQ, NMI and SWI) or not (FIRQ). More details are given in Section 6.1.

REGISTER ARRAY

The 6809 has two Data registers, termed **Accumulators A and B**. These Data registers are normally targeted by the ALU as the source and destination for at least one of its operands. Thus ADDA #50 adds 50 to the contents of Accumulator A (in register transfer language, RTL, this is symbolized as [A] <- [A] + 50, which reads 'the contents of register A become the original contents of A plus 50'). Operations requiring one operand can seemingly be done directly on external memory; for example, INC 6000h which increments the contents of location 6000h ([6000] <- [6000] + 1). The suffix h indicates the hexadecimal number base, whilst b denotes binary. However, in reality the MPU executes this by bringing down the contents of 6000h (written as [6000]), uses the ALU to add one and returns it. Whilst this fetch and execute process is invisible to the programmer, the penalty is space and time; INC M (3 bytes length) takes 7 μs and INCA or INCB

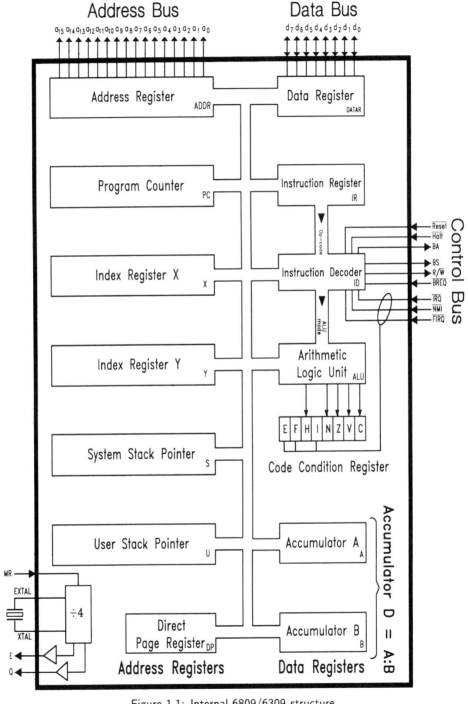

Figure 1.1: Internal 6809/6309 structure.

(1 byte length) takes 2 μs (at a 1 MHz clock rate). Thus while it is always better to use the Data registers for operands, this is difficult in practice because there are only two such registers. Unlike the older 6800 MPU, the 6809's two 8-bit Data registers can be concatenated to one 16-bit double register A:B; the **D Accumulator**. A few operations such as Add (e.g. ADDD #4567) can directly handle this. But although the 6809 has pretensions to be a 16-bit MPU, the ALU is only 8-bits wide and instructions such as this require two passes; but they are nevertheless faster than two single operations.

Six dedicated **Address registers** are accessible to the programmer and are associated with generating addresses of program and operand bytes external to the processor. The **Program Counter (PC)** always points to the current program byte in memory, and is automatically incremented by the number of operation bytes during the fetch. It normally advances monotonically from its start (reset) value, with discontinuities occurring only at Jump or Branch operations, and internal and external interrupts.

Two **Index registers** are primarily used when a computed address facility is desired. For example an Index register may be set up to address or *point to* the first element of a byte array. At any time after this, the nth element of this array can be fetched by augmenting the contents of the Index register by n. Thus the instruction LDA 6,X brings down array[6] to Accumulator_A ([A] <- [[X]+6]). Index registers can also be automatically or manually incremented or decremented and thus can systematically step through a table or array. The 6809 does not have a separate ALU for computed address generation, and this can make the execution of such operations rather lengthy. Sometimes Index registers are used, rather surreptitiously, to perform simple 16-bit arithmetic, for example counting loop passes. An example is given in the listing of Table 2.9.

The **System Stack Pointer (SSP)** register (also known as Hardware Stack Pointer) is normally used to identify an area of RAM used as a temporary storage area, to facilitate the implementation of subroutines and interrupts. These techniques are discussed in Chapters 5 and 6. Rather unusually the 6809 also has a **User Stack Pointer (USP)**, which can be usefully employed to point to an area of RAM which can be used by the programmer to place data for retrieval later and will not get mixed in with the automatic action of the **SSP**. Both Stack Pointers can also be used as Index registers.

The address size of most 8-bit MPUs is 16-bits wide, allowing direct access to 65,536 (2^{16}) bytes. With a data bus of only 8-bits width, instructions which specify absolute addresses will be at least three bytes long (one or more bytes for the operation code and two for the address). As well as needing space, the three fetches take time. To reduce this problem, the 6800 and 6502 processors use the concept of zero page addressing. This is a shortform absolute address mode which assumes that the upper address byte is 00h. Thus in 6800 code, loading data from location 005Fh (LDAA 005F) can be coded as: B6-00-5F (4 cycles) using the 3-byte Extended Direct address mode or 96-5F (3 cycles) with the 2-byte Direct address mode. In the 6809 MPU this concept has been extended in that the direct page can be moved to any 256-byte segment based at 00 to FFh, the segment number being held in the **Direct Page register** (DP). Thus, supposing locations 8000–80FFh hold peripheral interface devices which are frequently being accessed, then transferring the segment number 80h into the DP means that the instruction LDA 5F, coded as 96-5F, actually moves data from 805Fh into Accumulator_A. When the 6809 is Reset, the DP is set to 00h and, unless its value is changed, direct addressing is equivalent to zero page addressing. The DP can be changed dynamically as the program progresses, but this is worthwhile only if more than eight accesses within a page are to be made.

CONTROL CIRCUITRY

The remaining registers shown in Fig. 1.1 are invisible to the programmer, in that there is no direct access to their contents. Of these, the **Instruction decoder** represents the 'intelligence' of the MPU. In essence its job is to marshal all available resources in response to the operation code word fetched from memory. This sequential control function is the most complex internal process undertaken by the MPU; however, its design is beyond the scope of this text. References [9, 10] are useful background reading in this regard. Suffice to say that the 6809, like its earlier relatives, uses a random logic circuit for its decoder implementation. This provides for the highest implementation speed but at the expense of a less structured set of programming operations.

1.2 Outside the 6809

The 6809 MPU is available in a 40-pin package, whose pinout is shown in Fig. 1.2. The 40 signals can be conveniently divided into three functional groups, data, address and control. Unlike the 808x family, all signals are non-multiplexed, that is they retain the same function throughout the clock cycle, see Fig. 1.3. Signals are all Transistor-Transistor Logic (TTL) voltage-level compatible.

DATA BUS d(n)

A single bidirectional 8-bit **data bus** carries both instruction and operand data to and from the MPU (Read and Write respectively). When enabled, data lines can drive up to four 74 LS loads and a capacitive loading of 130 pF without external buffering. Data lines are high-impedance (turned off) when the processor is halted or in a direct memory access (DMA) mode.

ADDRESS BUS a(n)

Sixteen address lines can be externally decoded to activate directly up to 2^{16} byte locations which can be placed on the common data bus. During cycles when the MPU is internally processing, the **address bus** is set to all ones (FFFFh) and the data bus to Read. When enabled, up to four 74 LS loads and 90 pF can be driven. Activating $\overline{\text{Halt}}$ or $\overline{\text{DMA/BREQ}}$ turns off (or floats) these bus lines.

Outside the 6809 7

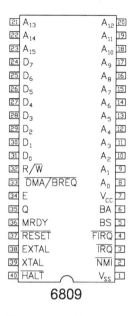

Figure 1.2: 6809 pinout.

CONTROL BUS

All MPUs have similar data and address buses, but differ considerably in the miscellany of functions conveniently lumped together as the **control bus**. These indicate to the outside world the status of the processor, or allow these external circuits control over the processor operation.

Power (V_{cc}, V_{ss})

A single $5\,V \pm 5\%$ supply dissipating a maximum of $1.0\,W$ ($200\,mA$). The analogous Hitachi 6309 CMOS MPU dissipates $60\,mW$ during normal operation and $10\,mW$ in its sleep mode.

Read/$\overline{\text{Write}}$ (R/$\overline{\text{W}}$)

Used to indicate the status of the data bus, high for Read and low for Write. $\overline{\text{Halt}}$ and $\overline{\text{DMA/BREQ}}$ float this signal.

$\overline{\text{Halt}}$

A low level here causes the MPU to stop running at the end of the present instruction. Both data and address buses are floated, as is R/$\overline{\text{W}}$. While halted, the MPU does not respond to external interrupt requests. The system clocks (E and Q) continue running.

DMA/BREQ

This is similar to $\overline{\text{Halt}}$ in that data, address and R/$\overline{\text{W}}$ signals are floated. However, the MPU does not wait until the end of the current instruction execution. This gives a response delay (sometimes called a latency) of $1\frac{1}{2}$ cycles, as opposed to a worst-case Halt latency of 21 cycles [5]. The payback is that because the processor clock is frozen, the internal dynamic registers will lose data unless periodically refreshed. Thus the MPU automatically pulls out of this mode every 14 clock cycles for an internal refresh before resuming (cycle stealing).

Reset

A low level at this input will reset the MPU. As long as this pin is held low, the vector address FFFEh will be presented on the address bus. On release, the 16-bit data stored at FFFEh and FFFFh will be moved to the **Program Counter**; thus the Reset vector FFFE:Fh should always hold the restart address (see Fig. 6.4).

$\overline{\text{Reset}}$ should be held low for not less than 100 ms to permit the internal clock generator to stabilize after a power switch on. As the $\overline{\text{Reset}}$ pin has a Schmitt-trigger input with a threshold (4 V minimum) higher than that of standard TTL-compatible peripherals (2 V maximum), a simple capacitor/resistor network may be used to reset the 6809. As the threshold is high, other peripherals should be out of their reset state before the MPU is ready to run.

Non-Maskable Interrupt ($\overline{\text{NMI}}$)

A negative *edge* (pulse width one clock cycle minimum) at this pin forces the MPU to complete its current instruction, save all internal registers (except the **System Stack Pointer, SSP**) on the System stack and vector to a program whose start address is held in the NMI vector FFFC:Dh. The **E** flag in the **CCR** is set to indicate that the Entire group of MPU registers (known as the **machine state**) has been saved. The **I** and **F** mask bits are set to preclude further lower priority interrupts (i.e. $\overline{\text{IRQ}}$ and $\overline{\text{FIRQ}}$). If the NMI program service routine is terminated by the RETURN FROM INTERRUPT (RTI) instruction, the machine state is restored and the interrupted program continues. After Reset, $\overline{\text{NMI}}$ will not be recognized until the **SSP** is set up (e.g. LDS #TOS+1 points the **System Stack Pointer** to just over the top of the stack, **TOS**). More details are given in Section 6.1.

Fast Interrupt Request ($\overline{\text{FIRQ}}$)

A low *level* at this pin causes an interrupt in a similar manner to $\overline{\text{NMI}}$. However, this time the interrupt will be locked out if the **F** mask in the **CCR** is set (as it is automatically on Reset). If **F** is clear, then the MPU will vector via FFF6:7h after saving *only* the **PC** and **CCR** on the System stack. The F and I masks are set to lock out any further interrupts, except NMI, and the E flag cleared to show that the Entire machine state has not been saved.

As $\overline{\text{FIRQ}}$ is level sensitive, the source of this signal must go back high before the end of the service routine.

Interrupt Request ($\overline{\text{IRQ}}$)

A low *level* at this pin causes the MPU to vector via FFF8:9h to the start of the IRQ service routine, provided that the I mask bit is cleared (it is set automatically at Reset). The entire machine state is saved on the System stack and I mask set to prevent any further IRQ interrupts (but not FIRQ or NMI). As in $\overline{\text{FIRQ}}$, the $\overline{\text{IRQ}}$ signal must be removed before the end of the service routine. On RTI the machine state will be restored, and as this includes the **CCR**, the I mask will return low automatically.

Bus Available, Bus Status (BA, BS)

These are status signals which may be decoded for external control purposes. Their four states (BA, BS) are:

        00 : Normally running
        01 : Interrupt or Reset in progress
        10 : A software SYNC is in progress (see Section 6.2)
        11 : MPU halted or has granted its bus to $\overline{\text{DMA/BREQ}}$

Clock (XTAL, EXTAL)

An on-chip oscillator requires an external parallel-resonant crystal between the **XTAL** and **EXTAL** pins and two small capacitors to ground (see Fig. 13.1). The internal oscillator provides a processor clocking rate of one quarter of the crystal resonant frequency. The basic 6809 MPU is a 1 MHz device requiring a 4 MHz crystal, whilst the 68A09 and 68B09 1.5 and 2 MHz versions need 6 and 8 MHz crystals respectively. The Hitachi 6309 MPU is available in a 3 MHz version. In all cases there is a lower frequency limit at 100 kHz, due to the need to keep the internal dynamic registers constantly refreshed. If desired, an external TTL-level oscillator may be used to drive **EXTAL**, with **XTAL** grounded.

The 6809E/6309E MPUs do not have an integral clock generator, but provide additional control functions suitable for multi-processor configurations.

Enable, Quadrature (E, Q)

These are buffered clock signals from the internal (or external) clock generator. They are used to synchronize devices taking data from or putting data on the data bus. We will look at the timing relationship between these signals and the main buses in the following section. E is sometimes labelled ϕ_2 after the second phase clock signal needed for the 6800 MPU, which fulfilled a similar role.

10 *The 6809 Microprocessor: Its Hardware*

Memory Ready (MRDY)

This is a control input to the internal clock oscillator. By activating $\overline{\text{MRDY}}$, a slow external memory or peripheral device can freeze the oscillator until its data is ready. This is subject to a maximum of 10 ms, in order to keep the MPU's dynamic registers refreshed.

1.3 Making the Connection

A microprocessor monitors and controls external events by sending and receiving information via its data bus through interface circuitry. In order to interface to a MPU, it is necessary to understand the interplay between the relevant buses and control signals. These involve sequences of events, and are usually presented as timing or flow diagrams.

Consider the execution of the instruction LDA 6000h ([A] <- [6000h]). This instruction takes four clock cycles to implement; three to fetch down the 3-byte instruction (B6-60-00) and one to send out the peripheral (memory or otherwise) address and put the resulting data into **Accumulator_A**. Figure 1.3 shows a somewhat simplified state of affairs during that last cycle, with the assumption of a

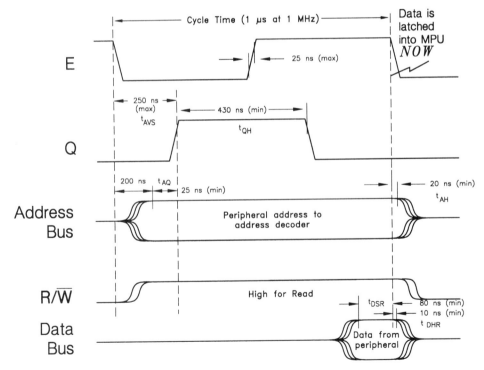

Figure 1.3: A snapshot of the 6809 MPU reading data from a peripheral device. Worst-case 1 MHz device times are shown.

1 MHz clock frequency. The address will be out and stable by not later than 25 ns before Q goes high (t_{AQ}). The external device (at 6000h in our example) must then respond and set up its data on the bus by no later than 80 ns (t_{DSR}) before the falling edge of E, which signals the cycle end. Such data must remain held for at least 10 ns (t_{DHR}) to ensure successful latching into the internal data register. t_{AQ}, t_{DSR}, t_{DHR} for the 68B09 2 MHz processor are 15, 40 and 10 ns respectively.

Writing data to an external device or memory cell is broadly similar, as illustrated in Fig. 1.4, which shows the waveforms associated with, for example, the last cycle of a STA 8000h (Store) instruction.

Once again the Address and R/$\overline{W}$ signals appear just before the rising edge of Q, t_{AQ}. This time it is the MPU which places the data on the bus, which will be stable well before the falling edge of Q. This data will disappear within 30 ns after the cycle end t_{DHW}; the corresponding address hold time t_{AH} is 20 ns.

Earlier members of the 6800 family did not provide a Q clock signal. In these cases the end of the E signal had to be used to turn off or trigger the external device when writing. As there are only 30 ns after this edge before the data collapses, care had to be taken to ensure that the sum of the address decoder propagation delay plus the time data must be held at the peripheral interface device after the trigger event (hold time) satisfies this criterion. Because of this tight timing requirement, the E clock is normally routed directly to the interface circuitry, rather than be

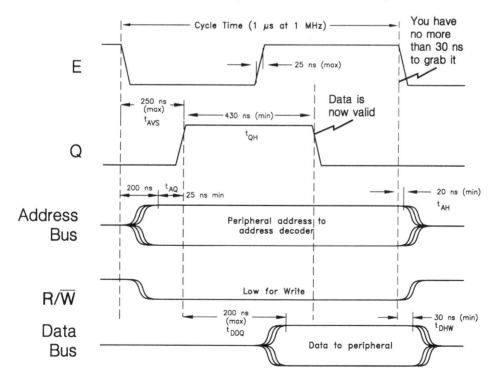

Figure 1.4: Sending data to the outside world.

12 *The 6809 Microprocessor: Its Hardware*

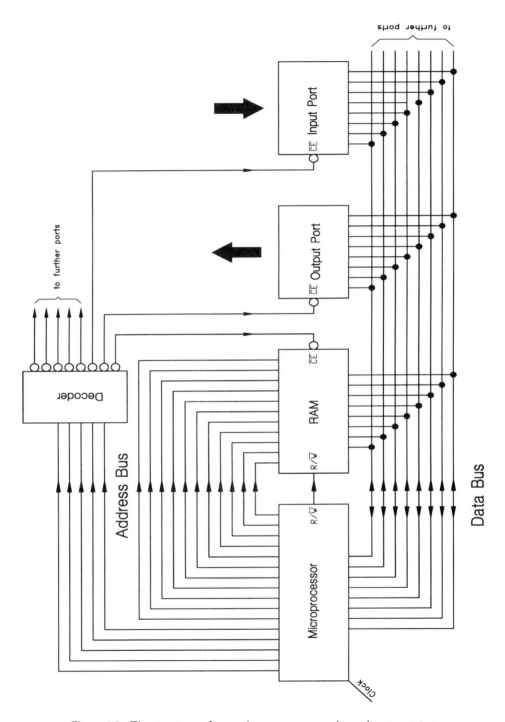

Figure 1.5: The structure of a synchronous common-bus microcomputer.

delayed by the address decoder (e.g. see Fig. 1.9). With the 6809, it is preferable to use the falling edge of the Q clock for this purpose when writing. While reading of course, the peripheral interface must be enabled up to (and a little beyond) the end of the E cycle, at which point the MPU captures the proffered data.

The basic structure of a synchronous common data bus MPU-based system is shown in Fig. 1.5. The term synchronous is used to denote that normal communication between peripheral device and MPU is open loop, with the latter having no knowledge of whether data is available or will be accepted at the end of a clock cycle. If a peripheral responds too slowly, its garbled data will be read at the end of the cycle irrespective of its validity. In such cases $\overline{\text{MRDY}}$ can be used to slow things down, although this is considered an abnormal transition. The alternative closed-loop architecture is discussed on page 72.

As all external devices communicate to the master through a single *common* data highway, it is necessary to ensure that only *one* is active on any exchange. All microprocessors use an address bus for this purpose. Taken together with external decoding circuitry, each target can be assigned a specific **address** and thus enabled uniquely. As depicted in Fig. 1.5, only one decoder is used, but in a larger system there is likely to be one central decoder dividing the available memory space into zones or pages, and local decoders providing the 'fine print'. Memory chips of course are not single devices, but comprise a multitude of addressable cells: they have their secondary decoder on-board. The 808x family use separate address buses for memory and peripheral selection. As well as requiring additional pins on the package, special instructions must be provided to use them.

There is nothing special about **address decoder** design [11, 12]. Implementation techniques range through gates, comparators, decoders, PROMs and PALs. Figure 1.6(a) shows a very simple page decoder which splits up the available 64 kbyte memory space into eight 4 kbyte zones. The decoder output of Fig. 1.6(b)(i) assumes that the 74138 is permanently enabled. Notice that the signal does not begin to go back high until after the address collapses, that is 10 ns after the cycle end. There is no problem during a Read, as the MPU will already have latched in the data; but during a Write, the data will collapse in 30 ns, leaving only 10 ns for decoder propagation delay and peripheral hold time. Using the E clock to enable the decoder (e.g. E to G1 in Fig. 1.6(a)) extends the permissible propagation delay plus hold time to 30 ns. For example, if we take the 74LS377 of Fig. 1.7 used as an 8-bit output port, then its hold time is 5 ns minimum and the propagation delay time for the 74LS138 from G1 is 26 ns worst-case. Clearly a hazardous race.

To avoid such races we can directly qualify each device which can be written to by either E, or preferably Q. The 74LS377 octal D flip flop array used as an 8-bit output port is selected at the appropriate address, $6000h$ in Fig. 1.7, by the decoder, but the data is only clocked in at the falling edge of Q. This leaves around $\frac{1}{4}$ cycle before the data collapses. Where separate enable and clock controls are not provided, the decoder signal may be gated by a derivative of Q.

RAM chips are more problematical as they need to be enabled until the end of the cycle when being read from, but cut off early when writing to. This differenti-

14 The 6809 Microprocessor: Its Hardware

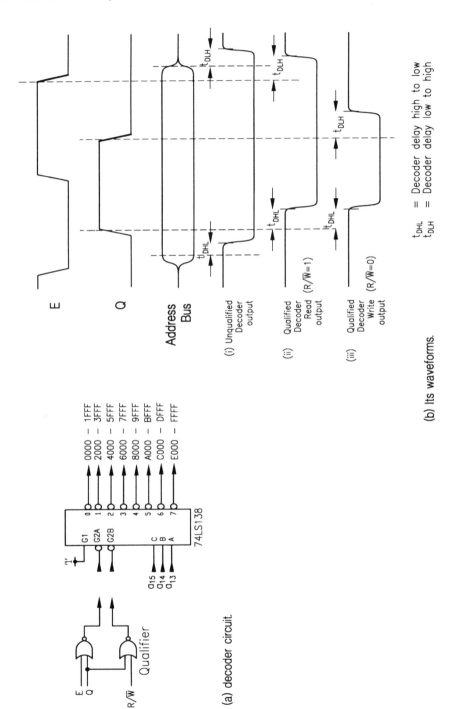

Figure 1.6: An elementary address decoding scheme.

Making the Connection 15

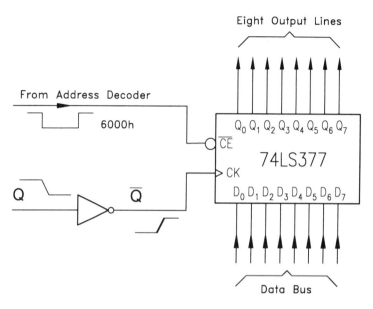

Figure 1.7: A simple byte-sized output port.

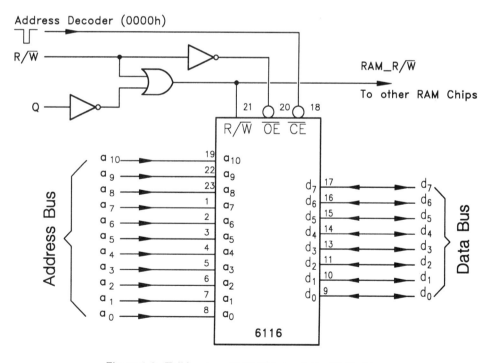

Figure 1.8: Talking to a 6116 2 kbyte static RAM chip.

ation can be accomplished by qualifying the R/$\overline{\text{W}}$ signal by Q, producing:

$$\text{RAM\_R}/\overline{\text{W}} = \overline{\text{R}/\overline{\text{W}}+\text{Q}}$$

which is high irrespective of Q during a Read, and is just Q when writing. As shown in Fig. 1.8, it is normal to ensure that the RAM will not output data during a Write-to operation, by driving the RAM's Output_Enable with the complement of R/$\overline{\text{W}}$. The 'doctored' RAM_R/$\overline{\text{W}}$ signal may of course be used for as many RAM chips as are present in the system. It may also be used to replace Q in Fig. 1.7, having the advantage that the output port cannot be erroneously read.

Care must be taken when interfacing memory chips to choose a device with a suitable access time. This is especially true for more recent MPUs, which can run at higher clock rates. The access time for a memory chip is normally given as the duration from the application of a stable address or chip enable until the activation of the cell to be read from or written to. In the 6116 RAM, this internal decoding occurs irrespective of the state of the chip enables. Looking first at RAM interfacing and taking Fig. 1.8 as an example, it is clear that the writing action is the more critical as this will end earlier at the falling edge of Q. From Fig. 1.4 we see that we have $t_{AQ} + t_{QH}$ less the RAM data setup time. The Hitachi HM6116AP-20 has a setup requirement of 50 ns and a 200 ns access time, so:

$$t_{AQ} + t_{QH} - 50 \geq 200$$
$$t_{AQ} + t_{QH} \geq 250 \text{ ns}$$

At 1 MHz, $t_{AQ} + t_{QH}$ is 455 ns, but this shrinks to 230 ns for a 2 MHz clock. Thus a 150 ns access time RAM chip must be used in the latter instance; for example the Hitachi HM6116AP-15. The 6264 RAM has an access time measured from the chip select. In this case the address decoder delay must be part of the calculation. An example of this is given in Section 3.3.

ROM chips are interfaced in a similar fashion, but of course they are read-only. Referring to the timing diagram of Fig. 1.3, we see that as data from the ROM must be present t_{DSR} before the end of the cycle, we have the relationship $t_{cyc} - t_{AVS} - t_{DSR} \geq t_{access}$. At 1 MHz this sums to 720 ns, and 380 ns at 2 MHz. Most of the smaller EPROMs, for example the 2 kbyte Texas Instruments TMSD2516JL, have 450 ns access times. The TMS2764-25JL is an 8 kbyte 250 ns device and is therefore suitable for the higher-speed processor.

Rather than qualifying each write-to peripheral by Q, it is possible to enable the address decoder directly. Thus the decoder should have a lengthy output pulse when a read is in operation, but be cut short (at the end of Q) when a write is in progress. This relationship can be written as:

$$\text{Enable} = (\overline{\text{R}/\overline{\text{W}}} \cdot (\text{E}+\text{Q})) + (\overline{\overline{\text{R}/\overline{\text{W}}} \cdot \text{Q}})$$

giving the decoder output waveforms shown in Fig. 1.6(b)(ii) and (iii). To make use of the two active low $\overline{\text{G2A}}$ and $\overline{\text{G2B}}$ 74138 inputs, a little Boolean algebra yields:

$$(R/\overline{W}\cdot E) + (R/\overline{W}\cdot Q) + \overline{(R/\overline{W}\cdot Q)}$$
$$(R/\overline{W}\cdot E) + Q\cdot(R/\overline{W} + \overline{R/\overline{W}})$$
$$(R/\overline{W}\cdot E) + Q$$
$$(R/\overline{W} + Q)\cdot(Q + E) = (G2A)\cdot(G2B)$$

giving the qualifying network of Fig. 1.6(a).

Special-purpose 6800 family peripheral interface devices, such as the PIA of Fig. 1.9 [13], are designed to work in harmony with older MPU types which only provide an E signal. They all have an enable input designed to be directly driven by E, and have data hold time requirements within the 30 ns limit. They must

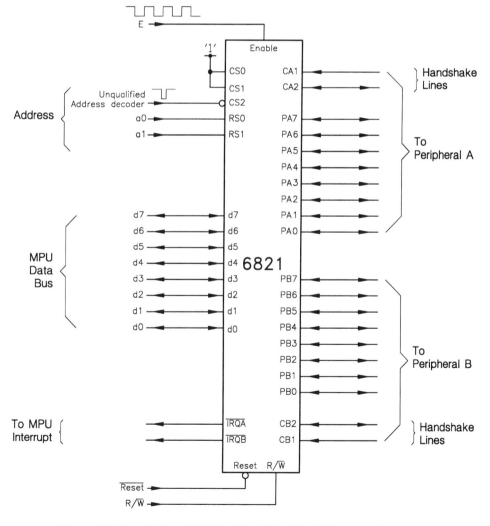

Figure 1.9: Interfacing a 6821 Peripheral Interface Adapter to the 6809.

not be disabled early in the cycle by a **Q** related signal. This means that 68xx peripherals cannot be selected by a modified decoder, such as in Fig. 1.6(a). However, it is permissible to mix the two kinds of peripheral devices, each enabled by the appropriate address decoder. For example, a primary address decoder could enable a simple secondary decoder for 68xx peripheral devices, and a more complex **Q** related secondary decoder for simple interface circuitry.

References

[1] Noyce, R.N. and Marcian, E. H.; A History of Microprocessor Development at Intel, *IEEE Micro*, Feb. 1981, pp. 8–21.

[2] Cahill, S.J.; *Designing Microprocessor-Based Digital Circuitry*, Prentice-Hall, 1985, Chapters 8 and 9.

[3] Frazer, D.A. et al.; *Introduction to Microcomputer Engineering*, Ellis Horwood/Halsted Press, 1985, Chapter 3.

[4] Cahill, S.J.; *The Single-Chip Microcomputer*, Prentice-Hall, 1988.

[5] Ritter, T. and Boney, J.; A Microprocessor for the Revolution: The 6809, *BYTE*, **4**, part 1, Jan. 1979, pp. 14–42; part 2, Feb. 1979, pp. 32–42; part 3, Mar. 1979, pp. 46–52.

[6] Wakerly, J.F.; *Microcomputer Architecture and Programming: The 68000 Family*, Wiley, 1989, Chapter 16.

[7] Horvath, R.; *Introduction to Microprocessors using the MC6809 or the MC68000*, McGraw-Hill, 1992.

[8] Hyman, A; *Charles Babbage: Pioneer of the Computer*, Princeton University Press/Oxford University Press, 1982, Chapter 16.

[9] Agrawala, A.K. and Rauscher, T.G.; *Foundations of Microprogramming*, Academic Press, 1976.

[10] Encegovac, M.D. and Larg, T.; *Digital Systems and Hardware/Software Algorithms*, Wiley, 1985, Chapter 11.

[11] Monolithic Memories; *PAL Handbook*, 3rd ed., 1983, pp. 6.27–6.39 and 8.40–8.43.

[12] Cahill, S.J.; *Digital and Microprocessor Engineering*, 2nd. ed., Ellis Horwood/Prentice-Hall, 1993, Chapter 5.3.

[13] Cahill, S.J.; *Digital and Microprocessor Engineering*, 2nd. ed., Ellis Horwood/Prentice-Hall, 1993, Chapter 5.3.4.

CHAPTER 2

The 6809 Microprocessor: Its Software

The 6809 processor's instruction set was designed to be upwardly compatible with its predecessor, the 6800. Indeed many of the common instructions even have the same machine code; for example the operation to clear location 2000h (CLR 2000h) is coded as 7F-20-00 in both cases. Notwithstanding, many new instructions were introduced giving greater flexibility and subsuming several older instructions. Thus the older 6800 device could only push its Accumulators into the stack (i.e. PSHA and PSHB; the equivalent 6809 instruction can push any or all its registers in one go: for example PSHS A,B,CC,DP,X,Y,U,PC.

As we shall see, enhancing stack-based operations facilitates the production of efficient high-level language code. To this end, the 6809 also features an extended arithmetic functionality and a limited repertoire of 16-bit operations. Additionally, the number of available address modes was considerably enlarged, in particular those involving computed effective addresses.

In this chapter I will overview the instruction set and address modes. Some example program subroutines will tie these together, and give us a base to compare with the 68000 MPU software introduced in Chapter 4. Detailed consideration of subroutines and interrupts are left to Chapters 5 and 6.

2.1 Its Instruction Set

Although the 6809 instruction set was designed to be upwardly compatible with that of the 6800, in fact the number of distinct operations was reduced from 72 to only 59. Its increased power, of the order of 260% [1], comes instead from the additional functionality of these instructions, the capability of using more registers and the extra address modes. First and second generation 8-bit MPUs, such as the 8080/8085 and 6800 devices, encoded all instructions as a byte-sized **operation code (op-code)**. Thus no more than 256 operation–register–address mode combinations were possible. Third generation devices such as the Z80 and 6809 MPUs can use two bytes for this function. Whereas the 6800 MPU has only 197 op-codes (out of a maximum of 256), the 6809 has 1464 op-codes. As an example, the primary op-code for PUSH ONTO THE SYSTEM STACK is 34h, the complete code for PSHS A,B,X is 34-16h. In binary this is 0011 0100 0001 0110, where each bit of the post-byte represents a register to be saved according to the format shown in Fig. 2.1. Of course the programmer normally need not be concerned with detail at this level; the assembler will take care of such matters.

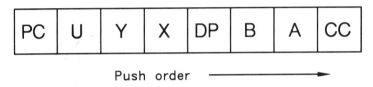

Push order ⟶

Figure 2.1: Postbyte for pushing and pulling.

Typically around 40% of instructions at machine-code level involve shuffling data in-between registers and out to memory [2], so we will look first of all at data movement instructions, as summarized in Table 2.1. The Load and Store operations copy data between memory and register. Both 8- and 16-bit moves are possible, but as memory is addressable only one byte at a time, the latter move involves two consecutive transfers. Thus the instruction LDX 0C100h will perform as shown in Fig. 2.2(a). Note how the most significant byte (MSB) of **X** comes from the least significant memory location C100h and the least significant byte (LSB) from the next highest location C101h, thus | MSB ^{C100h} | LSB ^{C101h} |. The same order is observed when sending out multiple-byte data, for example STX 0C000. In general, data structures in the 6800/68000 family are ordered with the MSB in the lowest consecutive memory location. Some other processors, such as the 808x family, are ordered with the MSB as the lowest successive memory location.

Notice that no Store to Direct Page register operation exists. To set up this register to, say, 80h, the sequence:

LDA   #80h
TFR   A,DP

first places the number 80h in Accumulator_A (it could equally be **B**) and then transfers this to the **DP register**. This overhead is justified as the **DP register** is (or should be) rarely altered. The TRANSFER instruction can move the contents of any 8-bit register (**A,B,DP,CC**) to any other, or any 16-bit register contents (**X,Y,U,S,D,PC**) to any other. The upper and lower nybbles (four bits) of the postbyte determine the source and destination register respectively, according to the code:

```
0000 = D     0001 = X     0010 = Y     0011 = U     0100 = S
0101 = PC    1000 = A     1010 = B     1010 = CCR   1011 = DP
```

thus TFR A,DP is coded as 1F-8Bh (post-byte 1000 1011b). EXCHANGE works in a similar way between like-sized registers with the same post-byte construction.

The programmer can easily keep two separate stacks using the **System Stack Pointer** and **User Stack Pointer** registers. These stacks are normally set up at the

Table 2.1: Move instructions.

Operation	Mnemonic	V	N	Z	C	Description
Exchange						Exchanges two like-sized
R1↔R2[1]	EXG R1,R2	•	•	•	•	register contents
e.g.	EXG A,B	•	•	•	•	[A]<-->[B]
Load						Moves data to register
to A; to B	LDA; LDB	0	√	√	•	[A]<-[M]; [B]<-[M]
to D	LDD	0	√	√	•	[D]<-[M:M+1]
to X; to Y	LDX; LDY	0	√	√	•	[X]<-[M:M+1]; [Y]<-[M:M+1]
to S; to U	LDS; LDU	0	√	√	•	[S]<-[M:M+1]; [U]<-[M:M+1]
Push						Moves registers onto Stack
to System stack	PSHS regs	•	•	•	•	Listed registers to S stack
to User stack	PSHU regs	•	•	•	•	Listed registers to U stack
e.g.	PSHS A,B,X	•	•	•	•	A,B and X to S stack
Pull						Moves stack data to registers
from System stack	PULS regs	•	•	•	•	S stack to listed registers
from User stack	PULU regs	•	•	•	•	U stack to listed registers
e.g.	PULS A,B,X	•	•	•	•	S stack to A,B and X
Store						Moves data from register
from A; from B	STA; STB	0	√	√	•	[M]<-[A]; [M]<-[B]
from D	STD	0	√	√	•	[M:M+1]<-[D]
from X; from Y	STX; STY	0	√	√	•	[M:M+1]<-[X]; [M:M+1]<-[Y]
from S; from U	STS; STU	0	√	√	•	[M:M+1]<-[S]; [M:M+1]<-[U]
Transfer						Transfers two like-sized
						register contents
R1↔R2[1]	TFR R1,R2	•	•	•	•	
e.g.	TFR A,DP	•	•	•	•	[DP]<-[A]

0   Flag always reset
1   Flag always set
•   Flag not affected
√   Flag operates in the normal way

Note 1: Register pairs must either be 8-bit A,B,CC,DP or 16-bit X,Y,S,U,PC.

beginning of the program, simply by using the relevant Load operation. Thus if we wish to define RAM from 1FFFh downwards as the System stack and 18FFh downwards as a User stack, the sequence:

LDS #02000h
LDU #01900h

will accomplish this. Notice that the Top Of Stack (TOS) in both cases is one

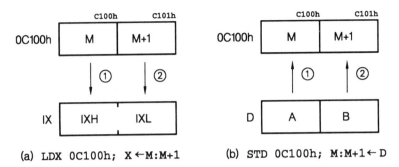

Figure 2.2: Moving 16-bit data at 'one go'.

above physical memory. This is because the Push and Pull operations, as well as the system operations of jumping to a subroutine and implementing an interrupt, decrement the relevant Stack Pointer *before* moving data. As mentioned earlier, the Push and Pull operations allow any register or set of registers to be pushed or pulled into or out of a stack at one go. This facilitates the passing of arguments to and from subroutines, and allows called subroutines to use registers without corrupting register-held data in the calling program (see Section 5.2).

Figure 2.1 shows how the post-byte is calculated for a Push or a Pull. Specifically the System stack is shown; if the User stack is being employed then **U** is replaced by **S**. Figure 2.3 shows a snapshot of memory after a Push onto the System stack. If only a subset of registers are saved, then the same order is preserved as in the diagram. The time-taken for a Push or Pull is five cycles plus one cycle per byte moved. In Fig. 2.3 this adds up to 17 cycles.

The 6809 implements the normal Add and Subtract operations, as shown in Table 2.2, both with and without carry, targeted on an 8-bit Accumulator. An Accumulator_D-based 16-bit Add and Subtract instruction is also provided, but unfortunately not with a carry. An unsigned addition of Accumulator_B to the 16-bit **X** Index register can also be classed as double, but the 8-bit addend is promoted to 16-bit at addition time, by assuming an upper byte of zero, hence the terminology unsigned. Thus for example, ABX #56h actually adds the constant $0056h$ to **X**.

It is possible to **promote** a signed number in Accumulator_B to its 16-bit equivalent in Accumulator_D by using the SIGN EXTENSION instruction. This zeros Accumulator_A if bit 7 of **B** is 0 and fills **A** with ones (A <- FFh) otherwise; for example [B] = $10110011b$ (-83) becomes [D] = 11111111 $10110011b$ (-83). The SIGN EXTENSION (SEX) instruction makes the 6809 unique as the only MPU offering sex appeal!

Any 16-bit Index or Stack register can be summed with an 8-bit Accumulator (which is automatically sign extended), Accumulator_D or a constant by means of the LOAD EFFECTIVE ADDRESS (LEA) instruction. This makes use of the arithmetic provision which computes effective addresses in the Indexed address mode. We will discuss this in the next section, but as an example the instruction:

```
LEAX 1,X     ; Coded as 30-01h
```

Its Instruction Set 23

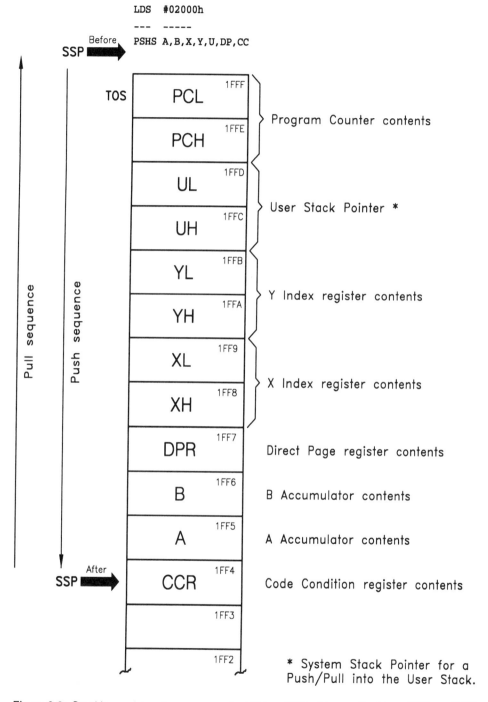

Figure 2.3: Stacking registers in memory using PSH and PUL. Also applicable to $\overline{IRQ}$ and $\overline{NMI}$ interrupts.

Table 2.2: Arithmetic operations

Operation	Mnemonic	V	N	Z	C	Description
Add						Binary addition
to A; to B	ADDA; ADDB	√	√	√	√	[A]<-[A]+[M]; [B]<-[B]+[M]
to D	ADDD	√	√	√	√	[D]<-[D]+[M:M+1]
B to X	ABX	•	•	•	•	[X]<-[X]+[00|B]
Add with Carry						Includes carry
to A; to B	ADCA; ADCB	√	√	√	√	[A]<-[A]+[M]+C; [B]<-[B]+[M]+C
Clear						Destination contents zeroed
memory	CLR	0	0	1	0	[M]<-00
A; B	CLRA; CLRB	0	0	1	0	[A]<-00; [B]<-00
Decrement						Subtract one, produce no carry
memory	DEC	1	√	√	•	[M]<-[M]−1
A; B	DECA; DECB	1	√	√	•	[A]<-[A]−1; [B]<-[B]−1
Increment						Add one, produce no carry
memory	INC	2	√	√	•	[M]<-[M]+1
A; B	INCA; INCB	2	√	√	•	[A]<-[A]+1; [B]<-[B]+1
Load Effective Address						Effective Address to register
X; Y	LEAX; LEAY	•	•	√	•	[X]<-EA; [Y]<-EA
S; U	LEAS; LEAU	•	•	•	•	[S]<-EA; [U]<-EA
Multiply						Multiplies [A] by [B]
	MUL	•	•	√	3	[D]<-[A]× [B]
Negate						Reverses 2's complement sign
memory	NEG	4	√	√	5	[M]<- −[M]
A; B	NEGA; NEGB	4	√	√	5	[A]<- −[A]; [B]<- −[B]
Sign Extend						Promotes signed B to signed D
	SEX	•	√	√	•	[D]<-00|[B] or [D]<-FF|[B]
Subtract						Binary subtraction
from A; from B	SUBA; SUBB	√	√	√	√	[A]<-[A]−[M]; [B]<-[B]−[M]
from D	SUBD	√	√	√	√	[D]<-[D]−[M:M+1]
Subt with Carry						Includes carry (borrow)
from A; from B	SBCA; SBCB	√	√	√	√	[A]<-[A]−[M]−C; [B]<-[B]−[M]−C

Note 1: Overflow set when passes from 10000000 to 01111111, i.e. an apparent sign change.
Note 2: Overflow set when passes from 01111111 to 10000000, i.e. an apparent sign change.
Note 3: Carry set to state of bit 7 product, i.e. MSB of lower byte; for rounding off.
Note 4: Overflow set if original data is 10000000 (−128), as there is no +128.
Note 5: Carry set if original data is 00000000; for multiple-byte negation.

calculates the effective address as [X] + 1 and loads it into the X Index register ([X] <- [X] + 1); thus it is the equivalent to an INCREMENT X (INX) instruction, which is missing from the 6809's repertoire. Much more powerful permutations of LEA exist, thus:

```
LEAY A,X    ; Coded as 31-96h
```

promotes a signed number in Accumulator_A to 16-bits, adds this to the contents of the X Index register and puts the result in the Y Index register ([Y] <- SEX|[A] + [X])!

The contents of any read–write memory location, or any 8-bit Accumulator can be directly incremented or decremented by using the INC or DEC instruction. As noted, the X,Y,S,U registers can be similarly augmented by using the LEA instruction. Notice that INC and DEC do not set the Carry flag, which makes multiple-byte Increment and Decrement operations awkward (use ADD #1 and SUB #1 instead). Increment sets the oVerflow flag when the target goes from 0,1111111b through to 1,0000000b (seemingly from + to −) and Decrement likewise when going from 1,0000000b through to 0,1111111b (− to +). INC and DEC on memory are classified as **read–modify–write** operations, as during execution, data is fetched from memory, modified and then sent back. Clearing (CLR) memory strangely works in the same way — although the original value is irrelevant.

It is possible to multiply the two 8-bit Accumulator contents using the MUL instruction, giving a 16-bit product overwriting the original contents of Accumulator_D; thus $\boxed{\text{A}}^A \times \boxed{\text{B}}^B$ leads to $\boxed{\text{A} \quad \times \quad \text{B}}^D$. For this purpose the multiplier and multiplicand are treated as unsigned. The 16-bit product may be truncated by using only the contents of Accumulator_A as the outcome, effectively dividing by 256 (equivalent to moving the binary point left eight places). Instead of truncating, this 8-bit product may be rounded off by adding the MSB of Accumulator_B to Accumulator_A, in effect adding the $\frac{1}{2}$ bit. To facilitate this, MUL sets the C flag to the state of bit 7 of **B**. Thus the sequence:

```
MUL         ; Multiply [A] and [B] giving a 16-bit product as [D]
ADCA #0     ; Add Carry to [A] (now can disregard contents of B)
```

would give the required rounded 8-bit product in Accumulator_A.

It is of course possible to multiply or divide by powers of two by shifting left or right as appropriate. Also a combination of shift and add or shift and subtract can be used to multiply or divide by any number [3]. Table 2.3 gives the range of Shift instructions available. All of these operate on an 8-bit Accumulator or on any read/write memory location through the read–modify–write mechanism.

Linear Arithmetic Shift instructions move the 8-bit operand left or right with the Carry flag catching the emerging bit. In the case of ASR, the sign bit propagates right; thus *1,1110100b* (−12) becomes *1,1111010b* (−6) → *1,1111101b* (−3) etc. and *0,0001100b* (+12) becomes *0,0000110* (+6) → *0,0000011b* (+3) etc. The LOGIC SHIFT RIGHT equivalent *always* shifts in zeros from the left. LOGIC SHIFT LEFT and ARITHMETIC SHIFT LEFT are equivalent, and some assemblers permit the use of the alternative LSL mnemonic.

Table 2.3: Shifting Instructions.

Operation	Mnemonic	Flags V N Z C	Description
Shift left, arithmetic or logic			Linear shift left into carry
memory	ASL	1 √ √ b7	
A; B	ASLA; ASLB	1 √ √ b7	C ← [ ] ← 0
Shift right, logic			Linear shift right into carry
memory	LSR	• √ √ b0	
A; B	LSRA; LSRB	• √ √ b0	0 → [ ] → C
Shift right, arithmetic			As above but keeps sign bit
memory	ASR	• √ √ b0	
A; B	ASRA; ASRB	• √ √ b0	b7 → [ ] → C
Rotate left			Circular shift left into carry
memory	ROL	1 √ √ b7	
A; B	ROLA; ROLB	1 √ √ b7	C ← [ ] ← C
Rotate right			Circular shift right into carry
memory	ROR	• √ √ b0	
A; B	RORA; RORB	• √ √ b0	C → [ ] → C

Note 1: $V = b_7 \oplus b_6$ before shift.

Circular or Rotate Shift instructions are similar to Add with Carry, in that they can be used for multiple-precision operations. A Rotate takes in the Carry from any previous Shift and in turn saves its ejected bit in the C flag. As an example, a 24-bit word stored in $\boxed{_{24}\ M\ _{16}\ |\ _{15}\ M{+}1\ _{8}\ |\ _{7}\ M{+}2\ _{0}}$ can be shifted right once by the sequence [4]:

```
LSR   M      ; 0 → [M]           b16 → C
ROR   M+1    ; b16/C → [M+1]     b8  → C
ROR   M+2    ; b8/C  → [M+2]     b0  → C
```

In all types of Left Shifts, the oVerflow flag is set when bits 7 and 6 differ *before* the shift (i.e. $b_7 \oplus b_6$), meaning that the (apparent) sign will change *after* the shift.

The logic operations of AND, OR, Exclusive-OR and NOT (Complement) are provided, as shown in Table 2.4. The only unusual feature here is the special instructions of ANDCC and ORCC for clearing or setting flags in the **Code Condition** register. Thus to clear the I mask (see Fig. 1.1) we have:

```
ANDCC  #11101111b   ; Coded as 1C-EFh (equivalent to CLI)
```

and to set it:

```
ORCC   #00010000b   ; Coded as 1A-10h (eqivalent to SEI)
```

This saves having to provide a series of separate instructions targeted at each of the **CCR** flags and masks, such as the 6800's CLI and SEI (CLEAR and SET INTERRUPT MASK), and also allows more than one flag to be set or cleared in a single instruction.

Table 2.4: Logic instructions.

Operation		Mnemonic	Flags V N Z C	Description
AND				Logic bitwise AND
	A; B	ASL	0 √ √ •	[A]<-[A]·[M]; [B]<-[B]·[M]
	CC	ANDCC #nn	Can clear	[CCR]<-[CCR]·#nn
Complement				Invert (1's complement)
	memory	COM	0 √ √ 1	[M]<-[$\overline{M}$]
	A; B	COMA; COMB	0 √ √ 1	[A]<-[$\overline{A}$]; [B]<-[$\overline{B}$]
Exclusive-OR				Logic bitwise Exclusive-OR
	A; B	EORA; EORB	0 √ √ •	[A]<-[A]⊕[M]; [B]<-[B]⊕[M]
OR				Logic bitwise Inclusive-OR
	A; B	ORA; ORB	0 √ √ •	[A]<-[A]+[M]; [B]<-[B]+[M]
	CC	ORCC #nn	Can set	[CCR]<-[CCR]+#nn

The setting of the **CCR** flags can be used after an operation to make some deduction about, and hence act on, the state of the operand data. Thus, to determine if the value of a port located at, say, $8080h$ is zero, then:

```
LDA  8080h       ; Move in data & set Z & N flags as appropriate {86-80-80h}
BEQ  SOMEWHERE   ; Go somewhere if Z flag EQuals zero {27-xxh}
```

will bring its contents into Accumulator_A and set the Z flag if it is zero. BRANCH IF EQUAL TO ZERO will then cause the program to skip to another place. The N flag is also set if bit 7 is logic 1, and thus a Load operation can enable us to test the state of this bit. The problem is, loading destroys the old contents of the Accumulator, and the new data is probably of little interest. A non-destructive equivalent of loading is TEST, as shown in Table 2.5. The sequence now becomes:

```
TST  8080h       ; Check data & set Z & N flags as appropriate {7D-80-80h}
BEQ  SOMEWHERE   ; Go somewhere if Z flag EQuals zero {27-xxh}
```

but the Accumulator contents are not overwritten. However, 16-bit tests must be carried out using a 16-bit Load operation as only 8-bit TeST instructions are provided.

TeST can only check for all bits zero or the state of bit 7. For data already in an 8-bit Accumulator, ANDing can check the state of any bit; thus:

```
ANDB    #00100000b    ; Clear all Accumulator B bits except 5 {C4-20h}
```

will set the Z flag if bit 5 is 0, otherwise Z will be cleared. Once again this is a destructive examination, and the equivalent from Table 2.5 is BIT TEST; thus:

```
BITB    #00100000b    ; Coded as C5-20h
```

does the same thing, but with the contents of Accumulator_B remaining unchanged; and more tests can subsequently be carried out without reloading.

Comparison of the magnitude of data in an Accumulator with either a constant or data in memory requires a different approach. Mathematically this can be done by subtracting [M] from [A] and checking the state of the flags. Which flags are relevant depend on whether the numbers are to be treated as unsigned (magnitude only) or signed. Taking the former first gives:

[A] *Higher than* [M] : [A]−[M] gives no Carry and non-Zero C=0, Z=0 $(\overline{C+Z}=1)$
[A] *Equal to* [M] : [A]−[M] gives Zero (Z=1)
[A] *Lower than* [M] : [A]−[M] gives a Carry (C=1)

The signed situation is more complex, involving both the Negative and oVerflow flag. Where a subtraction occurs and the difference is *positive*, then either bit 7 will be 0 and there will be no overflow (both N and V are 0) or else an overflow will occur with bit 7 at logic 1 (both N and V are 1). Logically, this is detected by the function $\overline{N \oplus V}$. A *negative* difference is signalled whenever there is no overflow

Table 2.5: Data test operations.

Operation	Mnemonic	V	N	Z	C	Description
Bit Test						Non-destructive AND
A; B	BITA; BITB	0	√	√	•	[A]·[M]; [B]·[M]
Compare						Non-destructive subtract
with A; B	CMPA; CMPB	√	√	√	√	[A]−[M]; [B]−[M]
with D	CMPD	√	√	√	√	[D]−[M:M+1]
with X; Y	CMPX; CMPY	√	√	√	√	[X]−[M:M+1]; [Y]−[M:M+1]
with S; U	CMPS; CMPU	√	√	√	√	[S]−[M:M+1]; [U]−[M:M+1]
Test for Zero or Minus						Non-destructive subtract from zero
memory	TST	0	√	√	•	[M]−00
A; B	TSTA; TSTB	0	√	√	•	[A]−00; [B]−00

and the sign bit is 1 (**N** is 1 and **V** is 0) or else an overflow occurs together with a positive sign bit (**N** is 0 and **V** is 1). Logically, this is N⊕V. Based on these outcomes we have:

[**A**] *Greater than* [**M**] : [**A**]−[**M**] → non-zero +ve result ($\overline{N \oplus V} \cdot \overline{Z} = 1$ or $N \oplus V + Z = 0$)
[**A**] *Equal to*   [**M**] : [**A**]−[**M**] → zero   (Z=1)
[**A**] *Less than*   [**M**] : [**A**]−[**M**] → a negative result   (N⊕V = 1)

Subtraction is a destructive test operation and Comparison is its non-destructive counterpart. It is the most powerful of the Data Testing operations, as it can be applied to both Index and Stack Pointer registers as well as 8- and 16-bit Accumulators.

All Conditional operations in the 6809 are in the form of a Branch instruction. These cause the **Program Counter** to skip *xx* places forward or backwards; usually based on the state of the **CCR** flags. Excluding BRANCH TO SUBROUTINE (see Section 5.1), there are 16 Branches provided, which can be considered as the True or False outcome of eight flag combinations. Thus BRANCH IF CARRY SET (BCS) and BRANCH IF CARRY CLEAR (BCC) are based on the one test (**C** =?).

If the test is True, the offset following the Branch op-code is added to the **Program Counter**. Thus if the **Carry flag** is zero:

```
E100:1     BCC-08      ; Coded as 24-08h
```

will add *0008h* to the **Program Counter** state E102*h* to give PC = E10A*h*. Note that the **PC** is already pointing to the following instruction when execution occurs, giving an effective destination of ten places on from the Branch location. The Branch offset is sign extended before addition to the **Program Counter**; thus if the N flag is zero:

```
E100:1     BPL-F8      ; Coded as 24-F8h
```

gives PC<−E102*h* + FFF8*h* = E0FA*h*, which is eight places back (six places back from the Branch itself). With such a single signed-byte offset, the maximum range is only +125 and −129 bytes.

Each 6809 Branch has a long equivalent which uses a double-byte offset. Thus the Conditional Branch:

```
E100:1:2:3 BCC-100F    ; Coded as 10-24-10-0Fh
```

if true forces **PC** to E104*h* + 100F*h* = F113*h*.

Long Branches can skip to anywhere in the 64 kbyte memory space, but occupy more room and take longer to execute. A normal Branch requires 3 cycles, whereas a Long Branch takes 6 cycles if carried out and 5 if not. Except for LONG BRANCH ALWAYS (LBRA), the op-code has a 10*h* byte fronting the normal Branch op-code; thus occupying four memory bytes. LBRA is exceptional, in that it has a special op-code of 16*h*, giving a 3-byte instruction always taking 5 cycles. Using a LONG BRANCH ALWAYS instead of a Jump is useful for **position independent code** (**PIC**); as by definition, the offset is relative to the **Program Counter**, the absolute

Table 2.6: Operations which affect the Program Counter.

Operation	Mnemonic	Description
Bcc **LBcc**		cc is the logical condition tested
Always (True) Never (False)	BRA; LBRA BRN; LBRN	Always affirmed regardless of flags Never carried out
Equal not Equal	BEQ; LBEQ BNE; LBNE	Z flag set (Zero result) Z flag clear (Non-zero result)
Carry Set Carry Clear	BCS; LBCS[1] BCC; LBCC[2]	[Acc] Lower Than (Carry = 1) [Acc] Higher or Same as (Carry = 0)
Lower or Same Higher Than	BLS; LBLS BHI; LBHI	[Acc] Lower or Same as (C+Z=1) [Acc] Higher Than (C+Z=0)
Minus Plus	BMI; LBMI BPL; LBPL	N flag set (Bit 7 = 1) N flag clear (Bit 7 = 0)
Overflow Set Overflow Clear	BVS; LBVS BVC; LBVC	V flag set V flag clear
Greater Than* Less Than or Equal*	BGT; LBGT BLE; LBLE	[Acc] Greater Than $(\overline{N \oplus V} \cdot \overline{Z} = 1)$ [Acc] Less Than or Equal $(N \oplus V \cdot Z = 0)$
Greater Than or Equal* Less Than*	BGE; LBGE BLT; LBLT	[Acc] Greater Than or Equal $(\overline{N \oplus V} = 1)$ [Acc] Less Than $(N \oplus V = 0)$
Jump	JMP	Absolute unconditional goto
No Operation	NOP	Only increments Program Counter

* 2's complement Branch

Note 1: Some assemblers allow the alternative BLO.
Note 2: Some assemblers allow the alternative BHS.

destination being irrelevant. This is convenient where the program is to run in ROM which may be based anywhere in memory space. A plain Jump can only be made to an absolute location, which by defination cannot be altered unless the ROM is reprogrammed.

Although Long Branches will cope with all destinations, where possible Short Branches should be used for efficiency. However, it can be difficult sometimes to predict whether a destination is within range. Some assemblers will choose for you at assembly time if advised accordingly, although they are unlikely to choose the Short Branch in all legal situations.

Table 2.7: (a) The M6809 instruction set (*continued next page*).

Instruction	Forms	Immediate Op / ~ / #	Direct Op / ~ / #	Indexed Op / ~ / #	Extended Op / ~ / #	Inherent Op / ~ / #	Description	5 H	3 N	2 Z	1 V	0 C
ABX						3A 3 1	B + X → X (Unsigned)	•	•	•	•	•
ADC	ADCA	89 2 2	99 4 2	A9 4+ 2+	B9 5 3		A + M + C → A	↕	↕	↕	↕	↕
	ADCB	C9 2 2	D9 4 2	E9 4+ 2+	F9 5 3		B + M + C → B	↕	↕	↕	↕	↕
ADD	ADDA	8B 2 2	9B 4 2	AB 4+ 2+	BB 5 3		A + M → A	↕	↕	↕	↕	↕
	ADDB	CB 2 2	DB 4 2	EB 4+ 2+	FB 5 3		B + M → B	↕	↕	↕	↕	↕
	ADDD	C3 4 3	D3 6 2	E3 6+ 2+	F3 7 3		D + M:M + 1 → D	•	↕	↕	↕	↕
AND	ANDA	84 2 2	94 4 2	A4 4+ 2+	B4 5 3		A ∧ M → A	•	↕	↕	0	•
	ANDB	C4 2 2	D4 4 2	E4 4+ 2+	F4 5 3		B ∧ M → B	•	↕	↕	0	•
	ANDCC	1C 3 2					CC ∧ IMM → CC					7
ASL	ASLA					48 2 1		8	↕	↕	↕	↕
	ASLB					58 2 1	A, B, M ← 0	8	↕	↕	↕	↕
	ASL		08 6 2	68 6+ 2+	78 7 3			8	↕	↕	↕	↕
ASR	ASRA					47 2 1		8	↕	↕	•	↕
	ASRB					57 2 1	A, B, M → C	8	↕	↕	•	↕
	ASR		07 6 2	67 6+ 2+	77 7 3			8	↕	↕	•	↕
BIT	BITA	85 2 2	95 4 2	A5 4+ 2+	B5 5 3		Bit Test A (M ∧ A)	•	↕	↕	0	•
	BITB	C5 2 2	D5 4 2	E5 4+ 2+	F5 5 3		Bit Test B (M ∧ B)	•	↕	↕	0	•
CLR	CLRA					4F 2 1	0 → A	•	0	1	0	0
	CLRB					5F 2 1	0 → B	•	0	1	0	0
	CLR		0F 6 2	6F 6+ 2+	7F 7 3		0 → M	•	0	1	0	0
CMP	CMPA	81 2 2	91 4 2	A1 4+ 2+	B1 5 3		Compare M from A	8	↕	↕	↕	↕
	CMPB	C1 2 2	D1 4 2	E1 4+ 2+	F1 5 3		Compare M from B	8	↕	↕	↕	↕
	CMPD	10 5 4 / 83	10 7 3 / 93	10 7+ 3+ / A3	10 8 4 / B3		Compare M:M + 1 from D	•	↕	↕	↕	↕
	CMPS	11 5 4 / 8C	11 7 3 / 9C	11 7+ 3+ / AC	11 8 4 / BC		Compare M:M + 1 from S	•	↕	↕	↕	↕
	CMPU	11 5 4 / 83	11 7 3 / 93	11 7+ 3+ / A3	11 8 4 / B3		Compare M:M + 1 from U	•	↕	↕	↕	↕
	CMPX	8C 4 3	9C 6 2	AC 6+ 2+	BC 7 3		Compare M:M + 1 from X	•	↕	↕	↕	↕
	CMPY	10 5 4 / 8C	10 7 3 / 9C	10 7+ 3+ / AC	10 8 4 / BC		Compare M:M + 1 from Y	•	↕	↕	↕	↕
COM	COMA					43 2 1	$\overline{A}$ → A	•	↕	↕	0	1
	COMB					53 2 1	$\overline{B}$ → B	•	↕	↕	0	1
	COM		03 6 2	63 6+ 2+	73 7 3		$\overline{M}$ → M	•	↕	↕	0	1
CWAI		3C ≥20 2					CC ∧ IMM → CC Wait for Interrupt					7
DAA						19 2 1	Decimal Adjust A	•	↕	↕	0	↕
DEC	DECA					4A 2 1	A − 1 → A	•	↕	↕	↕	•
	DECB					5A 2 1	B − 1 → B	•	↕	↕	↕	•
	DEC		0A 6 2	6A 6+ 2+	7A 7 3		M − 1 → M	•	↕	↕	↕	•
EOR	EORA	88 2 2	98 4 2	A8 4+ 2+	B8 5 3		A ⊻ M → A	•	↕	↕	0	•
	EORB	C8 2 2	D8 4 2	E8 4+ 2+	F8 5 3		B ⊻ M → B	•	↕	↕	0	•
EXG	R1, R2	1E 8 2					R1 ↔ R2[2]	•	•	•	•	•
INC	INCA					4C 2 1	A + 1 → A	•	↕	↕	↕	•
	INCB					5C 2 1	B + 1 → B	•	↕	↕	↕	•
	INC		0C 6 2	6C 6+ 2+	7C 7 3		M + 1 → M	•	↕	↕	↕	•
JMP			0E 3 2	6E 3+ 2+	7E 4 3		EA[3] → PC	•	•	•	•	•
JSR			9D 7 2	AD 7+ 2+	BD 8 3		Jump to Subroutine	•	•	•	•	•
LD	LDA	86 2 2	96 4 2	A6 4+ 2+	B6 5 3		M → A	•	↕	↕	0	•
	LDB	C6 2 2	D6 4 2	E6 4+ 2+	F6 5 3		M → B	•	↕	↕	0	•
	LDD	CC 3 3	DC 5 2	EC 5+ 2+	FC 6 3		M:M + 1 → D	•	↕	↕	0	•
	LDS	10 4 4 / CE	10 6 3 / DE	10 6+ 3+ / EE	10 7 4 / FE		M:M + 1 → S	•	↕	↕	0	•
	LDU	CE 3 3	DE 5 2	EE 5+ 2+	FE 6 3		M:M + 1 → U	•	↕	↕	0	•
	LDX	8E 3 3	9E 5 2	AE 5+ 2+	BE 6 3		M:M + 1 → X	•	↕	↕	0	•
	LDY	10 4 4 / 8E	10 6 3 / 9E	10 6+ 3+ / AE	10 7 4 / BE		M:M + 1 → Y	•	↕	↕	0	•
LEA	LEAS			32 4+ 2+			EA[3] → S	•	•	•	•	•
	LEAU			33 4+ 2+			EA[3] → U	•	•	•	•	•
	LEAX			30 4+ 2+			EA[3] → X	•	•	↕	•	•
	LEAY			31 4+ 2+			EA[3] → Y	•	•	↕	•	•

Legend:
- OP   Operation Code (Hexadecimal)
- ~   Number of MPU Cycles
- #   Number of Program Bytes
- +   Arithmetic Plus
- −   Arithmetic Minus
- •   Multiply
- $\overline{M}$ Complement of M
- →   Transfer Into
- H   Half-carry (from bit 3)
- N   Negative (sign bit)
- Z   Zero (Reset)
- V   Overflow, 2's complement
- C   Carry from ALU
- ↕   Test and set if true, cleared otherwise
- •   Not Affected
- CC   Condition Code Register
- :   Concatenation
- V   Logical or
- ∧   Logical and
- ⊻   Logical Exclusive or

Table 2.7: (b) The M6809 instruction set (*continued next page*).

Instruction	Forms	Immediate Op	~	#	Direct Op	~	#	Indexed[1] Op	~	#	Extended Op	~	#	Inherent Op	~	#	Description	5 H	3 N	2 Z	1 V	0 C
LSL	LSLA													48	2	1	$\begin{array}{c}A\\B\\M\end{array}\}$ C←b7...b0←0	•	↕	↕	↕	↕
	LSLB													58	2	1		•	↕	↕	↕	↕
	LSL				08	6	2	68	6+	2+	78	7	3					•	↕	↕	↕	↕
LSR	LSRA													44	2	1	$\begin{array}{c}A\\B\\M\end{array}\}$ 0→b7...b0→C	•	0	↕	•	↕
	LSRB													54	2	1		•	0	↕	•	↕
	LSR				04	6	2	64	6+	2+	74		3					•	0	↕	•	↕
MUL														3D	11	1	A × B → D (Unsigned)	•	•	↕	•	9
NEG	NEGA													40	2	1	$\overline{A}$ + 1 → A	8	↕	↕	↕	↕
	NEGB													50	2	1	$\overline{B}$ + 1 → B	8	↕	↕	↕	↕
	NEG				00	6	2	60	6+	2+	70	7	3				$\overline{M}$ + 1 → M	8	↕	↕	↕	↕
NOP														12	2	1	No Operation	•	•	•	•	•
OR	ORA	8A	2	2	9A	4	2	AA	4+	2+	BA	5	3				A V M → A	•	↕	↕	0	•
	ORB	CA	2	2	DA	4	2	EA	4+	2+	FA	5	3				B V M → B	•	↕	↕	0	•
	ORCC	1A	3	2													CC V IMM → CC	7				
PSH	PSHS	34	5+[4]	2													Push Registers on S Stack	•	•	•	•	•
	PSHU	36	5+[4]	2													Push Registers on U Stack	•	•	•	•	•
PUL	PULS	35	5+[4]	2													Pull Registers from S Stack	•	•	•	•	•
	PULU	37	5+[4]	2													Pull Registers from U Stack	•	•	•	•	•
ROL	ROLA													49	2	1	$\begin{array}{c}A\\B\\M\end{array}\}$ C←b7...b0	•	↕	↕	↕	↕
	ROLB													59	2	1		•	↕	↕	↕	↕
	ROL				09	6	2	69	6+	2+	79	7	3					•	↕	↕	↕	↕
ROR	RORA													46	2	1	$\begin{array}{c}A\\B\\M\end{array}\}$ C→b7...b0	•	↕	↕	•	↕
	RORB													56	2	1		•	↕	↕	•	↕
	ROR				06	6	2	66	6+	2+	76	7	3					•	↕	↕	•	↕
RTI														3B	6/15	1	Return From Interrupt					7
RTS														39	5	1	Return from Subroutine	•	•	•	•	•
SBC	SBCA	82	2	2	92	4	2	A2	4+	2+	B2	5	3				A − M − C → A	8	↕	↕	↕	↕
	SBCB	C2	2	2	D2	4	2	E2	4+	2+	F2	5	3				B − M − C → B	8	↕	↕	↕	↕
SEX														1D	2	1	Sign Extend B into A	•	↕	↕	0	•
ST	STA				97	4	2	A7	4+	2+	B7	5	3				A → M	•	↕	↕	0	•
	STB				D7	4	2	E7	4+	2+	F7	5	3				B → M	•	↕	↕	0	•
	STD				DD	5	2	ED	5+	2+	FD	6	3				D → M M+1	•	↕	↕	0	•
	STS				10 DF	6	3	10 EF	6+	3+	10 FF	7	4				S → M M+1	•	↕	↕	0	•
	STU				DF	5	2	EF	5+	2+	FF	6	3				U → M M+1	•	↕	↕	0	•
	STX				9F	5	2	AF	5+	2+	BF	6	3				X → M M+1	•	↕	↕	0	•
	STY				10 9F	6	3	10 AF	6+	3+	10 BF	7	4				Y → M M+1	•	↕	↕	0	•
SUB	SUBA	80	2	2	90	4	2	A0	4+	2+	B0	5	3				A − M → A	8	↕	↕	↕	↕
	SUBB	C0	2	2	D0	4	2	E0	4+	2+	F0	5	3				B − M → B	8	↕	↕	↕	↕
	SUBD	83	4	3	93	6	2	A3	6+	2+	B3	7	3				D − M M+1 → D	•	↕	↕	↕	↕
SWI	SWI[6]													3F	19	1	Software Interrupt 1	•	•	•	•	•
	SWI2[6]													10 3F	20	2	Software Interrupt 2	•	•	•	•	•
	SWI3[6]													11 3F	20	2	Software Interrupt 3	•	•	•	•	•
SYNC														13	≥4	1	Synchronize to Interrupt	•	•	•	•	•
TFR	R1, R2	1F	6	2													R1 → R2[2]	•	•	•	•	•
TST	TSTA													4D	2	1	Test A	•	↕	↕	0	•
	TSTB													5D	2	1	Test B	•	↕	↕	0	•
	TST				0D	6	2	6D	6+	2+	7D	7	3				Test M	•	↕	↕	0	•

Notes:
1. This column gives a base cycle and byte count. To obtain total count, add the values obtained from the INDEXED ADDRESSING MODE table.
2. R1 and R2 may be any pair of 8 bit or any pair of 16 bit registers.
   The 8 bit registers are: A, B, CC, DP
   The 16 bit registers are: X, Y, U, S, D, PC
3. EA is the effective address.
4. The PSH and PUL instructions require 5 cycles plus 1 cycle for each **byte** pushed or pulled.
5. 5(6) means: 5 cycles if branch not taken, 6 cycles if taken (Branch instructions).
6. SWI sets I and F bits. SWI2 and SWI3 do not affect I and F.
7. Conditions Codes set as a direct result of the instruction.
8. Value of half-carry flag is undefined.
9. Special Case — Carry set if b7 is SET.

Table 2.7: (c) (*continued*). The M6809 instruction set. Reproduced by courtesy of Motorola Semiconductor Products Ltd.

Indexed Addressing Mode Data

Type	Forms	Non Indirect Assembler Form	Non Indirect Postbyte OP Code	~	+ #	Indirect Assembler Form	Indirect Postbyte OP Code	+~	+#
Constant Offset From R (twos complement offset)	No Offset	,R	1RR00100	0	0	[,R]	1RR10100	3	0
	5 Bit Offset	n, R	0RRnnnnn	1	0	defaults to 8-bit			
	8 Bit Offset	n, R	1RR01000	1	1	[n, R]	1RR11000	4	1
	16 Bit Offset	n, R	1RR01001	4	2	[n, R]	1RR11001	7	2
Accumulator Offset From R (twos complement offset)	A — Register Offset	A, R	1RR00110	1	0	[A, R]	1RR10110	4	0
	B — Register Offset	B, R	1RR00101	1	0	[B, R]	1RR10101	4	0
	D — Register Offset	D, R	1RR01011	4	0	[D, R]	1RR11011	7	0
Auto Increment/Decrement R	Increment By 1	,R+	1RR00000	2	0	not allowed			
	Increment By 2	,R++	1RR00001	3	0	[,R++]	1RR10001	6	0
	Decrement By 1	,-R	1RR00010	2	0	not allowed			
	Decrement By 2	,--R	1RR00011	3	0	[,--R]	1RR10011	6	0
Constant Offset From PC (twos complement offset)	8 Bit Offset	n, PCR	1XX01100	1	1	[n, PCR]	1XX11100	4	1
	16 Bit Offset	n, PCR	1XX01101	5	2	[n, PCR]	1XX11101	8	2
Extended Indirect	16 Bit Address	—	—	—	—	[n]	10011111	5	2

R = X, Y, U or S X = 00 Y = 01
X = Don't Care   U = 10 S = 11

+~ and +# Indicate the number of additional cycles and bytes for the particular variation.

Branch Instructions

Instruction	Forms	OP	~	#	Description	5 H	3 N	2 Z	1 V	0 C
BCC	BCC	24	3	2	Branch C = 0	•	•	•	•	•
	LBCC	10 24	5(6)	4	Long Branch C = 0	•	•	•	•	•
BCS	BCS	25	3	2	Branch C = 1	•	•	•	•	•
	LBCS	10 25	5(6)	4	Long Branch C = 1	•	•	•	•	•
BEQ	BEQ	27	3	2	Branch Z = 0	•	•	•	•	•
	LBEQ	10 27	5(6)	4	Long Branch Z = 0	•	•	•	•	•
BGE	BGE	2C	3	2	Branch ≥ Zero	•	•	•	•	•
	LBGE	10 2C	5(6)	4	Long Branch ≥ Zero	•	•	•	•	•
BGT	BGT	2E	3	2	Branch > Zero	•	•	•	•	•
	LBGT	10 2E	5(6)	4	Long Branch > Zero	•	•	•	•	•
BHI	BHI	22	3	2	Branch Higher	•	•	•	•	•
	LBHI	10 22	5(6)	4	Long Branch Higher	•	•	•	•	•
BHS	BHS	24	3	2	Branch Higher or Same	•	•	•	•	•
	LBHS	10 24	5(6)	4	Long Branch Higher or Same	•	•	•	•	•
BLE	BLE	2F	3	2	Branch ≤ Zero	•	•	•	•	•
	LBLE	10 2F	5(6)	4	Long Branch ≤ Zero	•	•	•	•	•
BLO	BLO	25	3	2	Branch Lower	•	•	•	•	•
	LBLO	10 25	5(6)	4	Long Branch Lower	•	•	•	•	•
BLS	BLS	23	3	2	Branch Lower or Same	•	•	•	•	•
	LBLS	10 23	5(6)	4	Long Branch Lower or Same	•	•	•	•	•
BLT	BLT	2D	3	2	Branch < Zero	•	•	•	•	•
	LBLT	10 2D	5(6)	4	Long Branch < Zero	•	•	•	•	•
BMI	BMI	2B	3	2	Branch Minus	•	•	•	•	•
	LBMI	10 2B	5(6)	4	Long Branch Minus	•	•	•	•	•
BNE	BNE	26	3	2	Branch Z ≠ 0	•	•	•	•	•
	LBNE	10 26	5(6)	4	Long Branch Z ≠ 0	•	•	•	•	•
BPL	BPL	2A	3	2	Branch Plus	•	•	•	•	•
	LBPL	10 2A	5(6)	4	Long Branch Plus	•	•	•	•	•
BRA	BRA	20	3	2	Branch Always	•	•	•	•	•
	LBRA	16	5	3	Long Branch Always	•	•	•	•	•
BRN	BRN	21	3	2	Branch Never	•	•	•	•	•
	LBRN	10 21	5	4	Long Branch Never	•	•	•	•	•
BSR	BSR	8D	7	2	Branch to Subroutine	•	•	•	•	•
	LBSR	17	9	3	Long Branch to Subroutine	•	•	•	•	•
BVC	BVC	28	3	2	Branch V = 0	•	•	•	•	•
	LBVC	10 28	5(6)	4	Long Branch V = 0	•	•	•	•	•
BVS	BVS	29	3	2	Branch V = 1	•	•	•	•	•
	LBVS	10 29	5(6)	4	Long Branch V = 1	•	•	•	•	•

34 *The 6809 Microprocessor: Its Software*

The remaining instruction in Table 2.6 is NO OPERATION. NOP does just this, and as a consequence the fetch increments the **Program Counter**, taking 2 cycles to do it. NOPs are normally used in situations where a do-nothing delay is necessary. BRANCH NEVER (BRN) is effectively a 2-byte NOP with a 3-cycle delay and LBRN takes up 4 bytes for a 5-cycle delay.

Table 2.7 summarizes the instruction set and address modes of the 6809 family of microprocessors.

2.2 Address Modes

Virtually all instructions act on data; either outside the processor in its memory space, or in an internal register. Thus the op-code must include bits which inform the MPU's Control registers where this data is being held. There are a few exceptions to this, the so called Inherent operations, such as NOP (NO OPERATION) and RTS (RETURN FROM SUBROUTINE). Single-byte instructions whose operand is a single register, for example INCA (INCREMENT ACCUMULATOR A), are also sometimes classified as Inherent.

With the exception of Inherent instructions, the bytes following the op-code are either the (constant) operand itself, or more usually a pointer to where the operand can be found. We have already met the simplest of these, where the absolute address itself follows, as in:

```
LDA    2000h            ; [A] <- [2000] {Coded as B6-20-00h}
```

Table 2.8: Initializing a 256-byte array.

```
BEGIN: LDA   2000h ; Get array[0]
       ADDA  #30h  ; Add the constant (#) 30h
       STA   2000h ; Restore it
       LDA   2001h ; Get array[1]
       ADDA  #30h  ; Add the constant 30h
       STA   2001h ; Restore it
       LDA   2002h ; Get array[3]
        "      "   ; and so on
        "      "
        "      "
        "      "
       LDA   20FFh ; Get array[255]
       ADDA  #30h  ; Add the constant 30h
END:   STA   20FFh ; Restore it (phew!)
```

(a) Linear coding.

```
BEGIN: LDX  #2000h ; Point IX to array [0]
; While address less than 2100h add 30h to the contents of that address
LOOP:  LDA   ,X    ; Get array [IX]
       ADDA #30h   ; Add the constant 30h
       STA   ,X+   ; Put it away at [IX] and increment pointer
       CMPX #2100h ; Check for past array [256]
       BNE  LOOP   ; and repeat if not
END:
```

(b) Equivalent circular mode.

Absolute addressing is rather inflexible, as the address is fixed as part of the program, and this must be allocated by the programmer. One of the most important features of a processor is its range of **address modes**, that is different techniques for evaluating the operand address. To see why this is important, consider, say, the problem of adding the constant 30h to each element of an array of 256 data bytes stored consecutively between 2000h and 20FFh. If we had only absolute addressing, the routine would look something like the listing in Table 2.8(a), which is a pity because the same action is repeated 256 times, and takes 2048 bytes of program memory.

An alternative strategy is to use an address mode where the address is stored in a register which can be incremented, and fold our program into a **loop** as shown in Table 2.8(b). This only takes 16 bytes, less than 1% of the absolute version. Furthermore, the array can be of any length without increasing the size of the program. However, there is a penalty to pay for this flexibility. The more complex address modes take longer to execute (see Table 2.7(c) under ~), and the loop construct has the Test and Branch overhead. Thus, the absolute array program would take 3072 cycles, whilst the loop equivalent takes considerably longer at 4867 cycles to execute.

In the remainder of this section, we will look at the 6809 address modes. In this catalog, | op-code | may be one or two bytes.

Inherent

| op-code |

All the operand information is contained in the op-code, with no specific address-related bytes following. All of the 6809 inherent operations are one byte long except SOFTWARE INTERRUPT 2. An example is NOP (NO OPERATION). Motorola also classify most Register-Direct instructions as inherent, for example INCA (INCREMENT A). Table 2.7 gives the Inherent instructions.

Register Direct, $\sum$R

| op-code | post-byte |

Information concerning the source register(s) and/or destination register(s) are contained in a post-byte. For example TFR A,B (TRANSFER THE CONTENTS OF A TO B) is coded as 0001 1111 1000 1001b (1F-89h). The post-byte here is divided into two fields. The left field specifies the source register, and the right the destination. Each register is encoded as a bit in a 4-wide code. Thus 1000b is **A** and 1001b is **B**. A list of codes is given on page 20. The Transfer, Exchange, Push, and Pull operations come under this category. In Table 2.7 these are classified as Immediate.

Immediate, #kk

op-code	constant

8 bit

op-code	constant

16 bit

With Immediate addressing, the byte or bytes following the op-code are *constant* data and not a pointer to data. We have used this form of addressing before, in the array argument routine in Table 2.8. Some examples are:

```
ADDB #30h    ; Add the constant 30h to Acc. B       {Coded as CB-30h}
LDX  #2000h  ; Put the constant 2000h in X          {Coded as 8E-20-00h}
CMPY #21FFh  ; Compare [Y] with the constant 21FFh  {Coded as 10-8C-21-FFh}
```

The pound (hash) symbol # is commonly used to indicate a constant number.

Absolute, M

op-code	DP offset

Short (Direct)

op-code	Address

Long (Extended Direct)

In Absolute addressing, the address itself — either in whole or part — follows the op-code. Motorola terms the long 16-bit address version as Extended Direct. There is a short version just called Direct, where the effective address (ea) is the concatenation of the **Direct Page register** with the byte following the op-code. Thus if this register is set at, say, $80h$, then the instruction LDA 08h, coded as 96-08h, effectively brings down the byte from address $8008h$. Some assemblers have difficulty in deciding which of these forms to use. For example, in the fragment above, should the assembler generate the code B6-80-08 (LDA 8008) or **96-08** (LDA 08)? After all, the setting of the **DP register** may have been altered in a call to a subroutine yet to be linked in. There are ways around this, but none is entirely satisfactory.

Absolute Indirect, [M]

op-code	9Fh	Pointer to address

Here the op-code is followed by a post-byte 9Fh and then a 16-bit address. This is not the address of the operand but a pointer to where the operand address is stored in memory. Thus, if the locations 2000:2001h hold the address 80-08h, then the instruction:

```
LDA [2000h]      ; [A] <- [[2000:2001]]   {Coded as AF-9F-20-00h}
```

effectively fetches the data down from 2000h and then 2001h, puts them together as a 16-bit address and sends this address out on the address bus to fetch the data into **Accumulator_A**. Although the location in memory of this pointer address is absolute, the pointer residing there can be altered as the program progresses.

As an example, consider the problem of implementing a subroutine (see Chapter 5) which will process in some way the contents of an array of data. Rather than passing each element of the array to the subroutine it makes sense to send only the address or pointer to the first element. This can be done by using an absolute address, say 2000:2001h, to store the pointer prior to jumping to the subroutine. The subroutine can then use this pointer as a sort of base address to access any element of the array relative to this location.

As this indirect address is at an absolute location, this address mode is only slightly more flexible than the ordinary absolute modes. However, **indirection** can be used in conjunction with the Indexed addressing modes discussed below. As in the absolute case, the effective address is in fact only the address of a pointer to the data and not the data itself.

Branch Relative

op-code	offset

8-bit (Short)

op-code	offset

16-bit (Long)

We have already discussed this form of address mode in the previous section. Regular (or short) Branches *sign extend* the following 8-bit offset, and add this to the **Program Counter**. Effectively this means that offsets between 80h and FFh are treated as negative. For example the instruction BRA -06 is coded as 20-FAh (FAh is the 2's complement of 06h) when the **PC** is at E108h, is implemented as:

```
     1110 0001 0000 1000    (PC) = E108h
  +  1111 1111 1111 1010    (offset) = FFFAh = -6
  1̸ 1110 0001 0000 0010    (E102h, which is E108h - 0006h)
```

In calculating this offset, it must be remembered that the **PC** is already pointing to the next instruction. Thus the maximum forward point is $(00)7Fh + 2 = 127 + 2 = 129$ bytes from the op-code and $(FF)80h + 2 = -128 + 2 = 126$ bytes back. Long Branches have a 16-bit offset and can range from +32,767 and -32,768 bytes from the following op-code, effectively anywhere in the full 64 kbyte address space of memory that the processor can address at one time. Of course Long Branch code is bigger and slower to execute (see Table 2.7(c) under the column ~).

Indexed

The Absolute address modes are used where operands lie in *fixed* locations. In many cases, this places an unacceptable restriction on the data structures which can easily be processed. Compilers, for example, like to pass parameters in a stack, and these should then be capable of being retrieved in locations relative to the Stack Pointer. The 6800 MPU has a primitive form of computed effective address (ea), where this could be up to +FFh (+255) bytes from the contents of one Index register thus:

```
LDAA 8,X           ; [A] <- [X] + 8
```

means that if **X** is 8000h at the time of execution, then 8008h is the ea of the data brought down to Accumulator_A. The 6809 has an additional complement of Index registers (**X**, **Y**, **S**, **U** and sometimes the **PC**), as well as an extended repertoire of offsets. Constant offsets of up to $\pm 2^{15}$ are now possible, and Accumulator_A, _B or _D can act as a variable offset. In addition, automatic incrementation or decrementation submodes are possible. A level of indirection is also provided for most combinations. Table 2.7(c) summarizes the submodes, which are coded as an op-code followed by a post-byte. Notice that Absolute Indirect is part of this table, although strictly it is not an Indexed address mode.

Constant Offset from Register

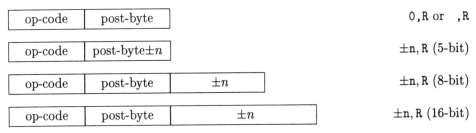

op-code	post-byte		0,R or ,R
op-code	post-byte$\pm n$		$\pm n$,R (5-bit)
op-code	post-byte	$\pm n$	$\pm n$,R (8-bit)
op-code	post-byte	$\pm n$	$\pm n$,R (16-bit)

Here the effective address is R$\pm$n where R is **X**, **Y**, **S** or **U**. The actual machine code produced depends on the size of n, with a single post-byte capable of integrally handling up to ± 15. This complex encoding scheme is worthwhile, as most offsets are small; for example, an analysis has shown that 40% of this type of indexing uses a zero offset [1]. Indirect Constant Offset Index does not have an 8-bit (± 127) offset version, the 16-bit variety being used. Fortunately the task of evaluating the post-byte and following bytes is handled automatically by the assembler.

Post-Auto-Increment / Pre-Auto-Decrement from Register

| op-code | post-byte | ,R+ / ,R++ / ,-R / ,--R

As we saw in the listing of Table 2.8(b), indexing comes into its own when stepping through blocks of memory, arrays and related structures. To avoid having to follow (or lead) the use of the Index register with an Increment or Decrement, this mode provides for automatic advance or retard; thus:

```
LDA  ,R+  ; Bring down data byte and then increment Index register R
LDA  ,-R  ; Bring down data byte and then increment Index register R twice
LDA  ,R++ ; Decrement Index register R and then bring down data byte
LDA  ,--R ; Decrement Index register R twice and then bring down data byte
```

where R is **X**, **Y**, **S** or **U**. Notice that incrementing is done *after* and decrementing *before* the Index register is used. Double Increment/Decrement modes are useful when the arrays contain addresses or other double-byte data. Indirection is only available for this double form, as by its nature addresses are likely to be being accessed.

As an example of these modes, consider the problem of multiplying two 256-byte arrays to give a 256 double-byte array. If array_1 begins at 2000h with the second array following directly, and the product array commences at 3000h, then we have:

```
      LDX  #2000h     ; Point IX to array_1[0]
      LDY  #3000h     ; Point IY to array_3[0]
LOOP: LDA  256,X      ; Get array_2[i]
      LDB  ,X+        ; Get array_1[i]; increment i
      MUL             ; Multiply them
      STD  ,Y++       ; Put it away and move on twice
      CMPX #21FFh     ; Last element yet?
      BLS  LOOP       ; IF not past it THEN repeat
      RTS             ; ELSE finished
```

Accumulator Offset from Register, A,R / B,R / D,R

op-code	post-byte

As an alternative to a constant offset, any Accumulator can hold a variable offset to an Index register, for example:

```
LDA  B,X        ;[A] <- [SEX|[B]+[X]]
LDB  A,Y        ;[B] <- [SEX|[A]+[Y]]
LDX  D,U        ;[X] <- [[D]+[U]]:[[D]+[U]+1]
```

Note that the value of the 8-bit Accumulator is *sign extended* before the addition, giving a range of +127 to −128. Thus if **B** is FEh, then FFFEh is added to the **X Index register** in the first example above to give the effective address. Of course, FFFEh is effectively −2, so the target memory location is actually **X** − 2. If this is not desirable, Accumulator_A may be cleared and **D** used as the offset, e.g.:

```
CLRA
LDA  D,X        ;[A] <- [00|[B]+[X]]
```

and this allows an offset of up to +255 (FFh) in Accumulator_B.

The use of an Accumulator allows the offset to be dynamically calculated as the program runs. A typical example is listed below, where we require access to one of a table (array) of ten elements, actually the 7-segment code. The requested element is already in the Accumulator_B (the decimal number 0 − 9), and it is to be replaced with the 7-segment equivalent code on exit. We are assuming that the subroutine starts at E200h.

```
E200/1/2    8E-E2-06              LDX #TABLE_BOT  ; Point X to table
E203/4      E6-85                 LDB B,X         ; Get element [B]
E205        39                    RTS             ; Exit

; Table of 7-segment codes begins here
E206-E20A   01-4F-12-06-4C  TABLE_BOT:  .BYTE 1,4Fh,12h,6,4Ch
E20B-E20F   24-20-0F-00-0C              .BYTE 24h,20h,0Fh,0,0Ch
```

The first instruction puts the absolute address of the first table element (E206h) in the X Index register. The effective address calculated in the following instruction is B + X. If, say, B is 04h on entry, then this gives $0004 + E206 = E20Ah$. The data in here is 4Ch, and this is the value loaded into Accumulator_B. Notice the assembler directive .BYTE, which states that the following bytes are to be put into memory verbatim; that is not to be interpreted as instruction mnemonics.

Constant Offset from Program Counter

op-code	post-byte	±n

± n,PC (8-bit)

op-code	post-byte	±n

± n,PC (16-bit)

One of the major advantages of the Relative address mode is that it produces **position independent code (PIC)**. Thus a Branch is relative to where the program is at the time the decision is taken. If the program is moved to a different part of memory, all the offsets move with it unchanged. This is what differentiates a Branch from a Jump operation. The Program Counter Offset mode extends the PIC capability to any instruction which has an Indexed address mode. This is similar to the Constant Offset from Register mode, but with the **Program Counter** being the Index register. For example in:

```
LDA     200h,PC      ;[A] <- [200+[PC]]
```

the data 200h bytes on from where the PC is on execution (pointing to the following instruction) is placed in Accumulator_A. Of course this is not an absolute address, as only the distance from the instruction is of interest. PIC is especially suitable for code in ROM (i.e. firmware) which can be placed anywhere in the address space. Thus a vendor could sell a ROM-based floating-point package with no a priori knowledge of where the customer will locate the firmware in memory.

As an example of this, consider the 7-segment decoder routine previously discussed. Line 1 of the actual code (shown second column from the left) contained the bytes E2-06h, which is the absolute location of the table bottom. If, say, the table of data was to start at C180h, then the ROM would have to be reprogrammed to make these two bytes C1-80h, the rest of the code remaining unaltered. Here is a PIC version of the same routine:

```
C102/3/4    30-8C-03              LEAX    3,PC
;Effective address PC+3 is loaded into X, which then points to the table
```

```
C105/6      E6-85                   LDB    B,X    ; Get element [B]
C107        39                      RTS

; Table of 7-segment codes begins here
C108-C10C   01-4F-12-06-4C  TABLE_BOT:  .BYTE 1,4Fh,12h,6,4Ch
C10D-C111   24-20-0F-00-0C              .BYTE 24h,20h,0Fh,0,0Ch
```

The only difference between the two programs is in line 1. In the first case, the *absolute* address of the table bottom is put into the X Index register. In the relocatable case, the X Index register is loaded with the contents of the Program Counter+3, which is again the address of the bottom of the table, but is the difference between the instruction following step 1 (i.e. at C105h) and the base of the table. If the program is bodily moved somewhere else, the offset of three bytes to the table remains the same. Thus the address of the table is calculated during each run rather than before (at load time).

As with Branch operations, assemblers save the programmer having to calculate this offset, by permitting the use of an absolute label in this type of address mode; thus assembling:

```
LEAX    TABLE_BOT,PC
```

still produces the same code 30-8C-03h; that is the label TABLE_BOT is interpreted by the assembler as the *distance* from the following instruction to the absolute address TABLE_BOT and not the absolute value C108h.

We first met the LOAD EFFECTIVE ADDRESS (LEA) instruction in Table 2.2. Here we observed that it could be used to perform simple arithmetic on the X, Y, U or S registers. Essentially, any effective address computed by any of the Direct Index address modes, except Post-Increment/Pre-Decrement, can be loaded into one of these four registers. A few examples are:

```
LEAX   +2,X  ; The EA of X+2 is put into X, effectively incrementing X by 2
LEAY    D,X  ; Adds [D] to [X] and puts sum in Y
LEAS  -20,S  ; Moves the Stack Pointer down 20 bytes
```

2.3 Example Programs

Previously we have used program fragments to illustrate various instruction/address mode combinations. Here we conclude our look at 6809 assembly-level software by developing three programs of a slightly more elaborate nature. This will serve to integrate at least some of the concepts we have discussed, and provide for a comparison with equivalent software using 68000 code in Chapter 4. Each program module is written in the form of a complete **subroutine**; that is data is assumed present on entry in some place, usually in a register, and is terminated by a RETURN FROM SUBROUTINE (RTS) instruction. Subroutine structure is the subject of Chapter 5.

Implementing a software function involves developing an appropriate algorithm, writing code in a suitable language, testing and debugging. There is little that can be done to mechanize the former, as algorithms are an expression of human creativity. Once this has been done, a range of software tools, such as assemblers, linkers, compilers and simulators, exist to aid in the production of the latter phases. We will look at these in some detail in Part 2.

The most fundamental software tool is the **assembler**. An assembler is a program that translates, on a line for line basis, symbolically-coded native language to machine code for the target processor. This saves the error-prone tedium of working out op-codes and relative offsets. Nearly as important is the use of mnemonics for instructions and names for locations (labels). These, together with the use of comments, provide superior documentation compared to strings of hexadecimal digits (see page 170).

At this point in the text, we are only concerned to provide sufficient background to allow the reader to follow program syntax as presented in the remainder of the text. Assemblers, like any other commercial package (such as a word processor), have their own peculiar rules and peccadilloes, which have to be learnt. One common denominator is the virtually unanimous use of the processor manufacturers' standard instruction mnemonics, with minor variations. Most of the variations lie in the layout of the source code and the directives (or pseudo operators) used to pass information from the programmer to the assembler.

A line of source code comprises four fields: an optional label, the essential instruction mnemonic, the operand (if any) and an optional comment. Some assemblers require all fields to be present in spirit, their absence being signalled by spaces or tabs. The Real Time Systems XA8 cross assembler[1] used here has a free format, where absent fields can simply be omitted. The only essential role of space is in separating the instruction mnemonic from its operand. However, as the following code fragment shows, spaces and tabs should be used for readability:

```
BCC NEXT;IF no Carry THEN don't add one to int X
ADDD #1
NEXT:RTS;and return
```

or

```
        BCC     NEXT        ; IF no Carry THEN don't add one to int X
        ADDD    #1
NEXT:   RTS                 ; and return
```

The latter source code is obviously more pleasing to the eye. Notice that lines 1 and 2 have no label, line 2 no comment and line 3 no operand field.

Looking at the syntax in more detail.

---

[1] Real Time Systems, M & G House, Head Road, Douglas, Isle of Man, British Isles; Intermetrics Microsystems Software Inc., 733 Concord Avenue, Cambridge MA 02138, USA.; Whitesmiths Australia Pty Ltd. PO Box 756, Suite 3, 47 Regent Street, Kogarah NSW 2217, Australia; COSMIC SARL, 33 rue Le Corbusier, EUROPARC CRETEIL, 94035 CRETEIL CEDEX, France and ADaC, Nihon Seimei Otsuka Bldg., No. 13-4 Kita Otsuka 1-chome, Toshima-Ku, Tokyo 170 Japan.

Labels

These are defined in the first field and should be delineated by a colon. The colon is omitted when the label is referred to in the operand field. The label takes on the value of the **Program Counter** pointing to the first instruction byte. Labels can be up to 15 meaningful alphanumeric (including _ and .) characters long, and should not start with a numeral.

Operator mnemonics

These are the standard manufacturer's mnemonics, with a few minor extensions. There *must* be an entry in this field.

Operand

These may be a label, defined name, address or data constant. Numbers may be in decimal, hexadecimal, octal, binary or ASCII. Thus the following all translate to the same:

```
LDA #43h         ; Codes as 86-43h. Use a 0 prefix if MSD is alpha, e.g. 0F6h
LDA #67          ; Codes as 86-43h. Decimal 67 is 43 hex
LDA #01000011b   ; Codes as 86-43h. Binary 01000011 is 43 hex
LDA #103o        ; Codes as 86-43h. Octal 103 is 43 hex
LDA #'C'         ; Codes as 86-43h. ASCII 'C' is 43 hex
```

but the use of the appropriate form aids in readability and thus documentation.

Mathematical expressions can be used to generate a constant at assembly time, thus:

```
LDA    MSD-1          ; Get data from address MSD less one
LDA    ARRAY+(i*5)+j  ; Get data from address ARRAY plus
                      ; i rows of 5 and j columns
BRA    .+3            ; Branch forward 3 places
```

Comment

The final field is simply a documentation comment, delimited by a semicolon ;. Whole-line comments are possible with an initial semicolon. Some assemblers use an asterisk * to delimit comments.

Some of the more common assembler directives, all of which are distinguished by a leading period, are:

.PROCESSOR

The first line of source code must indicate which processor is being targeted, e.g.:

```
.processor  m6809
```

for the 6809 MPU.

.END

The last line of source code must be .end.

.DEFINE

This gives a permanent value to a symbol. For example:

```
.DEFINE    ERROR    = 0FFh,
           TRUE     = 01,
           FALSE    = 0,
           PIA_BASE = 8080h
           --------------------
           --------------------
CMPA       #ERROR
BEQ        ABORT
CMPA       #FALSE
BEQ        REPEAT
CMPA       #TRUE
BNE        ABORT
LDB        PIA_BASE+2
           --------------------
           --------------------
```

This mechanism is useful in assigning names to absolute locations, such as those associated with hardware interface ports, and to constants which have a readily identifiable meaning. Placing definitions at the start of the source program means that such constant data and addresses can be altered throughout the source file by simply altering this **header**. The mnemonic EQU (EQUATE) is frequently used in other assemblers to perform the same function; see page 183.

.INCLUDE

Source code in separate files can be included for assembly by using this directive, for example:

```
.INCLUDE "stdio.h"   ; Insert the I/O header file at this point
```

.PSECT

A useful feature of this assembler is the ability to delineate sections of the source program to produce code in different memory areas. Thus program code and fixed constants can be assigned to area _text which the linker can place in memory occupied by ROM, whilst section _data can be used for variable data destined for RAM. An example of the use of .psect is given in Table 2.12.

.ORG

The assembler used here is configured to be relocatable, that is absolute addresses are not assigned until link time (see Section 7.2). The .ORG function is normally used in an absolute assembler (one in which absolute locations are assigned at assembly time) to denote where the code commences. In the RTS assembler .ORG can be used in a relocatable manner relative to a label, for example:

```
          .PSECT  _text               ; Program code
START:    LDA     MEMORY              ; Program start (e.g. 0E000h)
          --------------------
          --------------------
RVECTOR:  .ORG    START+1FFEh         ; Move on from start (e.g. 0FFFEh)
          .WORD   START               ; Put in Reset vector
          .end
```

Assuming that the section _text is linked to $0E000h$, then the code at RVECTOR is commanded to be placed in $0E000h + 1FFEh = 0FFFEh$.

.BYTE, .WORD, .DOUBLE, .TEXT

In the code fragment above, the assembler is commanded to place the double-byte constant $E0$-$00h$ in at RVECTOR:RVECTOR+1, using the .WORD directive. The directives .BYTE and .DOUBLE are similar, but allocate storage of 8 and 32 bits respectively. .TEXT allows series of bytes, entered as strings within quotes, to be stored in a similar manner. Other assemblers use FCB (FORM CONSTANT BYTE), RMB (RESERVE MEMORY BYTE), FDB (FORM DOUBLE BYTE), FCC (FORM CONSTANT CHARACTER) as equivalent directives.

We have already seen an example of .BYTE when we designed the 7-segment decoder subroutine on page 40. A simple example of .TEXT is:

```
.TEXT  "This is an example", 0
```

which is considerably more convenient than the equivalent:

```
.BYTE  54h,68h,69h,73h,20h,69h,73h,20h,61h,6Eh
.BYTE  20h,65h,78h,61h,60h,70h,6Ch,65h,0
```

Statements such as this have to be used with caution where the program is blasted into ROM. Constants can be located in ROM (e.g. .psect _text). but not in RAM (e.g. .psect _data). This is because there is no download of code prior to the run, and volatile memory is unpredictable on power up. Care must be taken when using a simulator to debug such programs, as this data is downloaded into RAM from the assembled machine code file and will then appear to be available at start-up.

Our first program generates the sum of all integers n up to a maximum of 255 (FFh). We assume that n is passed to the subroutine in Accumulator_B. The maximum possible total of 32,640 can comfortably fit into the 16-bit Accumulator_D for return.

Table 2.9: Source code for sum of n integers program.

```
        .processor m6809
; ***************************************************************
; * FUNCTION : Sums all unsigned byte numbers up to n            *
; * ENTRY    : n is passed in Accumulator B                      *
; * EXIT     : Sum is returned in D Accumulator                  *
; * EXIT     : Index X = sum                                     *
; ***************************************************************
;
        .psect _text        ; Direct code into text area
; for (sum=0;n>0;n--){
        ldx    #0           ; Sum = 0000
SLOOP:  tstb                ; n > 00?
        beq    SEND         ; IF not THEN end
        abx                 ; ELSE sum = sum + n
        decb                ; n--
        bra    SLOOP        ; }
SEND:   tfr    x,d          ; Put sum in D Accumulator as asked
S_EXIT: rts                 ; for return
        .end
```

The algorithm used in Table 2.9 simply clears the initial total, temporarily located in the X Index register, and adds to it the progressively decrementing integer, kept in Accumulator_B. When B reaches zero, the grand total is transferred to Accumulator_D for return. The instruction ADD B TO X (ABX) is a convenient vehicle to add the 8-bit integer to the 16-bit partial summation. Without it, n would have to be unsigned promoted to 16-bits by zeroing Accumulator_A and then the instruction LEAX D,X used for the addition.

The **source-code file** is translated by the assembler program to produce a **machine-code file**, which will eventually find its way into program memory. An absolute **listing file** is also generated, which documents the machine code and its location together with the original source code. The listing of Table 2.10 shows the outcome of the translation, with the line number, location and machine code occupying the leftmost three columns. This type of file is often referred to as object code. The absolute location of the machine code is decided by the linker-locator program, as described in Section 7.2. All 6809-based programs in this text assume ROM from E000h upwards for the program sections designated _text, and RAM from 0000h upwards for the _data sections. Only _text is needed in this case.

The program of Table 2.10 is 12 bytes long and takes $16 + 13n$ cycles (maximum 3331). An alternative algorithm recognizes that the total is given by the expression $n \times (n+1) \div 2$. In Table 2.11 this is implemented by copying n into Accumulator_A, incrementing it, multiplying the two Accumulators and doing a sin-

Table 2.10: Object code generated from Table 2.9.

```
1                         .processor m6809
2    ; ***************************************************************
3    ; * FUNCTION : Sums all unsigned byte numbers up to n            *
4    ; * ENTRY    : n is passed in Accumulator B                      *
5    ; * EXIT     : Sum is returned in D Accumulator                  *
6    ; * EXIT     : Index X = sum                                     *
7    ; ***************************************************************
8    ;
9                         .psect _text  ; Direct code into text area
10   ; for (sum=0;n>0;n--){
11 E000  8E0000  SUM_OF_N: ldx    #0      ; Sum = 0000
12 E003  5D      SLOOP:    tstb           ; n > 00?
13 E004  2704              beq    SEND    ; IF not THEN end
14 E006  3A                abx            ; ELSE sum = sum+n
15 E007  5A                decb           ; n--
16 E008  20F9              bra    SLOOP   ; }
17 E00A  1F10    SEND:     tfr    x,d     ; Put sum in D Accumulator as asked
18 E00C  39      S_EXIT:   rts            ; for return
19                         .end
```

gle double-byte shift right (i.e. ÷ 2). Only six bytes long and executing in a fixed 28 cycles, this illustrates that time taken in refining the problem algorithm can be profitable. However, there is a bug in this implementation, with one value of n giving an erroneous zero answer. Can you determine which, and recode to avoid this problem?

Table 2.11: A superior implementation.

```
1                         .processor m6809
2    ; ***************************************************************
3    ; * FUNCTION : Sums all unsigned byte numbers up to n            *
4    ; * ENTRY    : n is passed in Accumulator B                      *
5    ; * EXIT     : Sum is returned in D Accumulator                  *
6    ; * EXIT     : No other registers disturbed                      *
7    ; ***************************************************************
8                         .psect _text  ; Direct code into text area
9    ;
10   ; sum = n*(n+1)/2
11 E000  1F98    SUM_OF_N: tfr    b,a    ; Copy n into Acc.A
12 E002  4C                inca          ; which becomes n+1
13 E003  3D                mul           ; n*(n+1) now in Acc.D
14 E004  44                lsra          ; Divide by two
15 E005  56                rorb          ; by shifting right once
16 E006  39      S_EXIT:   rts           ; for return
17                         .end
```

Our second program is more elaborate. We are required to convert a 16-bit binary word to a string of ASCII-coded decimal digits, terminated with 00h (ASCII NULL). The more usual mathematical conversion algorithm requires that the base-M number be continually divided by ten, the series of remainder digits being the base-10 equivalent (see the listing of Table 4.14). Implementing this requires a lengthy division/remainder subroutine. If this is already present for use by another program module, the resulting code will be acceptably short. In any case, in the absence of a hardware divide operation in the 6809, execution time is likely to be long.

An alternative algorithm, which is especially suitable for small numbers, is illustrated in Fig. 2.4. Essentially the nth-decade digit is evaluated as the number of successful subtractions by 10^n, where n begins at the highest possible value, and is decremented towards zero after each decade evaluation. As the maximum value for a 16-bit binary number is 65,535, this requires subtraction by 10,000, 1000, 100, 10 and 1. With the procedure being the same for each decade, it is easier to store the constants as a table in ROM and use a loop with an advancing pointer to select the decade and its corresponding table entry. This look-up table is shown in the listing of Table 2.12 in line 43. Notice the additional zero word at the end of the table; this is used to provide an escape mechanism after the decade passes 10^0.

The actual subtraction of 10^n is performed in line 23, with the **X Index register** pointing into the table of powers. If no borrow is generated ($\mathbf{C} = 0$), the byte holding the nth string character (initialized to ASCII 0 = 30h in lines 18–21) is incremented and the process repeated (lines 25–28). On emerging from this inner (decade) loop, the 10^n constant is added back to compensate for the one subtraction too many. As line 30 uses the Double-Increment Index address mode (ADDD ,X++), the table pointer is simultaneously advanced one word. LEAY 1,Y then increments the string pointer (the **Y Index register**) one byte, and the scene is set for the next decade evaluation. Before returning to the top of this outer loop, the escape condition (i.e. NULL) must be tested. There is no instruction to test the zero state of a double memory location; instead an unused double register is loaded with the word data (LDU 0,X in line 34) and the **Z flag** will be set accordingly. An alternative escape procedure would be to decrement a count on each loop pass or simply to check the table pointer for 0E030h (e.g. CMPX #PWR_10+10). Using a special terminate character is better where the length of the table can vary, and is the normal approach to character strings, as is specified in this example (line 38).

None of the MPU's registers are altered by this subroutine, except the **Code Condition register**. A subroutine with this property is known as **transparent**. This is achieved by pushing the used registers onto the System stack at the beginning (line 13) and restoring them at the end (line 40). In general the number of Push and Pull operations should match to ensure that the **System Stack Pointer** is back up to the return **Program Counter**, which was shoved out automatically when the subroutine was called. Thus RETURN FROM SUBROUTINE (RTS) will then be able to retrieve the original **PC** as required. One trick sometimes seen is to add the **PC** to the last PULS, which of course does the same thing; thus:

Table 2.12: Object code for the conversion of 16-bit binary to an equivalent ASCII-coded decimal string.

```
1                              .processor m6809
2    ; *********************************************************************
3    ; * Converts 16-bit binary to a string of five ASCII-coded         *
4    ; * characters terminated by 00 (NULL)                             *
5    ; * EXAMPLE : FFFF -> '6''5''5''3''5''0'  (36/35/35/33/35/00h)     *
6    ; * ENTRY   : Binary word in D                                     *
7    ; * EXIT    : Decimal string in 6 RAM bytes starting from DEC_STRG *
8    ; * EXIT    : All register contents unchanged                      *
9    ; *********************************************************************
10                             .list     +.text
11                             .define   NUL = 0000
12                             .psect    _text
13   E000 3476      BIN_2_DEC: pshs a,b,x,y,u ; Save pointer registers used
14   ; N=4
15   E002 308C21              leax PWR_10,pc ; Point to table bottom (10^4)
16   E005 108E0000            ldy  #DEC_STRG ; Point to beginning of string in RAM
17   ; Nth decade = '0'
18   E009 1F03       NEW_N:   tfr  d,u      ; Put away binary for safekeeping
19   E00B 8630                lda  #'0'     ; Put ASCII '0' in nth decade of string
20   E00D A7A4                sta  0,y
21   E00F 1F30                tfr  u,d      ; Get binary back
22   ; Binary - 10**N
23   E011 A384     NEXT_SUBT: subd 0,x
24   ; Can do?
25   E013 2504                bcs  NEXT_DEC ; A Carry/borrow means No
26   ; IF Yes THEN increment Nth decade
27   E015 6CA4                inc  0,y
28   E017 20F8                bra  NEXT_SUBT
29   ; ELSE restore 10**N to binary
30   E019 E381     NEXT_DEC:  addd ,x++
31   ; N = N -1
32   E01B 3121                leay 1,y      ; Advance one decade
33   ; N < 0?
34   E01D EE84                ldu  0,x      ; Look for double-byte NULL in table
35   ; No
36   E01F 26E8                bne  NEW_N
37   ; Yes
38   E021 6FA4                clr  0,y      ; IF Yes terminate the string
39   ; End
40   E023 3576                puls a,b,x,y,u ; Return old register values
41   E025 39                  rts           ; Omit if above is puls a,b,x,y,u,pc!
     ; ***************************************************************
42   ; This is the table of powers of 10
43   E026 2710     PWR_10:    .word 10000,1000,100,10,1,NUL
          03E8
          0064
          000A
          0001
          0000
     ; ***************************************************************
44   ; This is the area of RAM where the number string is to be returned
45                            .psect   _data
46   0000          DEC_STRG:  .byte    [6]   ; Reserve six memory bytes for string
47                            .end
```

50 The 6809 Microprocessor: Its Software

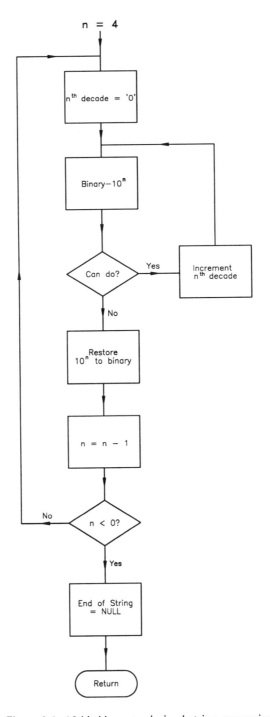

Figure 2.4: 16-bit binary to decimal string conversion.

```
PULS A,B,X,Y,U,PC
```

is the same as

```
PULS A,B,X,Y,U
RTS
```

The two pointers, **X** to the table and **Y** to the string, are set up just after the initial Push. The table pointer is set up in line 15 using the Program Counter Relative address mode, `LEAX PWR_10,PC`. Looking at the machine code produced (namely 30-8C-21h), shows an operand of 21h, being the distance between the **PC** (pointing at execution time to the following instruction at 0E005h) and the start of the table at 0E026h. This relative operand ensures that no matter where the program/table ROM is placed in address space, the code need not be altered. This code is strictly speaking not position independent, as the string is in a fixed location in the _data program section, that is in RAM. If DEC_STRG is the first occurrence of .psect _data, then our linker will place the string at locations 0000h to 0005h. Thus the code in line 16 for `LDY #DEC_STRG` is 108E-0000h. We could use the Program Counter Relative mode here (i.e. `LEAY DEC_STRG,PC`) but this would mean that the address distance between the ROM and RAM chips would have to remain constant, and they could not be independently relocated: not very convenient.

Our last example also has a mathematical flavor. We are required to calculate the factorial of an integer n passed in **Accumulator_B**. The factorial of n (represented as $n!$) is defined as $n \times (n-1) \times (n-2) \times \cdots \times 3 \times 2 \times 1$. By convention 0! is defined as 1 [5].

Superficially this appears to be the same as our first example, but with multiplication replacing addition, see Fig. 2.5. However, the product rapidly becomes very large, with $12! = 479,001,600$ being the largest factorial fitting into a 32-bit binary number. Thus we will restrict n to the range 0–12, and will have to return $n!$ in four memory bytes, as no 6809 register of this size is available (although it could be returned in two pieces using, for example, the **X** and **Y Index registers**). Furthermore, we will use **Accumulator_B** to return an error status byte of **FF**h if the programmer sent an out of range integer ($n > 12$), otherwise 00h indicating success.

Our first problem is that product generation is a 4-byte long-word process, whilst the 6809 can only perform an 8×8 multiplication. Thus our requirement for an 8×32 product will have to be met by four 8×8 operations. Hence we will require four memory bytes to hold the product (after each total multiplication) and at least four memory bytes to act as a temporary workspace, where the four multiplications will be summed as they happen.

The initial value of the product is set to 0001h in lines 21–25 of the listing in Table 2.13, and the 7-byte temporary workspace cleared in lines 30–33. The actual

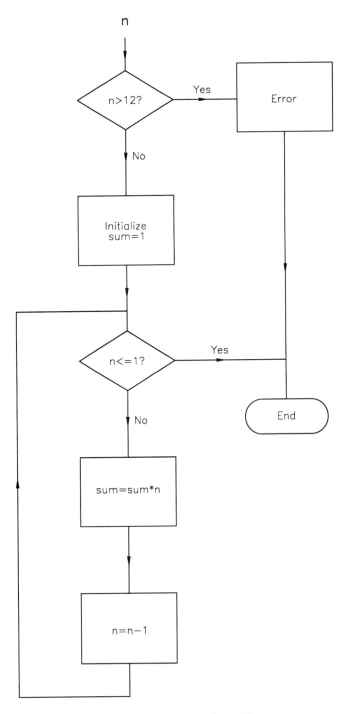

Figure 2.5: Evaluating factorial n.

Table 2.13: Fundamental factorial-n code.

```
 1                          .processor m6809
 2   ;************************************************************
 3   ; * Subroutine calculates the factorial of n (n!)            *
 4   ; * EXAMPLE : n = 12; n! = 479,001,600                       *
 5   ; * ENTRY   : n in Acc.B; maximum value 12                   *
 6   ; * EXIT    : n! in 4 bytes PROD -> PROD+3                   *
 7   ; * EXIT    : Acc.B = -1 (FFh) if error (n>12) ELSE 00       *
 8   ;************************************************************
 9                          .define  ERROR = -1
10   ; Initialize
11                          .psect   _text
12 E000 3412    FACTORIAL:  pshs   a,x        ; Save these registers
13 E002 3404                pshs   b          ; Put n away for safekeeping
14   ; Error condition
15 E004 C10C                cmpb   #12        ; IF >12 THEN an error condition
16 E006 2306                bls    CONTINUE   ; ELSE continue
17 E008 C6FF                ldb    #ERROR     ; Put FFh in B to signal error
18 E00A E7E4                stb    0,s        ; and into where it is in the stack
19 E00C 204B                bra    FEXIT      ; and exit with it
20   ;
21 E00E 7F0003 CONTINUE:    clr    PROD+3     ; Initialize product to 0001h
22 E011 7C0003              inc    PROD+3
23 E014 7F0002              clr    PROD+2
24 E017 7F0001              clr    PROD+1
25 E01A 7F0000              clr    PROD
26   ; N <=1?
27 E01D E6E4    OUTER_LOOP:ldb    0,s        ; Get factor n (or residue) back
28 E01F C101                cmpb   #1
29 E021 2336                bls    FEXIT      ; IF <=1 then answer is in PROD
30 E023 8E0004              ldx    #TEMP      ; Now clear temporary product area
31 E026 6F80    CLOOP:      clr    ,x+        ; all five 7 bytes
32 E028 8C000B              cmpx   #TEMP+7
33 E02B 26F9                bne    CLOOP
34   ; Now begin the multiple multiplication (PROD = PROD*n)
35 E02D 8E0004              ldx    #PROD+4    ; Point to just past LSB product
36 E030 E6E4    MUL_LOOP:   ldb    0,s        ; Get residue of n from stack
37 E032 A682                lda    ,-x        ; and ith byte of product, i++
38 E034 3D                  mul               ; [D] holds the product
39 E035 E306                addd   6,x        ; Add it to temporary product
40 E037 ED06                std    6,x        ; 6,x points into temp product area
41 E039 A605                lda    5,x        ; Now add any carry to the third byte
42 E03B 8900                adca   #0
43 E03D A705                sta    5,x
44 E03F A604                lda    4,x        ; and to the next higher byte
45 E041 8900                adca   #0
46 E043 A704                sta    4,x
47 E045 8CFFFF              cmpx   #PROD-1    ; i over the MSB product
48 E048 26E6                bne    MUL_LOOP   ; IF not then again once left
49   ;
50 E04A 3001                leax   1,x        ; Increment pointer to MSD product
51 E04C A607    MOVE_LOOP:  lda    7,x        ; Moving temporary product bytes
52 E04E A780                sta    ,x+        ; which is the new product
53 E050 8C0004              cmpx   #PROD+4    ; to its rightful place
54 E053 26F7                bne    MOVE_LOOP
55   ; n=n-1
56 E055 6AE4                dec    0,s
57 E057 20C4                bra    OUTER_LOOP
58 E059 6FE4    FEXIT:      clr    0,s        ; Zero (no error), to B in stack
59   ; End
60 E05B 3504    ERR_EXIT:   puls   b          ; Gets error condition from stack
61 E05D 3512                puls   a,x        ; Retrieve used registers
62 E05F 39                  rts               ; n! is in the four PROD locations
63   ;
64                          .psect  _data     ; Define the data area
65 0000        PROD:         .byte [4]        ; The area holding the product
66 0004        TEMP:         .byte [7]        ; The temporary product area
67                          .end
```

54   *The 6809 Microprocessor: Its Software*

4-stage multiplication of the partial product to the integer byte n takes place in the following lines 35–48. This is shown diagrammatically at the right of Fig. 2.6, from where it can be seen that each process is similar, but with the addition shifted left once each move towards the MSB of PROD. Thus the word $n \times$ [PROD+3] is added to TEMP+5:TEMP+6, with any carries up to TEMP+3 (no more, as we know the result will never exceed four bytes). The second product of $n \times$ [PROD+2] is added to TEMP+4:TEMP+5, with any carry to TEMP+3. The word $n \times$ [PROD+1] is summed to TEMP+3:TEMP+4, whilst the same 4-byte restriction means that only the lower-byte of the final product $n \times$ [PROD] (i.e. [**B**]) need be added to the temporary store.

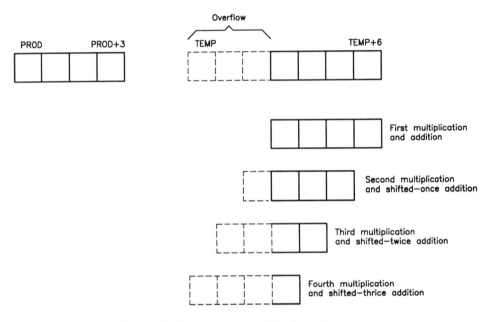

Figure 2.6: A memory map of the factorial process.

From the above discussion, we see that the addition process is different at each position, as the 16-bit result from the multiplication 'slides' from right to left. This is a pity, as otherwise the four multiply/add steps are the same. This inefficiency can be circumvented by allowing three buffer temporary bytes, as shown dashed at the top of Fig. 2.6. This allows us to put the multiply/add process in a loop ensuring that no other data object is inadvertently altered by the slide leftward. In lines 36–48 of this loop, the **X Index register** is used to point to both the relevant product byte (line 37) and, with offset, to the temporary addition target bytes (lines 39–46). When the multiplication is over, the result becomes the new product (lines 51–56). n is decremented in situ on the System stack, using the **System Stack Pointer** as an Index register (line 56), and the process continued until $n = 1$ (line 28). On exit Accumulator_B is cleared to indicate success (line 58), unless n is >12 on entry, in which case FFh is put into **B** (line 17), and an immediate exit made. Notice how four bytes for the product and seven temporary locations are reserved in the data

Table 2.14: Factorial using a look-up table.

```
1                          .processor m6809
2     ; **********************************************************************
3     ; * Subroutine calculates the factorial of n (n!)                      *
4     ; * EXAMPLE : n = 12; n! = 479,001,600                                 *
5     ; * ENTRY   : n in Acc.B; maximum value 12                             *
6     ; * EXIT    : n! in 4 bytes PROD -> PROD+3                             *
7     ; * EXIT    : Acc.B = -1 (FFh) if error (n>12) ELSE 00                 *
8     ; **********************************************************************
9                          .list +.text
10                         .define ERROR = -1
11    ; Initialize
12                         .psect _text
13 E000 3430     FACTORIAL: pshs   x,y          ; Save registers
14    ; Error condition
15 E002 C10C              cmpb    #12          ; IF >12 THEN an error condition
16 E004 2304              bls     CONTINUE     ; ELSE continue
17 E006 C6FF              ldb     #ERROR       ; Put FFh in B to signal error
18 E008 2016              bra     FEXIT        ; and exit with it
19    ; Get factorial out of table
20 E00A 58       CONTINUE: lslb                ; Multiply n by four
21 E00B 58                lslb                 ; as table is 4-wide
22 E00C 8EE023            ldx     #TABLE       ; Point to bottom of table
23 E00F 3085              leax    b,x          ; Point to relevant table entry
24 E011 10AE81            ldy     ,x++         ; Get top two bytes, and advance pointer
25 E014 10BF0000          sty     PROD         ; and put away
26 E018 10AE84            ldy     0,x          ; Get lower two bytes
27 E01B 10BF0002          sty     PROD+2       ; and put these away
28 E01F 5F                clrb                 ; Signal no error state
29 E020 3530     FEXIT:   puls    x,y          ; Retrieve used registers
30 E022 39                rts                  ; n! is in the four PROD locations
31    ;
32    ; Now the table which is in the text (ROM) area
33 E023 00000001 TABLE:   .double 1,1,2,6,24,120,720,5040
        00000001
        00000002
        00000006
        00000018
        00000078
        000002D0
        000013B0
34 E043 0000              .word   0,9d80h,5,8980h,37h,5f00h,261h,1500h,1c8ch,0fc00h
        9D80
        0005
        8980
        0037
        5F00
        0261
        1500
        1C8C
        FC00
35    ;
36                        .psect _data      ; Define the data area
37 0000         PROD:     .byte   [4]       ; The area holding the product
38                        .end
```

program section (RAM) in lines 65 and 66.

As there are only 13 legitimate outcomes of the program for $n = 0 \to 12$, a more efficient technique is to use a **look-up table**. The coding for this approach is shown in Table 2.14. Basically, the X Index register is pointed to the bottom of TABLE (line 22) and n (stored in Accumulator_B) is used as an offset to point into the relevant area. As each table entry occupies four bytes, B must be multiplied by four (by shifting twice left in lines 20 and 21), so that it goes up in 4-byte steps. The operation LOAD EFFECTIVE ADDRESS INTO X with the address mode B,X points X to the entry in line 23 (the maximum value of B is 48, thus its sign extension inherent with this address mode will have no deleterious effect). Now the high word can be moved from the table to 2 bytes of memory via Index register_Y (lines 24 and 25). As the Indexed with Post Double Increment address mode is used, X will automatically point to the lower word, for a repeat performance (lines 26 and 27).

The coding shows the assembler directive .DOUBLE being used for the first eight table entries and .WORD twice for each of the remaining entries. This is deliberate, as the assembler used here has a bug which gives incorrect values for .DOUBLE above 32,767 (00007FFFh). Assemblers, as all other software, are not immune to bugs! See Table 4.14 for a look-up table using .DOUBLE for this situation.

It is interesting to compare the performance of the two implementations. The former mathematical algorithm requires 96 bytes of ROM and 11 of RAM. Its operation time varies with n, from 53 cycles with $n = 0$ or 1 to 1724 cycles with $n = 12$. The tabular approach takes 84 bytes of ROM and 4 of RAM, and takes a fixed 42 cycles for n between 0 and 12. In both cases an error situation requires 30 cycles. The conclusion is obvious.

References

[1] Ritter, T and Boney, J.; Preliminary Detailed Description MC6809, *Motorola Bulletin 055*, March 1978.

[2] Ritter, T and Boney, J.; A Microprocessor for the Revolution: The 6809, *BYTE*, **4**, part 1, no. 1, Jan. 1979, pp. 14–42.

[3] Bartee, T.C.; *Digital Computer Fundamentals*, 5th ed., McGraw-Hill, 1981, Section 6.16.

[4] Cahill, S.J.; *The Single-Chip Microcomputer*, Prentice-Hall, 1987, Section 1.2.

[5] Dorn, W.S. and McCracken, D.D.; *Introductory Finite Mathematics with Computing*, Wiley, 1976, Section 8.3.

CHAPTER 3

The 68000/8 Microprocessor: Its Hardware

At its inception, the microprocessor was perceived as a replacement for many applications then implemented by standard logic circuitry. The considerable enhancement of facilities offered by second generation MPUs led to their use as the engine of a number of simple general purpose computers, such as the APPLE II. Whilst these were initially targeted at the home and education markets, the evolution of affordable magnetic disk technology quickly created an explosive growth in their use in the business and scientific communities.

The large potential market thus opened up was the impetus in the development of a new generation of more powerful MPUs. Although, as we have seen, there was some movement in that direction by 8-bit devices, in the main the opportunity was taken to expand the internal architecture to use 16 and 32-bit registers and ALUs. As well as increasing the data throughput, especially where floating-point computation is being used, this makes it easier to support larger external buses. Along with enhanced power, a larger memory space and support for data structures targeted to high-level languages were provided. Typically an increase in execution speed of around ten was achieved by this strategy.

The Intel 8086/8088 16-bit MPU released in 1978 was designed to be compatible with the older 8-bit 8080/8085 MPUs, and as such perpetuated many of their limitations. Internal registers were dedicated, rather than general purpose, and the address range of 1 Mbyte was fractured into 64 kbyte segments. Later members of the family increased the register sizes and address capacity, with the 32-bit 80386/486 being able to address 2^{32} bytes. This family was popularized by their use in the IBM series of personal computers.

First devices from the Motorola 68000 family were released in 1979 [1, 2, 3]. In contrast these took the chance of breaking completely with the past generation. The 68000 MPU offered a 32-bit register structure from the beginning, although the 16-bit data bus and ALU really marks it as 16-bit with 32-bit pretensions. A non-multiplexed address bus with effectively 24 lines gives a 16 Mbyte directly addressable memory capacity. This was later extended to 32 lines in the 68020/30/40 devices, giving a potential 4 Gbyte memory size. All the 8086 family, as well as the 68020 up, provide the capability to easily ride tandem with a floating-point hardware co-processor, which considerably extends their capabilities in mathematics intensive computing, such as computer-aided design graphics. In general the 68000 family is found in the more powerful personal computers, such as the APPLE Macintosh, as well as graphic workstations such as the Hewlett Packard Apollo DN series.

All this growth in raw power has made the microcomputer at least as powerful as a minicomputer from the last decade, but there has also been a spin-off into the area of embedded microprocessor circuitry, with which we are concerned in this text. Although the current 8-bit microprocessors are adequate in the majority of embedded applications, either singly on in multiple-processor configurations, many of the more powerful tasks are being implemented using these newer devices. This is not necessarily due to their virtues, but because more aids to hardware and software design, which have appeared in the last decade, have been targeted in this direction. This is especially true in the field of compiler and simulator work.

The 68000/8 is the second of our MPUs we have chosen to illustrate high-level language techniques. This and the following chapter overviews its hardware and software features.

3.1 Inside the 68000/8

Here we look at the register model of the 68000 and 68008 MPUs. A highly simplified equivalent circuit of the former is shown in Fig. 3.1. Following the classification developed in Section 1.1, we will discuss the internal attributes of the device in terms of the mill, the register array and the control unit.

THE MILL

A 16-bit ALU implements in hardware the arithmetic operations of Addition, Subtraction, Multiplication and Division; the former with and without carry/borrow and the latter in signed and unsigned representations. The logic operations of AND, OR, Exclusive-OR, NOT and Shift are also provided.

Five flags in the associated **Code Condition register** (**CCR**) provide a status report on ALU activity. The Carry, Negative, Zero and 2's complement oVerflow semaphores are standard, but the eXtend flag needs some explanation. The **X flag** is similar to Carry, but is only affected by Addition, Subtraction, Negate and certain Shift operations. Multiple-precision versions of these instructions use the **X flag** for their carry; thus the familiar ADD WITH CARRY (ADC) instruction appears here as ADD WITH EXTEND (ADDX). For example, this means that a Compare operation, which of course affects the C flag, can be done in-between multiple-precision operations without affecting the 'true' carry information (which is in **X**).

We shall see that the 68000 MPU directly operates on byte (8-bit), word (16-bit) or long-word (32-bit) data. All **CCR** flags operate correctly (e.g. Carry from bit 7, bit 15 or bit 31 respectively), automatically reflecting the operand size.

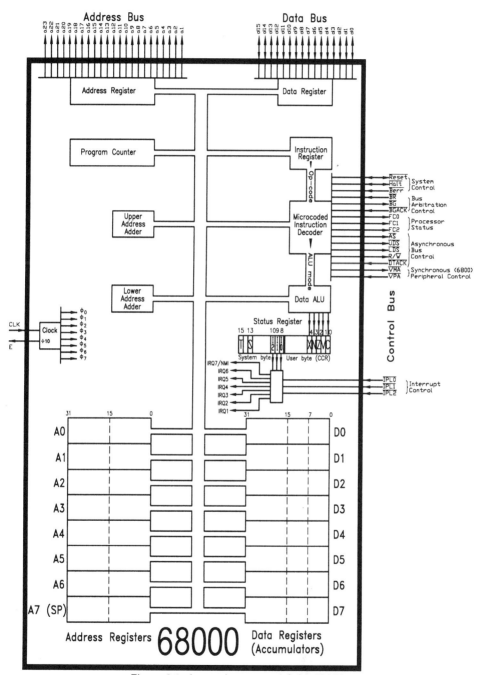

Figure 3.1: Internal structure of the 68000.

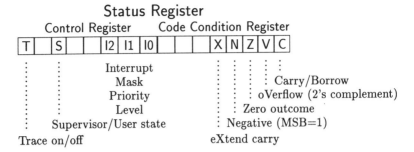

As shown, the Code Condition register occupies the lower byte of the 16-bit Status register, the upper field containing masks and bits which control the operating mode of the processor. The three bits I2 I1 I0 represent the Interrupt mask. The MPU will only respond to an interrupt request signalled externally on pins $\overline{\text{IPL2}}\,\overline{\text{IPL1}}\,\overline{\text{IPL0}}$ (IPL stands for Interrupt Priority Level) if this active-low IPL number is *above* the mask number. For example an IPL number Low High Low (active-low 5) will trigger a level-5 request (IRQ5 in Fig. 3.1) if the mask is set at between 000 and 100. The exception is a level-7 request, which is non-maskable. More details are given in Section 6.1. The mask is set to level 7 at Reset, thus inhibiting all but a non-maskable interrupt.

The 68000 MPU leads a Jekyll and Hyde existence, in that it has two states of existence, which are virtually independent of each other. These are somewhat more prosaically termed the **Supervisor** and **User states**. When the MPU is Reset, the S bit in the Status register is automatically set to 1, the Supervisor state. Certain so called **privileged instructions** can only be executed in this state. These instructions generally deal with the overall operation of the processor. Thus it is only possible to change the Interrupt mask in the supervisor state, for example:

```
MOVE    #00100 100 00000000b,SR
```

sets the mask to level 4. Moving data into the Status register is a privileged instruction (*but not reading it*).

The *only* way to exit the Supervisor state is to clear the S bit, for example:

```
ANDI    #11 0 1111111111111b,SR
```

will clear bit 13 of the Status register and leave all else unchanged. As you might expect AND IMMEDIATELY TO STATUS REGISTER is privileged, as is ORI #data,SR (to set individual bits) and EORI #data,SR (to toggle individual bits).

Once in the User state you cannot return to the Supervisor state by simply setting the S bit as the MOVE, ANDI, ORI and EORI #data,SR instructions are illegal in this situation, but note that the same instructions targeted to the **CCR** part of the Status register are perfectly legal; for instance:

```
ORI #00000001b,CCR
```

sets the **Carry flag**). The only way back to the Supervisor state is when a Hardware interrupt or Trap occurs (a Trap is a type of Software interrupt, and is described in Chapter 6).

What is the point in having two distinct states? In a multitasking environment (more than one program running concurrently on the same machine) it is usual to have a master program, known as the **operating system**. The operating system provides resources to the user program, such as an interface to a magnetic disk store. Where more than one user program appears to run simultaneously, it may switch between these programs in a time-slice manner in a fairly complex way [4]. As a simple example, consider a microprocessor development system to which software can be downloaded into RAM, whence it can be run and tested. The operating system, here called a monitor, usually resides in ROM. Once control is passed from the monitor to the user program running in real time, the only way back to the operating system is via a Software interrupt, Hardware interrupt or Reset. In all these situations it is important to ensure that user programs do not corrupt memory or other resources used by the operating system.

In the 68000 MPU, this operating system runs in the Supervisor state, into which it enters automatically on Reset. The MPU informs the outside world which mode it is running by using the three **Function Code** pins FC2 FC1 FC0, as detailed in Section 3.2. Thus the hardware engineer can design the address decoder to access Supervisor ROM and RAM chips in an entirely separate address space than that accessible to the User program. Furthermore, the Supervisor and User modes have separate System Stack Pointers, the **Supervisor Stack Pointer** (**SSP**) and **User Stack Pointer** (**USP**). Thus, in reality there are two **A7 registers**, only one of which is active in any mode. Both separate and mutually exclusive memory spaces and System Stack Pointers make it difficult for the user program to accidentally corrupt the operating software.

In small dedicated embedded systems there is often no operating system as a separate entity. In such naked cases, it is normal to stay in the Supervisor state and ignore the existence of the User state. We will do this for our project in Part 3. However, the security of two distinct states is important for the reliable operation of more sophisticated embedded systems, especially where an extensive interrupt-driven configuration is being used.

Finally, bit 15 of the **Status register** is the **Trace bit**. When set to 1, a Software interrupt/Trap will occur at the end of each instruction execution. This can be used in conjunction with a suitable operating system routine to print out information, such as the register contents after each step of the program [5]. The **Trace bit** is turned off on Reset.

REGISTER ARRAY

As in all microprocessors, the 68000 has a **Program Counter** (**PC**) which essentially points to the next instruction to be fetched. With this MPU, the situation is a little more complex. This is because use is made of time when the external buses would normally be idle, to bring down words from program memory into a 2-

word prefetch queue buffer [1]. For example, when a Branch is executed, both the next instruction and the Branch-to op-code will already be in the buffer. Which one executed will depend on the outcome of the condition test. Like most of the registers, the **PC** is 32-bit wide, but in the basic 68000 MPU only the lower 24 bits have any connection to the external address bus.

Two arrays of eight 32-bit registers are of major concern to the programmer. These are functionally divided into Data and Address registers. Data registers provide the source or/and the destination data for most operations. The Address registers hold pointers to data stored outside in the MPU's memory. Motorola have made a considerable effort to ensure that these registers behave in a consistent and regular manner (they use the term orthogonal), for example anything that can be done on **D0** can also be done in exactly the same manner on, say, **D7**. However, they have made a clear distinction between registers holding operational data (Data registers) and those used to compute addresses (Address registers).

The eight Data registers are the equivalent to the one or two Accumulator registers found in most 8-bit MPUs. Most instructions need to hold a source and/or destination operand in a Data register, for example:

```
ADD.L   [ea],D0      ; [D0] <- [D0] + [ea]
```

adds the 32-bit long operand at some effective address (ea) to **D0**, answer in **D0**.

```
ADD.L   D1,[ea]      ; [ea] <- [ea] + [D1]
```

adds the long operand at some ea to **D1**, answer in ea.

Any Data register can be treated as an 8-bit, 16-bit or 32-bit Accumulator, for example:

```
CLR.L   D2                 ; [D2(31:0)] <- 00000000 00000000 00000000 00000000
MOVE.B  #0FFh,D2           ; [D2(7:0)]  <- 00000000 00000000 00000000 11111111
MOVE.W  #0FFFFh,D2         ; [D2(15:0)] <- 00000000 00000000 11111111 11111111
MOVE.L  #0FFFFFFFFL,D2     ; [D2(31:0)] <- 11111111 11111111 11111111 11111111
```

Any bits outside the target field remain *unchanged*. I have used the notation D2(n:m) as meaning bits n to m of **Data register_2**. Most instructions acting on Data registers come in all three size varieties, indicated to the assembler by using the extensions .B (for byte), .W (default word) and .L (for long-word). Two bits in the op-code word are used to represent the size, as shown in Fig. 4.4. There are also a few instructions which can affect any bit in a Data register, for example:

```
BSET    #12,D4       ; Sets bit 12 of D4.L high, the rest unchanged.
```

In order to make it difficult to use an Address register for anything other than its legitimate role, only a small range of special instructions can be used to alter their contents. For example, to set up **A0** to address $0000C000h$ we have:

```
MOVEA.L #0C000L,A0   ; [A0(31:0)] <- 0C000h
```

An ordinary MOVE cannot target an Address register, although it is possible to copy the data in an Address register to a Data register, for example:

```
MOVE.L   A1,D4         ; [D4(31:0)] <- [A1(31:0)]
```

Other Address register modification instructions include (ADD TO ADDRESS REGISTER) ADDA, (COMPARE WITH ADDRESS REGISTER) CMPA and (SUBTRACT FROM ADDRESS REGISTER) SUBA. Except for CMPA, such operations do not affect the **CCR** flags. Only long and word-sized operations are allowed. The full 32-bits are *always* affected, even where word-sized operations are used. In this case, bit 15 is sign-extended to 32 bits, for example:

```
MOVEA.W  #0C000h,A0    ; [A0(31:0)] <- FFFFC000h
```

There are no byte-sized operations on Address resisters.

Like the Data register array, all Address registers behave in the same way, except that **A7** is special in that it is used as the **System Stack Pointer** for subroutines and interrupts. The MOVE MULTIPLE (MOVEM) instruction, which when targeted to **A7** is equivalent to Push and Pull in other MPUs, can also be used with any other Address register (see Section 4.1).

The Address registers have their own arithmetic circuitry, allowing effective addresses to be calculated in parallel with any data calculation. Like the 6809 MPU, the 68000 has an extensive range of Indexed addressing modes; for example:

```
MOVE.B   64(A0,D7.L),D0 ; [D0(7:0)] <- [64+[A0(31:0)]+[D7(31:0)]]
```

copies the contents of the data byte located in wherever **A0** points to plus the 32-bit variable in **D7** plus the constant 64 into the lower byte of **D0**! Incidentally, if there is going to be lots of activity around this area of memory, the instruction:

```
LEA      64(A0,D7.L),A1 ; [A1(31:0)] <- 64+[A0(31:0)]+[D7(31:0)]
```

puts the effective address (**A0.L** plus **D0.L** plus 64) in **A1**; and future accesses can be made without further calculation using **A1** as a pointer. More about LOAD EFFECTIVE ADDRESS and address modes is given in Chapter 4.

CONTROL CIRCUITRY

The 68000's Instruction decoder uses a microcoded design [6] as opposed to the random logic employed by 8-bit processors, such as the 6809. The order of magnitude increase in complexity exhibited in 4th generation devices makes the design and testing of the more efficient random logic circuitry difficult. Thus the disadvantages of larger and slower circuitry are considered more than offset by the advantages of simplicity of design and testing, as well as the flexibility of an easier change or enhancement of operation. In a microcoded design, the sequence of steps in implementing an instruction is stored in integral ROMs [2].

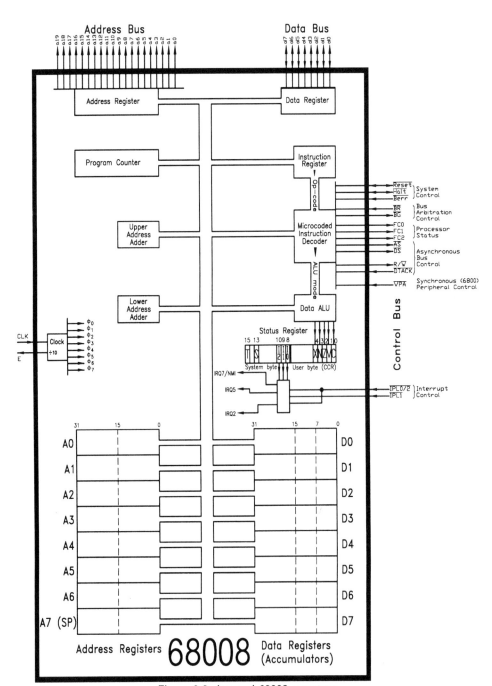

Figure 3.2: Internal 68008 structure.

The 68008 MPU is an 8-bit data bus version of the 68000. Despite the reduced external functionality, as can be seen from Fig. 3.2, internally the two processors are the same. Software is identical for both processors, although execution times are typically 40% longer, due to the larger numbers of 8-bit fetches, as opposed to 16-bit equivalents [7]. This still makes the 68008 a powerful alternative to a purely 8-bit MPU, and it is often used for this purpose in embedded MPU circuitry. Although the device itself is similar in price to its bigger brother, the smaller package, bus width and number of memory chips (see Figs 3.11, 3.12 and 13.3) considerably reduces board space and hence costs.

3.2 Outside the 68000/8

The 68000 MPU is available in a 64-pin package, which is shown in Fig. 3.3 together with the 48-pin 68008. Unlike the 8086 family, all bus signals are non-multiplexed. All signals are TTL voltage-level compatible. The 68HC000 is a CMOS version with slightly different electrical and timing specifications. Unless otherwise stated, figures are given for the normal HMOS version.

ADDRESS BUS and $\overline{\text{ADDRESS STROBE}}$ (a_n and $\overline{\text{AS}}$)

The term 'address' is normally used in a rather careless way without qualification. Address of what? In an 8-bit processor, at the hardware level it can be taken as the bit pattern on the address bus, which is externally decoded to physically enable the target 8-bit byte in memory or port onto the 8-bit data bus. Thus it is a byte address. In a 16-bit processor, it is a word address, that is, points to a word in memory space. In a high-level language, what meaning do we attach to the address of, say, an object comprising an array of ten byte-sized elements stored in consecutive memory locations? The general convention is to specify the lowest byte address of the object. This is mainly for historic reasons, as MPU technology came of age with 8-bit devices. Thus, if the array fred[ ] is stored in memory between byte addresses 01C030h and 01C03Ah, then its address is 01C030h. In the 68000 MPU this base address is used for word and longword sized objects. Thus the instruction MOVE.W 01C030h,D0 will bring the object

| MSB ^{1C030h} | LSB ^{1C031h} |

down into D0(15:0).

The physical address bus reflects this natural word size by omitting line a_0. Thus each pattern on the bus $a_{23}-a_1$ spans two internal byte addresses $a_{23}-a_0$, one even and one odd. As we shall see, the missing a_0 line is implicitly available in the guise of the two Data Strobe lines. Up to 8 Mwords or 16 Mbytes are directly accessible on this address bus. The 68008 MPU has a natural byte-sized word, as reflected in its byte-organized address bus, which does explicitly provide an a_0 line. This 68008 has 20 address pins, from a_{19} to a_0, giving a 1 Mbyte address space (there is a 52-pin version with 22 lines).

66 The 68000/8 Microprocessor: Its Hardware

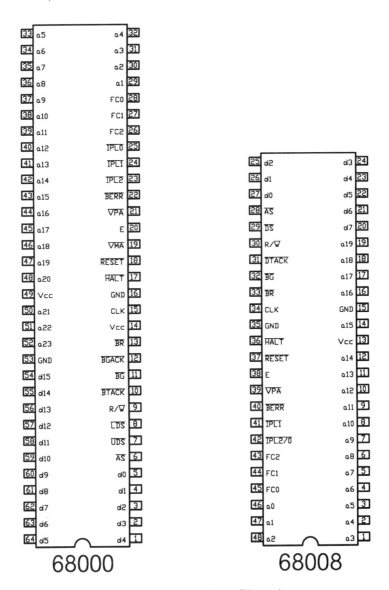

Figure 3.3: 68000 and 68008 DIL packages.

$\overline{\text{Address\_Strobe}}$ ($\overline{\text{AS}}$) is asserted when the state of the address bus is valid, see Figs 3.6 and 3.7. When enabled, the address lines can drive up to four LSTTL loads into a 130 pF capacitive load. $\overline{\text{AS}}$ can similarly drive six LSTTL loads. Both sets of lines are off when in a direct memory access (DMA) mode, whilst only the address bus is off when halted.

DATA BUS and DATA STROBES (d_n and $\overline{\text{UDS}}$ / $\overline{\text{LDS}}$ / $\overline{\text{DS}}$)

The 68000 MPU uses a single bidirectional 16-bit data bus to carry both instruction and operand data to the MPU (Read) and from (Write). There is a problem here, in that the 68000 sees a byte-organized world out there through a word-sized eye. Figure 3.4 shows the execution cycle of a MOVE instruction in byte, word and long-word versions. In the case of a Read-byte action, the actual data lines used for the transfer depend on whether an even address (upper eight lines) or odd address (lower eight lines) is specified. Data as considered in byte-sized lumps is organized as | UDS ^{EVEN} | LDS ^{ODD} |. Thus the $\overline{\text{Upper\_Data\_Strobe}}$ is seen to be equivalent to the missing $\overline{a}_0$ (active when a_0 is 0, that is on even-byte addresses) and $\overline{\text{Lower\_Data\_Strobe}}$ is active when a_0 is 1 (odd-byte address). Thus the two Data Strobes have a dual role. Firstly they signal when data is valid during a Write action, as shown in Fig. 3.7. Secondly they can enable either the upper or lower byte of an addressed word, effectively enabling the 16-bit data bus to carry a single byte from a word-organized memory space.

A word transfer is signalled with both $\overline{\text{UDS}}$ and $\overline{\text{LDS}}$ being asserted together, and the two bytes feeding the bus simultaneously. Notice that the most significant byte (MSB) is always in the lower byte address (even address) in common with all Motorola MPUs (see page 20). A long-word transfer simply involves two word transfers in sequence. As can be seen, the execution time here is longer by four clock periods (see Fig. 3.5) due to the extra transfer cycle. Byte and word execution takes the same time. In both word and long-word cases the data has to be organized starting with an *even address* (MSB). Attempts to do an odd-address word or long-word access, for example:

```
MOVE.W   0C101h,D0     ; This is erroneous
```

are an error, and the 68000 will terminate execution by returning to the Supervisor state via an **Address Error Trap** (see Section 6.2).

The 68008 has only a byte-sized data bus and a single Data Strobe ($\overline{\text{DS}}$). There is no problem here, as address line a_0 is provided explicitly to reflect the natural byte size of the data bus, and thus each target memory byte is individually enabled. This is exactly the same as an 8-bit MPU seeing the world through an 8-bit eye. Nevertheless, the even boundaries restriction for word and long-word memory data are retained for compatibility with the 68000 processor. Execution times for the 68008 are shortest for a byte operand, word and long-word operands taking one and three extra access cycles respectively. Fetching the op-code also takes twice as long. At a clock frequency of 8 MHz, the 68000 moves a word to a data register in $2\,\mu$s, whilst a 68008 takes $4\,\mu$s. However moving between registers, for example:

```
MOVE.W   D7,D0     ; A register to register move
```

takes $\frac{1}{2}\,\mu$s in both cases. The moral being to keep as much in the Data registers as possible.

When the data lines and $\overline{\text{DS}}$ signals are enabled, they can drive up to six 74LS loads and 130 pF without external buffering. Data lines are high impedance when the processor is halted or in a DMA mode. $\overline{\text{DS}}$ signals are off only during DMA.

$\overline{\text{Reset}}$

Asserting both $\overline{\text{Reset}}$ and $\overline{\text{Halt}}$ together initiates a total Reset of the processor.

68 The 68000/8 Microprocessor: Its Hardware

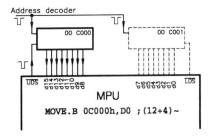

(a) Reading an even-eddress byte.

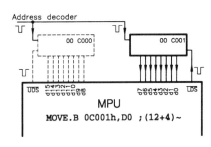

(b) Reading an odd-address byte.

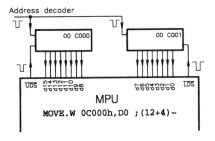

(c) Reading a word.

Key:
LSB Least Significant Byte
MSB Most Significant Byte
Execution time (fetch + execute)
~ = Clock periods
For the 68008
(a) and (b) = (24+4)~
(c)   = (24+16)~
(d)   = (24+16)~

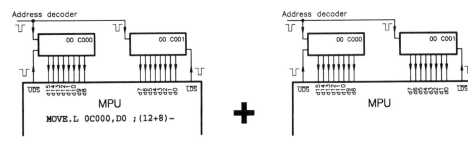

(d) Reading a long-word.

Figure 3.4: Memory Organization for the 68000.

This must be held for at least 100 ms when the power is initially applied. This ensures stabilization of the internal bias voltage generator and external clock source. Otherwise a duration of ten clock cycles is sufficient.

A total Reset causes the contents of the long-word at $000000-3h$ to be moved into the **Supervisor Stack Pointer** (its initial setting) and long-word $000004-7h$ to be moved into the **Program Counter** (the Restart address, see Fig. 6.5). The **Status register** is also set to Supervisor state (**S** = 1), the **Interrupt** mask to 7 (I2 I1 I0 = 111) and Trace is turned off (**T** = 0). No other registers are affected.

$\overline{\text{Reset}}$ can also act as an output signal, activated by the privileged instruction RESET. This drives the $\overline{\text{Reset}}$ pin low for 124 clock periods, which can be used to reset peripheral devices. Because of this bidirectional action, external restart circuitry must be more complex than a simple switch. An example of a typical circuit [8] is shown in Fig. 13.3. $\overline{\text{Reset}}$ will also be driven low, together with $\overline{\text{Halt}}$ when a Double-Bus fault occurs, as described in the next paragraph.

$\overline{\text{Halt}}$

Like $\overline{\text{Reset}}$, this is also a bidirectional line. As an input it can be used in conjunction with $\overline{\text{Reset}}$ or alone. When asserted alone, it will stop the processor after the current instruction is finished. The address and data buses will then be floated, and other Control outputs negated. If $\overline{\text{Halt}}$ is then released for one cycle, the processor will execute the next instruction and then stop. So, $\overline{\text{Halt}}$ can be used to single-step the processor for debug purposes [9].

As an output, $\overline{\text{Halt}}$ is driven low (together with $\overline{\text{Reset}}$) when the initial **Supervisor Stack Pointer** setting obtained from the Vector table on Reset is odd or a Bus Error is active in an exceptional event (see page 162). This is known as a Double-Bus fault. Halting the MPU is the obvious thing to do in these cases, as such events are unrecoverable.

Read/$\overline{\text{Write}}$ (R/$\overline{\text{W}}$)

This is low during a Write cycle, otherwise high. It is floated during DMA and, as a precaution, normally has a pull-up resistor to prevent erroneous writes during this situation. It can drive up to six LSTTL loads into 130 pF.

Data_Transfer_ACKnowledge ($\overline{\text{DTACK}}$)

This is a signal sent back by the addressed device to indicate that the peripheral's data is valid during a Read cycle and that the peripheral is ready to accept the data during a Write cycle. This asynchronous handshake protocol is discussed in detail in the next section.

Interrupt_Priority_Level ($\overline{\text{IPL0}}$ $\overline{\text{IPL1}}$ $\overline{\text{IPL2}}$)

These input pins are driven from external devices requesting an interrupt. The 3-bit active-low code thus placed is its priority level, ranging from zero (*111*) for no interrupt (quiescent state) to seven (*000*) for a non-maskable top-priority request. The $\overline{\text{IPL}}$ pins are constantly monitored, and any change lasting a minimum of two successive clock periods is internally latched. At the end of each instruction, the latched request level is compared with the **Interrupt mask** bits setting in the **Status register** and acted upon if higher. If masked to level 7, a *change* from a lower level to level 7 request will trigger an edge-triggered non-maskable interrupt response. More details in Section 6.1.

The 68008 MPU (except in its 52-pin version) internally connects the $\overline{\text{IPL0}}$ and $\overline{\text{IPL2}}$ lines as shown in Fig. 3.2. This means that only levels where bits 0 and 2 are the same (*111* = 0, *101* = 2, *010* = 5 and *000* = 7) are available.

Bus_ERRor (BERR)

This input acts as a special type of interrupt used to inform the processor that something has gone wrong out there. As an example of what can go awry, the addressed peripheral may not send back its $\overline{\text{DTACK}}$ acknowledge signal. If this continues indefinitely, the processor will hang up forever waiting for the peripheral to respond. Using a re-triggerable monostable activated by $\overline{\text{DTACK}}$ to drive $\overline{\text{BERR}}$ would ensure that in the absence of a correct response, say within 10 ms, the monostable will relax and alarm the processor. The use of a 'watch-dog' timer like this can be extended to ensure that the veracity of the program in high-noise situations, which can corrupt data and address lines, causing the processor to go off to some illegal memory space and do its own thing. By using a few lines of the legitimate program to trigger a watch-dog at some regular interval, a Bus Error can be signalled if this area of program is not entered. See Section 6.2 for more details. If a Bus Error occurs during the Restart process, signalling that the Reset vectors cannot be accessed, then the MPU stops with the $\overline{\text{HALT}}$ pin low.

During normal execution, if the external error-detection circuitry also drives the $\overline{\text{Halt}}$ line in the correct fashion, the processor can be persuaded to rerun the cycle which caused the error [10].

When a Bus Error occurs, the processor pushes data onto the Supervisor stack, which can then be used by the operating system for diagnostic purposes. If a Bus Error continues to be signalled, then a Double-Bus fault is said to have occurred. The processor signals this catastrophe by bringing $\overline{\text{Halt}}$ low and stopping.

Function_Code (FC2 FC1 FC0)

These three outputs inform the outside world concerning the state of the processor according to the codes:

FC2	FC1	FC0	
0	0	1	User state accessing Data memory
0	1	0	User state accessing Program memory
1	0	1	Supervisor state accessing Data memory
1	1	0	Supervisor state accessing Program memory
1	1	1	Interrupt acknowledge

Being able to distinguish between User and Supervisor states allows the hardware engineer to design address decoding circuitry which accesses separate memory devices for the two operating states. Knowing that an interrupt is being serviced is useful in cancelling the request, as discussed in Section 6.1.

Function_Code outputs can drive up to four LSTTL loads into 130 pF. They go high impedance on DMA.

Bus_Request ($\overline{\text{BR}}$)

External devices that wish to take over the buses for direct memory access (DMA) do so by asserting $\overline{\text{BR}}$ for as long as necessary. Tied high if not being used.

Bus_Grant ($\overline{\text{BG}}$)

The 68000 asserts $\overline{\text{BG}}$ in response to a Bus Request. Once the $\overline{\text{Address\_Strobe}}$ is negated, the DMA device can take over the buses.

Bus_Grant_ACKnowledge ($\overline{\text{BGACK}}$)

Before taking over the buses, the DMA device checks that no other DMA device is asserting $\overline{\text{BGACK}}$. If it is, the new device waits until $\overline{\text{BGACK}}$ is negated before asserting its own $\overline{\text{BGACK}}$ and proceeding. All DMA devices have their $\overline{\text{BGACK}}$ outputs wire-ORed together. The 68008 does not have this handshake input (except for the 52-pin version) and so can only handle systems where only one DMA device is present.

CLocK (CLK)

This must be driven by an external TTL compatible oscillator. Small crystal controlled DIL packaged circuits are readily available for this purpose. Rise and fall times should be 10 ns or better (8 and 10 MHz). Maximum frequency versions of 8 and 10 MHz are readily available with 12.5 and 16 MHz (not 68008) variants obtainable. The 68040/68060 can run up to 50 MHz. A typical Read or Write cycle needs four clock pulses (see Figs 3.5 and 3.6), thus taking between 500 ns (8 MHZ) through to 80 ns (50 MHz). The 68000/8 has internal dynamic circuitry, and so has a lower clock frequency bound (2 MHz for the HMOS devices, 4 MHz for CMOS versions).

E

This output is CLK frequency divided by ten (six low, four high). It is equivalent to the same-named signal in the older 6800 and 6809 MPUs (see Figs 1.3 and 1.4), and is used when interfacing to the older-style specialized 6800-oriented peripheral devices. It can drive up to six LSTTL loads at 130 pF.

Valid_Memory_Access (VMA)

This is also an 'old-style' 6800 type signal (not 6809). It indicates that the address bus data is valid, and is used as an Address Strobe synchronized to E for 'old-style' peripheral devices, such as the 6821 PIA (see Fig. 3.14). This is not available on the 68008 MPU, but can be generated with external circuitry [11]. It is only generated when external circuitry asserts the MPU's $\overline{\text{VPA}}$ pin, and then will take some time to lock into the E signal.

Valid_Peripheral_Address ($\overline{\text{VPA}}$)

This input, which is usually driven from the address decoder, indicates that the location the MPU wishes to communicate with is populated with a 6800-style peripheral, and that a special 6800-type data transfer cycle (using E and $\overline{\text{VMA}}$) should be used. $\overline{\text{VPA}}$ is also used to indicate that the processor should use automatic vectoring to respond to an interrupt, as described in Section 6.1.

Power (V_{CC} and GND)

The HMOS 68000/8 MPU dissipates 1.5 W maximum at a V_{cc} of 5 ± 0.25 V and a mean current of 300 mA. However, current peaks of as high as 1.5 A can be expected. The CMOS 68HC000 uses a maximum average current of 25 mA at 8 MHz (35 mA at 12.5 MHz), but may still require peaks of 1.5 A. These figures do not include that current taken by any loads.

3.3 Making the Connection

Like all microprocessors, the 68000/8 communicates with the outside world via its data bus through interface circuitry. The sequence of events during a transaction is a consequence of the interplay of the various control signals. However, unlike most 8-bit MPUs, the 68000/8 is controlled in an asynchronous manner, where the completion of a Read or Write cycle is dependent on the source or destination responding with a handshake when ready to go ahead. In the simple open-loop synchronous situation, as shown in Fig. 1.5, the transaction is completed at the end of the clock cycle irrespective of the state of readiness of the peripheral. Although it is certainly possible to extend the cycle by freezing the clock (in the 6809 by using $\overline{\text{MRDY}}$), this is very much the exceptional way of acting.

The closed-loop nature of asynchronous data transfer is clearly shown in Fig. 3.5, where feedback lines exist between each peripheral in the system and the microprocessor. When contacted (i.e. enabled by the address decoder) the external device responds when ready with a Data_Transfer_ACKnowledge ($\overline{\text{DTACK}}$) signal. Only then will the MPU complete the transaction.

We will use timing diagrams to look at this sequence of events in more detail, both when doing a Read and doing a Write to the outside world. In both cases the clock is internally split into eight phases (see Fig. 3.1), each of which initiates some micro-action. Based on this division, the sequence of events can be illustrated.

The Read cycle of Fig. 3.6 shows the address stabilizing early in the cycle, with the $\overline{\text{AS}}$ and $\overline{\text{DS}}$ Strobes then being asserted. $\overline{\text{DS}}$ is used as the generic term for $\overline{\text{UDS}}$ and $\overline{\text{LDS}}$, one or both of which are asserted according to the rules of Fig. 3.2.

When the *peripheral is ready*, it responds by putting its data on the bus and asserting its handshake, $\overline{\text{DTACK}}$. The MPU then proceeds by latching in the data. The MPU then terminates the cycle by negating its Strobes. The peripheral then responds by removing its data and raising its $\overline{\text{DTACK}}$.

In more detail:

1. The address bus's data will be valid within t_{CLAV} (Clock Low to Address Valid) of the beginning of phase 1 (ϕ_1).

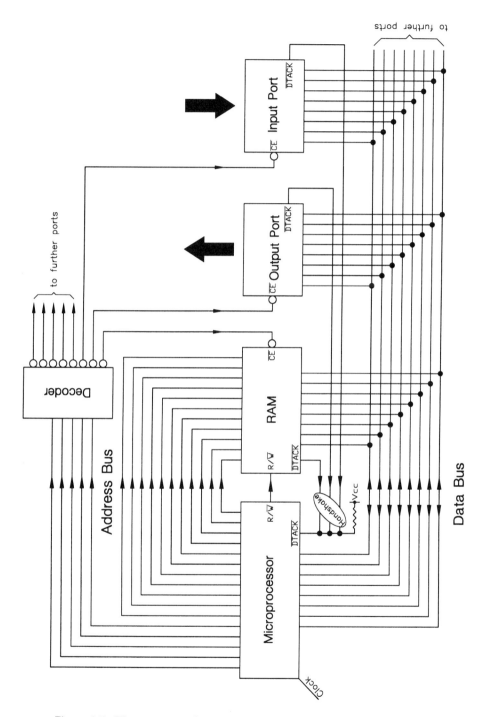

Figure 3.5: The structure of an asynchronous common-bus microcomputer.

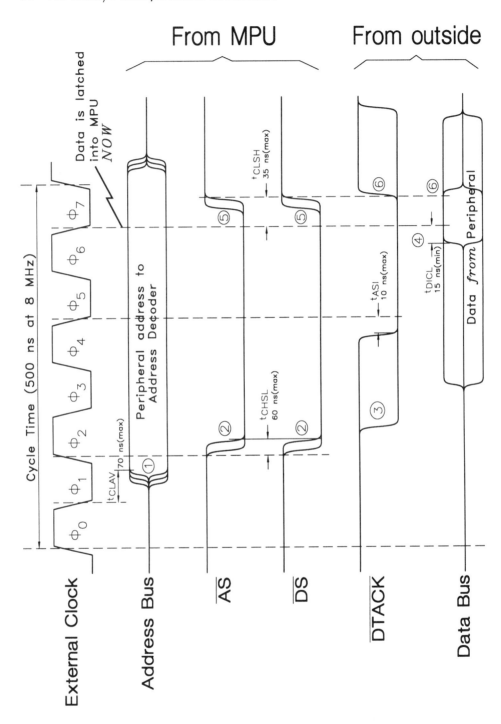

Figure 3.6: The 68000/8 Read cycle. Times given are for the 8 MHz HMOS version.

2. The $\overline{AS}$ and $\overline{DS}$ strobes are asserted by t_{CHSL} (Clock High to Strobe Low) following the start of ϕ_2.

3. The peripheral device responds when ready by asserting its $\overline{DTACK}$ line. If this can be done by t_{ASI} (Asynchronous Setup Input) preceding the end of ϕ_4, then the cycle will go ahead. Otherwise, the MPU will insert wait states of one clock period each (two phases) until $\overline{DTACK}$ is recognized on the falling edge.

4. The peripheral must set up its data on the bus no less than t_{DICL} (Data In to Clock Low) before the ⎤⎣ of ϕ_6, to ensure a successful read by the processor.

5. The $\overline{AS}$ and $\overline{DS}$ Strobes are then negated by no more than t_{CLSH} (Clock Low to Strobe High) following ϕ_6.

6. The peripheral has up to two clock periods from this point to negate its $\overline{DTACK}$ and remove its data.

Function_Code values, not shown in the diagram, are stable for the duration of the asserted Strobe signals, as is R/$\overline{W}$ (high for Read.).

The Write cycle time sequence shown in Fig. 3.7 is broadly the same as for reading. This time data is put on the bus by the MPU, and it is the job of the peripheral device to capture this and acknowledge with the $\overline{DTACK}$ handshake. The Data_Strobes are not asserted until the outgoing data is valid, somewhat later in this situation than the Address_Strobe, which indicates a valid address. After $\overline{UDS}/\overline{LDS}$ is negated, the data is taken off the bus, and the peripheral should now terminate its handshake.

In more detail:

1. The address bus will be valid within t_{CLAV} (Clock Low to Address Valid) of the beginning of phase 1 (ϕ_1).

2. $\overline{AS}$ is asserted by t_{CHSL} (Clock High to Strobe Low) following the start of ϕ_2.

3. The MPU sends out data on the bus by no later than t_{CLDO} (Clock Low to Data Out) following ϕ_3.

4. The $\overline{UDS}/\overline{LDS}$ Strobes are asserted by t_{CHSL} following the start of ϕ_4.

5. The peripheral device responds when ready by asserting its $\overline{DTACK}$ line. If this can be done by t_{ASI} (Asynchronous Setup Input) preceding the end of ϕ_4, then the cycle will go ahead. Otherwise the MPU will insert wait states of one clock period each (two phases) until $\overline{DTACK}$ is recognized on the clock ⎤⎣ .

6. All Strobes are negated by no more than t_{CLSH} (Clock Low to Strobe High) following ϕ_6.

7. Anytime after this, the peripheral can lift its $\overline{DTACK}$ handshake.

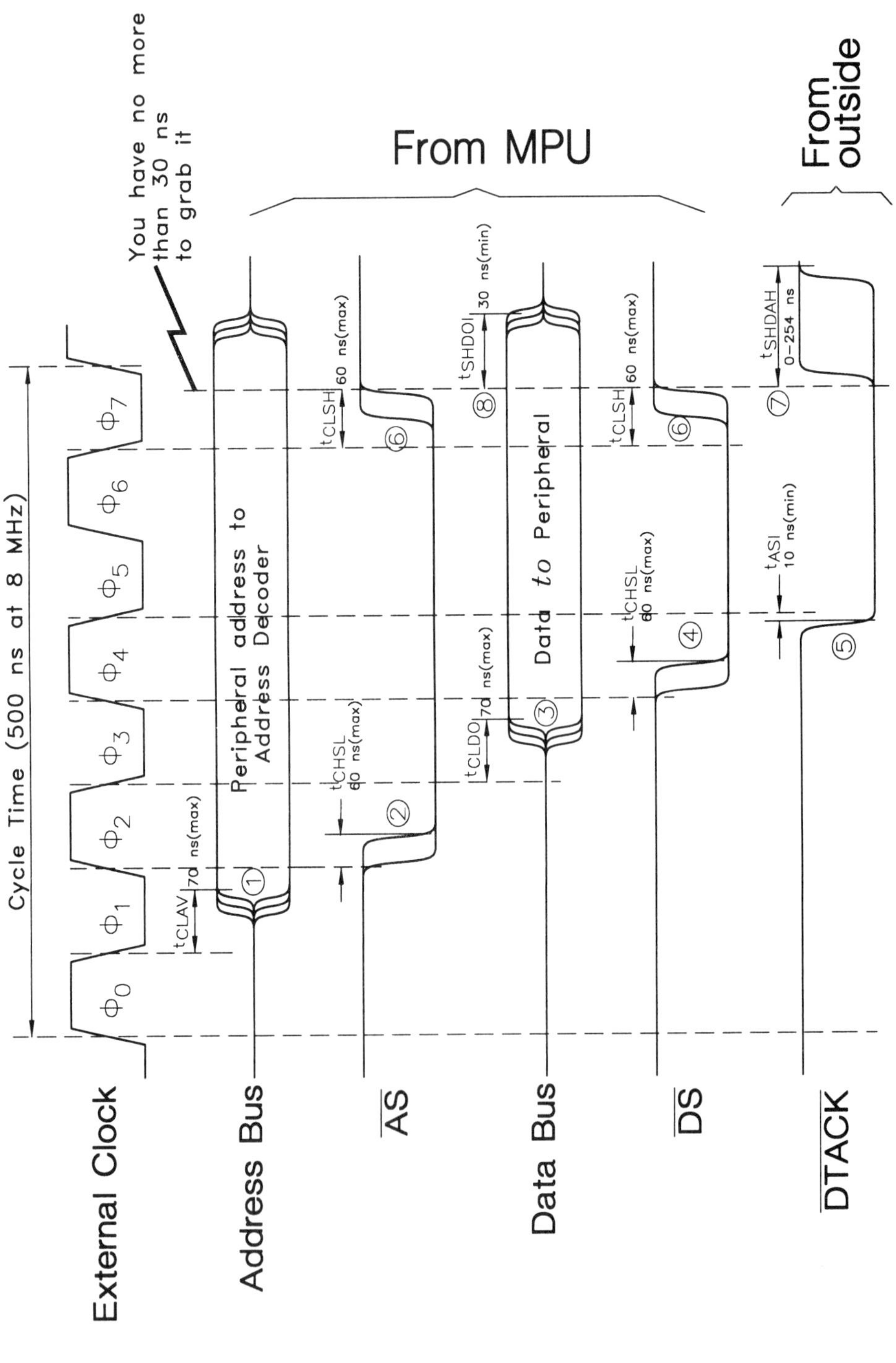

Figure 3.7: The 68000/8 Write cycle. Times given are for the 8 MHz HMOS version.

8. The MPU lifts its data off the bus by no less than t_{SHDOI} (Strobe High to Data Out Invalid) after the Strobes negate. This is the time a peripheral has to grab the data (including its setup time) after a ⎍ Strobe edge (30 ns for the 8 MHz device, 20 and 15 ns for the 10 and 12.5 MHz devices respectively).

Not shown are the Function_Code settings, which are valid for the duration of the $\overline{AS}$ Strobe, whilst R/$\overline{W}$ is low for Write as long as the $\overline{DS}$ is active.

Designing an address decoder involves the definition of logic which will implement the Boolean equations describing which combinations (addresses) of input variables (address lines) are to select the various peripheral devices. In this regard the 68000/8 does not differ from that for an 8-bit processor (see Section 2.3), although the larger number of variables is a further inducement to use more sophisticated implementations, such as programmable array logic [12]. This is especially the case where high speed versions demand small propagation delays. It is beyond the scope of this book to discuss the merits and features of the various circuitry, reference [13] gives a good review for the interested reader.

A rather unlikely, but nevertheless working circuit, is shown in Fig. 3.8. Here the 16 Mbyte address map can be considered split into four quarters using a_{23} and a_{22}. A 74LS154 4-to-16-line decoder further splits the quarter defined by $a_{23} a_{22} = 00$ into 16 pages of 256 kbytes each. Page 0 is again subdivided into eight 'paragraphs' of 16 kbytes, which are assumed to directly enable the labelled devices. In the cases where only a single peripheral interface is indicated, further levels of decoding may be used. $\overline{EPROM\_EN}$ combines two of these paragraphs using a 74LS08 AND gate, as 27128 EPROM pairs have a 16 kword (32 kbyte) capacity.

The secondary decoder is qualified by $\overline{AS}$. As $\overline{AS}$ is only asserted when the address signals have stabilized, this ensures that there are no spurious outputs during times when the address bus is in transition. With $\overline{AS}$ being asserted approximately one clock phase after the address is valid, it should be applied to the last decoder stage. This allows primary stages to 'get on with it' as soon as possible, and hence reduce the decoder's overall propagation delay. When high clock-speed versions of the 68000 are used, $\overline{AS}$ is commonly fed directly to the peripheral or memory's Chip Enable to further reduce this delay; an example of this is shown in Fig. 3.11(b).

Address decoding for the 68008 is identical to that for the 68000, but only address lines up to a_{19} are available. Thus, a functionally equivalent page 0 split could be obtained by replacing the 74LS154 decoder by a 74LS08 AND gate acting on a_{19} and a_{18}.

As we have seen, each peripheral addressed by a 68000 family MPU must reply by asserting the $\overline{DTACK}$ line when ready. All 68xxx peripheral devices specifically designed to function in an asynchronous manner automatically provide this handshake signal. An example of this is the 68230 Parallel Interface/Timer (PI/T) shown in Fig. 3.13. However, memory chips and elementary interface devices such as 3-state buffers and latches do not generate this information.

In the simplest of situations the 68000/8 MPU will run with its $\overline{DTACK}$ input permanently asserted. No wait states will be inserted into its Read or Write cycle, so all memory and peripheral interface must be fast enough to function correctly

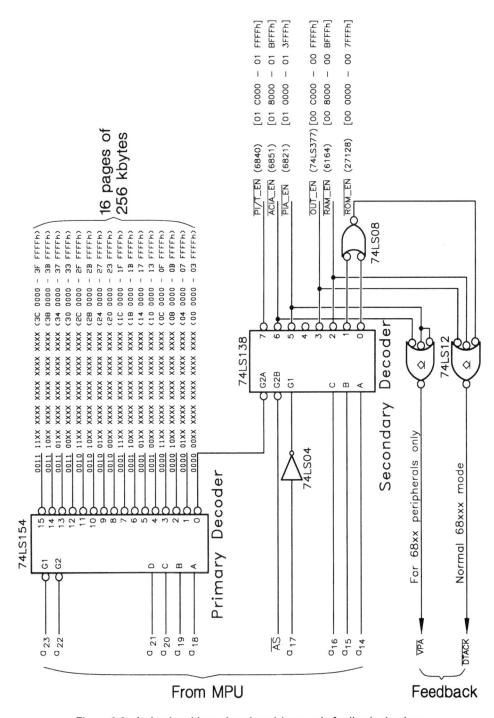

Figure 3.8: A simple address decoder with no-wait feedback circuitry.

in the allowed time. Figure 15.6 shows an example of this treatment of $\overline{\text{DTACK}}$.

A slightly more sophisticated approach is depicted at the bottom of Fig. 3.8. Here the pulse actually enabling the relevant device is also fed back to acknowledge readiness. This will activate shortly after $\overline{\text{AS}}$ is asserted, and will thus appear well before the end of clock phase 4, and no wait states will be introduced. The AND gate used to sum the Enable signals to the relevant interfaces and memory is open-collector. Thus other similar signals from elsewhere in the memory space can be wire-ORed to the one $\overline{\text{DTACK}}$ pin, see Fig. 3.9. The $\overline{\text{PI/T\_EN}}$ of Fig. 3.8 does not take any part in this scheme, as the 68230 provides its own open-collector $\overline{\text{DTACK}}$ handshake output (see Fig. 3.12).

Although this approach is more flexible than simply grounding $\overline{\text{DTACK}}$, it still assumes that the addressed device is fast enough not to require wait states. Where fast 68000 MPUs are used, this is not likely to be the case for all peripherals. Devices such as EPROMs and LCD interfaces tend to be rather slow. In such situations a delay circuit is needed for each such $\overline{\text{DTACK}}$ reply. This may take the form of a monostable, counter or shift register. An example of the latter is given in Fig. 3.9. Normally when the device in question is not being accessed, $\overline{\text{DEV\_EN}}$ is high and all eight flip flops are low. The 74LS05 open-collector buffer is then off. When the device is selected, $\overline{\text{DEV\_EN}}$ goes low trailing $\overline{\text{AS}}$ by the address decoder's propagation delay; thus releasing the register's $\overline{\text{CLR}}$. As the serial inputs are permanently held high, the flip flops will each in turn become logic 1, with an advance from Q_A to Q_H on the rising edge of the 68000's Clock. Assuming that the decoder's and 74LS05's propagation delay plus the 74LS164's setup time is less than the difference between $\overline{\text{AS}}$ being asserted and t_{ASI} before the end of clock phase 4 (approximately one clock cycle, see Figs 3.6 and 3.7), then wait states of between 0 and 7 clock periods are available according to the position of the link. Once the logic 1 reaches the link, the 74LS05's output goes low and $\overline{\text{DTACK}}$ is asserted.

Two 74LS377 octal flip flop registers are used in Fig. 3.10 to illustrate the implementation of an elementary 16-bit output port. The registers are both enabled by the address decoder, and the data clocked in by one or both Data Strobes, as appropriate (see Fig. 3.4). The rising edge of the Strobe is the active transition, $\boxed{6}$ in Fig. 3.7. There is a minimum of t_{SHDOI} between this point and the data becoming invalid. In determining the margin, the hold time (5 ns) for the 74LS377 must be subtracted. In the case of the 8 MHz 68000, this gives a worst-case margin of 25 ns, which shrinks to 10 ns for the 12.5 MHz version. There is no problem meeting the 25 ns 74LS377 setup requirement.

From these figures, it is clear that the Data Strobes should *directly* clock the registers and not be gated via additional logic. For example, it is tempting to use $\text{R}/\overline{\text{W}}$ ANDed with $\overline{\text{UDS}}/\overline{\text{LDS}}$ to ensure that an accidental read from this port does not latch in irrelevant data, but this will delay the active clock edge. The alternative of using $\text{R}/\overline{\text{W}}$ in conjunction with $\overline{\text{OUT\_EN}}$ is preferable for this purpose. The falling edge of $\overline{\text{UDS}}/\overline{\text{LDS}}$ via an inverter or gate cannot be reliably used as the clock, as it is just possible that if t_{CLDO} is a maximum and t_{CHSL} is a minimum, the

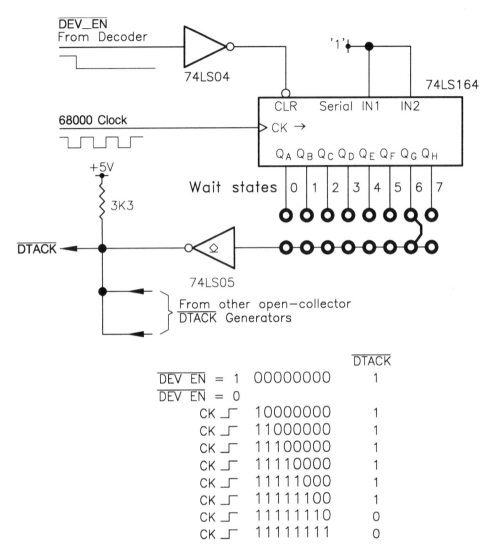

Figure 3.9: A DTACK generator for slow devices.

data will not be valid at this point.

In the case of the 68008 MPU, one 74LS377 will give an 8-bit output port, with $\overline{DS}$ acting as the clock (see Fig. 13.1). The same timing considerations hold.

The 6264 is a static CMOS 64 kbit RAM organized as an $8K \times 8$ array. It is commonly available in 100, 120 and 150 ns access time selections. Taking the Hitachi HM6264CP-10 as an example, the access time defining the minimum period from a stable address and device enabled ($\overline{CS1} = 0$, $CS2 = 1$) before data becomes valid during a Read is 100 ns. When writing, the address must be stable for the full 100 ns and for at least 80 ns of this time the device must be enabled and $R/\overline{W} = 0$

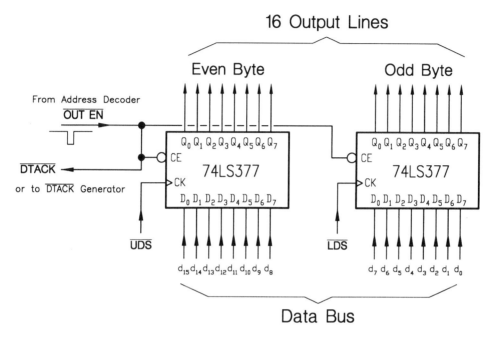

Figure 3.10: A simple word-sized output port.

for a successful Write-to action. The address must remain stable for at least 5 ns after $\overline{CS1}$ or R/$\overline{W}$ change state, or 15 ns after CS2 deactivates.

Referring to Fig. 3.11(a), we see that two broadside 6264s provide the 16 bits at each word address. As there is no a_0 byte address bit available from the 68000 MPU, address lines $a_1 - a_{13}$ drive the $A_0 - A_{12}$ RAM inputs, with $\overline{UDS}$ and $\overline{LDS}$ effectively providing the byte selection.

To determine whether wait states are required in using these devices, we need to analyze the timing constraints [14]. Essentially the RAM is enabled for the duration of the Data Strobes. As this is shortest during a Write cycle, we will use this as the determining factor. From Fig. 3.7, the worst-case width of $\overline{UDS}/\overline{LDS}$ is $\boxed{6} - \boxed{4}$, or three clock phases − t_{CHDL}; if we assume a minimum t_{CLSH} of zero (no figure is given). For the 8, 10 and 12.5 MHz MPUs, this is 120, 90 and 60 ns respectively. Thus the 80 ns HM6264LP-10 figure is suitable for up to 10 MHz systems. Actually we are being unduly pessimistic, as the 68000 data sheet gives t_{DSL} (Data Strobe Low) minimum as 80 ns for the 12.5 MHz MPU. For the Read cycle, 160 ns is the equivalent 12.5 MHz figure, rising to 240 ns for the 8 MHz version.

We have assumed that the propagation delay through the address decoder is such that $\overline{RAM\_EN}$ is asserted before the Data Strobes. During a Write cycle this is the time between $\boxed{4}$ and $\boxed{2}$ in Fig. 3.7, which is around one clock cycle. In the case of a Read cycle, the propagation delay must be subtracted from the t_{DSL} time for which the Data Strobes are low. In higher speed circuits, this propagation delay can be minimized by omitting $\overline{AS}$ from the address decoder and using it to qualify

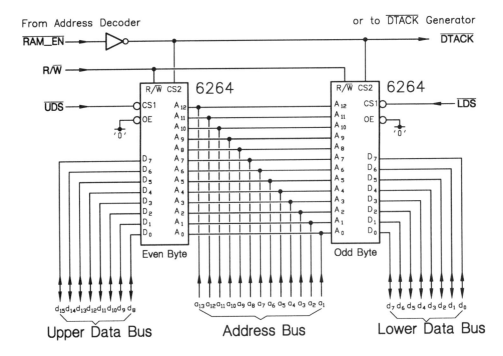

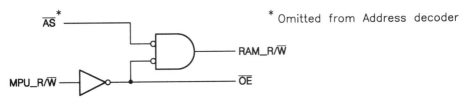

Figure 3.11: Interfacing 6264 RAM ICs to the 68000 MPU.

the R/$\overline{\text{W}}$ signal, as shown in Fig. 3.11(b). This is more economical than qualifying the $\overline{\text{RAM\_EN}}$ signal, as the modified R/$\overline{\text{W}}$ (i.e. RAM_R/$\overline{\text{W}}$) can be used for any number of RAM chips. The inverted MPU_R/$\overline{\text{W}}$ is normally used in this situation to turn off the output 3-state buffers during a Write, by activating $\overline{\text{Output\_Enable}}$ ($\overline{\text{OE}}$). Turn-off time is quicker from $\overline{\text{OE}}$ than from the RAM's Chip Select or R/$\overline{\text{W}}$.

EPROMs cause problems as they tend to be very much slower. A typical 27128 $16\,\text{K} \times 8$ EPROM has a 250 ns access time from stable address/asserted $\overline{\text{Chip\_Select}}$. Even at 8 MHz, there are only 235 ns from the falling edge of $\overline{\text{AS}}$ to the beginning of the setup time before the end of ϕ_6 ($5 \times \text{cycles} - t_{\text{CHSL}} - t_{\text{DOCL}}$). Fortunately, the time from $\overline{\text{Output\_Enable}}$ ($\overline{\text{OE}}$) to data valid is much less, for example 100 ns for the Hitachi HN4827128AG-25, and the circuit of Fig. 3.12 makes use of this means of

Making the Connection 83

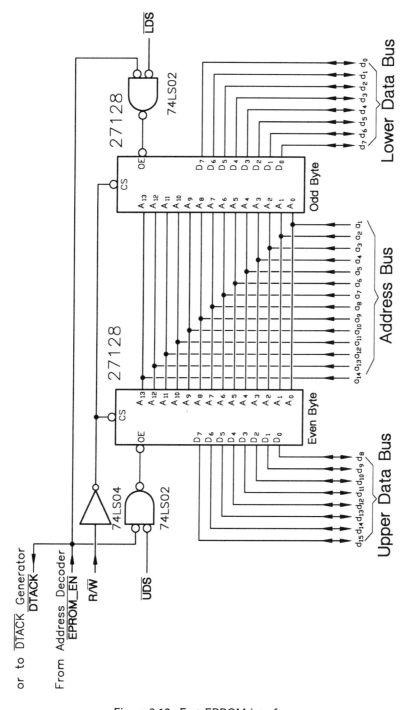

Figure 3.12: Fast EPROM interface.

access. Here $\overline{\text{CS}}$ is enabled whenever $\text{R}/\overline{\text{W}}$ is high, that is, during each Read. The $\text{R}/\overline{\text{W}}$ signal is valid no later than 70 ns after ϕ_0, which gives around 350 ns enabling time to the end of ϕ_6, less setup time t_{DICL} ($\boxed{4}$ in Fig. 3.6). Provided that the EPROM's $\overline{\text{OE}}$ is enabled at least 100 ns prior to this endpoint, a successful Read will occur. As the time between $\overline{\text{AS}}$ enabling the address decoder and this point is 235 ns, the remaining 135 ns will more than adequately cover this delay.

Faster CMOS EPROMs, such as the 150 ns National Semiconductor NMC27C64 (60 ns from $\overline{\text{OE}}$), facilitate no-wait state operations for faster processors. Alternatively the contents of slow EPROM could be transferred 'lock-stock and barrel' to fast RAM at the beginning of the program, and the EPROM henceforth ignored. This technique is frequently used in IBM PCs, where the BIOS is shadowed in RAM during the booting process.

RAM and ROM are interfaced to the 68008 MPU in the same way, but this time the MPU provides the byte-address bit a_0, and this goes to the memories' A_0 line. $\overline{\text{DS}}$ replaces $\overline{\text{UDS}}$ and $\overline{\text{LDS}}$, see Fig. 13.3.

The 68000 family are supported by a series of dedicated peripheral interface devices. The 68230 PI/T is typical of these, providing three 8-bit peripheral ports, two with handshake, and sharing functions with an internal timer together with interrupt facilities. As shown in Fig. 3.13, interfacing is straightforward, with a Data Strobe enabling the device together with the address decoder output. $\overline{\text{DTACK}}$ is internally generated and is connected directly to the MPU's $\overline{\text{DTACK}}$ node. Handshaking for the Interrupts (one for the parallel interface $\overline{\text{PIRQ}}/\overline{\text{PIACK}}$ and one for the timer $\overline{\text{TOUT}}/\overline{\text{TIACK}}$) is provided, as described in Chapter 6.

There are 25 internal registers addressed by the five Register Select inputs (RS1–RS5). As shown driven by address lines $a_1 - a_5$, they will appear at alternate byte addresses. Although this presents little inconvenience, a special instruction, MOVEP, can transfer two or four bytes at alternate addresses to suit this arrangement.

The two main peripheral ports can be set up to act as one 16-bit port, although the rather strange decision to use an 8-bit data bus means that two cycles are needed to transfer the data word. Programming the 68230 is complex and beyond the scope of this text; see reference [15] for a good description.

When the 68000 MPU was first released in 1979, the decision was taken to provide an operating mode to allow its use with the existing 68xx family of peripheral interface devices. This would ensure that the MPU was immediately useful without having to wait for further device introductions. We have already met the 6821 PIA in Fig. 1.9, and Fig. 3.14 shows this device in the alien environment of the 68000.

Essentially a 68xx device prompts the 68000 MPU about its special status by asserting the latter's $\overline{\text{VPA}}$ input, rather than $\overline{\text{DTACK}}$, as shown in Fig. 3.8. The Read and Write cycles are then synchronized to the E clock to give the normal 6800/6809-type synchronous data transfer sequence. The $\overline{\text{Valid\_Memory\_Address}}$ ($\overline{\text{VMA}}$) status output is used as an Address Strobe in this mode. $\overline{\text{DTACK}}$ should not be asserted during this time. As E is the 68000's clock divided by ten, the normal 1 MHz 6821 version is adequate for up to 10 MHz systems. The 1.5 MHz 68A21 is suitable for the 12.5 MHz 68000 MPU.

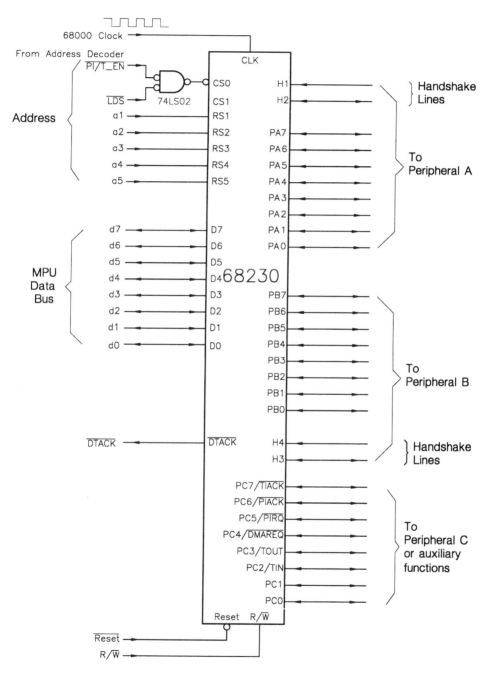

Figure 3.13: Interfacing the 68230 PI/T to the 68000's buses.

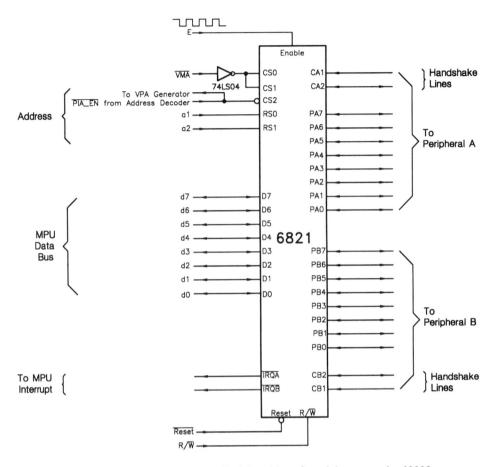

Figure 3.14: Interfacing a 6821 Peripheral Interface Adapter to the 68000.

References

[1] Starnes, T.W.; Design Philosophy Behind Motorola's MC68000; Part 1: A 16-bit Processor with Multiple 32-bit Registers, *BYTE*, **8**, no. 4, April 1983, pp. 70–92.

[2] Starnes, T.W.; Design Philosophy Behind Motorola's MC68000; Part 2: Data Movement, Arithmetic, and Logic Instructions, *BYTE*, **8**, no. 5, May 1983, pp. 342–367.

[3] Starnes, T.W.; Design Philosophy Behind Motorola's MC68000; Part 3: Advanced Instructions, *BYTE*, **8**, no. 6, June 1983, pp. 339–349.

[4] Lawrence, P.D. and Mauch, K.; *Real-Time Microcomputer System Design: An Introduction*, McGraw-Hill, 1987, Chapter 16.

[5] Kane, G et al.; *68000 Assembly Language Programming*, Osbourne/McGraw-Hill, 1981, Chapter 19.

[6] Stritter, S and Tredennic, N.; Microprogrammed Implementation of a Single Chip Microprocessor, *Prog. 11th Ann. Microprogramming Workshop*, Nov. 1978, IEEE, pp. 8–16.

[7] Browne, J.W.; µp Fits 16-bit Performance into 8-bit Systems, *Electronic Design*, **30**, April 15th, 1982, pp. 183–187.

[8] Wilcox, A.D.; *68000 Microcomputer Systems: Designing and Troubleshooting*, Prentice-Hall, 1987, Section 9.13.

[9] Starnes, T.W.; Handling Exceptions Gracefully Enhances Software Reliability, *Electronics*, 11th Sept. 1980, pp. 153–155.

[10] Clements, A.; *Microprocessor Systems Design: 68000 Hardware, Software, and Interfacing*, PWS-KENT, 2nd ed., 1992, Section 6.5.

[11] Barth, A.J.; Designing with the 68008 MPU, *Wireless World*, **90**, no. 1579, April 1984, pp. 30–33 and 41.

[12] Cahill, S.J.; *Digital and Microprocessor Engineering*, Ellis Horwood/Prentice-Hall, 2nd ed., 1993, Section 6.1.

[13] Clements, A.; *Microprocessor Systems Design: 68000 Hardware, Software, and Interfacing*, PWS-KENT, 2nd ed., 1992, Sections 5.1 and 5.2.

[14] Wilcox, A.D.; *68000 Microcomputer Systems: Designing and Troubleshooting*, Prentice-Hall, 1987, Section 10.6.

[15] Clements, A.; *Microprocessor Systems Design: 68000 Hardware, Software, and Interfacing*, PWS-KENT, 2nd ed., 1992, Section 8.3.

CHAPTER 4

The 68000/8 Microprocessor: Its Software

Although the 68000 architecture represents a complete break with its progenitor 6800 family; its software is in reality an evolution rather than a break from earlier implementations. Many of the characteristics exhibited by the 6809 instruction set (see Chapter 2) also appear in 68000 software, and indeed this is not surprising as they both support high-level language compilation, with extensive stack-oriented operations and a large repertoire of computed address modes.

The use of a full 16-bit op-code allows considerable scope in handling the many instruction:op-code:register combinations. Nevertheless, a special effort was made to make the assembly-level software user friendly. There are only 56 primary instructions [1], although variations on themes of several of these add another 29 mnemonics (eg. MOVE and MOVEQ for MOVE and MOVE QUICK). Most instructions are orthogonal, in that they apply to all registers within a group (Data or Address) in the same manner. The 'rules of grammar' are fairly consistent across the range of instructions with relatively minor quirks [2].

In this chapter we look at the more important of the instructions and their address modes. We will tie these together with the same example subroutines used to illustrate 6809 software in Section 2.3. The same assembler will be used here, details of which were given at that point. 68008 software is identical to that for the 68000 (except that only the lower twenty address bits are significant) and we will use the term 68000 as generic of the two.

It would take a complete book, rather than a single chapter, to do justice to assembly-level programming for such a complex processor. References [3, 4, 5, 6] are recommended to the interested reader.

4.1 Its Instruction Set

We will briefly look at the machine-code structure of 68000 instructions at the end of the next section. As far as assembly level is concerned, instructions may be classified as three kinds; that is, inherent, single- and dual-operand.

Inherent instructions have no operand, and are represented by mnemonic only, for instance the instruction RETURN FROM SUBROUTINE:

```
RTS    ; Program counter is pulled from System stack {Coded as 4E75h}
```

Single-operand (or monadic) instructions, such as CLEAR, have only one entry in the operand field, for example:

```
CLR.B  0E000h  ; [E000]     <- 00         {Coded as 4439-0000-E000h}
CLR.L  D0      ; [D0(31:0)] <- 00000000   {Coded as 4480h}
```

Dual-operand (diadic) operations such as Move have the form:
    Mnemonic   <Source operand>,<Destination operand>
For example:

```
MOVE.L  D0,D1         ; [D1(31:0)] <- [D0(31:0)] {Coded as 2200h}
MOVE.B  4000h,0E000h  ; [E000]     <- [4000]     {Coded as 03F9-4000-0000-E000h}
MOVE.W  D0,0E000h     ; [E000:1]   <- [D0(15:0)] {Coded as 33C0-0000-E000h}
```

Data Movement is the the most common operation executed. Reference [7] reports a frequency count of about 33% for MOVE, and it is with this in mind that we start with Table 4.1. Here we can see that only three mnemonics cover the range (see also LEA and PEA in Table 4.2). Of these the chief is MOVE, which subsumes the Load and Store operations of the 6809 MPU. MOVE is so frequently used that Motorola made it the most flexible of all the 68000 operations, a true 2-address instruction. Data in 8-, 16- or 32-bit packets can be copied from anywhere in memory, any register (except the **PC**) or immediately to any alterable memory or to any register (except **PC**). All other 2-operand instructions must specify a register as the source and/or destination, for instance ADD.B 0C000h,D0.

The MOVEA variation of the plain MOVE instruction must be used where an Address register is the destination. For example:

```
MOVEA.L #0C000h,A0 ; [A0(31:0)] <- 0000C000 {Coded as 207C-0000-C000h}
```

Like all specific Address register-destination operations, the **CCR** flags are not altered, and only word and long-word sizes are permitted. Word-sized operands are *sign extended* to 32 bits, for example:

```
MOVE.W  #0C000h,A0  ; [A0(31:0)] <- FFFFC000 {Coded as 307C-C000h}
```

The state of the **CCR** flags can be set up using the MOVE <ea>,CCR variant (some assemblers use the non-standard mnemonic MTCCR for MOVE TO CCR). Notice that its size is word only (the .W is usually omitted) although the **CCR** is byte sized. The **Status register** equivalent is MOVE <ea>,SR (or MTSR <ea>), and is only legal in the Supervisor state, that is privileged; but a MOVE FROM THE SR, MOVE SR,<ea> (or MFSR <ea>), can be made from anywhere. The MOVE FROM THE CCR is only available on the 68010 MPU and higher family members.

The MOVE QUICK (MOVEQ) instruction is targeted exclusively to the Data registers. It is used to set up a *32-bit* Data register to a fixed long number between +127 and −128 (signed 8-bit). Of course an ordinary MOVE can be used, but as the immediate data is included in the op-code for MOVEQ, the latter's execution is much faster, as shown here:

90   *The 68000/8 Microprocessor: Its Software*

```
MOVE.L #1,D0   ; [D0(31:0)] <- 00000001 (12~) {Coded as 223C-0000-0001h}
MOVEQ  #1,D0   ; [D0(31:0)] <- 00000001 (4~ ) {Coded as 7001h}
MOVEQ  #-1,D0  ; [D0(31:0)] <- FFFFFFFF (4~ ) {Coded as 70FFh}
```

where ~ indicates clock cycle. Thus the ordinary MOVE takes 1.5 μs at an 8 MHz clock rate against 0.5 μs for a MOVEQ. The timings for the 68008 MPU are 24~ (3 μs) and 8~ (1 μs) respectively. Note that *all* 32 bits of the Data register are affected. There is *no* MOVEQ.B or MOVEQ.W; an ordinary MOVE must be used in cases where only the lower 8 or 16 bits are to be setup.

Using a regular MOVE with the appropriate address mode gives the equivalent of a Push or Pull operation; for example:

```
MOVE.L D0,-(SP)  ; Same as PSHS D0 (14~) {Coded as 2F00h}
```

pushes all of **D0** out to the System stack, *after* the System Stack Pointer **A7** has been decremented four bytes, and

```
MOVE.L (SP)+,D0  ; Same as PULS D0 (12~) {Coded as 201Fh}
```

pulls four bytes off the System stack into **D0.L** and then increments the System Stack Pointer. The actual System stack used depends on whether the MPU is in

Table 4.1: Move instructions.

Operation	Mnemonic	X	N	Z	V	C	Description
Move							Data, source to destination
data	MOVE.s3 ea1,ea2	•	√	√	0	0	[ea2] <- [ea1]
to Address reg.	MOVEA.s2 ea,Dn	•	•	•	•	•	[An] <- [ea]
quick	MOVEQ #±d₈,Dn	•	√	√	0	0	[Dn] <- #±d_8
regs to memory	MOVEM.s2 ∑Rn,ea	•	•	•	•	•	[-ea] <- ∑Rn
memory to regs	MOVEM.s2 ea,∑Rn	•	•	•	•	•	∑Rn <- [ea+]
to CCR	MOVE.W ea,CCR	√	√	√	√	√	[CCR] <- [ea]
to SR	MOVE.W ea,SR	√	√	√	√	√	[SR] <- [ea], privileged
from SR	MOVE.W SR,ea	•	•	•	•	•	[ea] <- [SR]
Exchange							Switch two registers
	EXG.L R1,R2	•	•	•	•	•	[R2] <--> [R1]
Swap							Switch lower/upper words
	SWAP Dn	•	√	√	0	0	[D(31:16)] <--> [D(15:0)]

0	Flag always reset	Rn	Data or Address register n
1	Flag always set	An	Address register n
•	Flag not affected	Dn	Data register n
√	Flag operated in the expected way	Dn(x:y)	Data register n, bits x to y
s3	Three sizes, .B, .W, .L	#±d₈	Signed 8-bit value
s2	Two sizes, .W, .L	[ ]	Contents of
ea	Effective Address or immediate data	<-	Becomes

the Supervisor or User mode, the assembler allowing the use of the mnemonic SP or, indeed A7, for either System Stack Pointer. Note that a MOVE.B to/from the System stack always results in a word being transferred, to preserve the evenness of the System Stack Pointer (i.e. **A7**). Any of the other Address registers may be used in place of **A7**. Pre-Decrement and Post-Increment address modes are discussed in the next section.

As there are 16 registers which may have to be pushed or pulled, clearly a single instruction which can save or retrieve any or all Address and Data registers at one go will be more efficient. The MOVE MULTIPLE instruction fulfils this task; for example:

MOVEM.L D2/D3/D4/A2,-(SP) ; Same as PSHS D2,D3,D4,A2 (40~) {Coded as 48E7-3820h}

pushes all of **D2,D3**, **D4** and **A2** out to the System stack, the **System Stack Pointer** ending 16 bytes down; and

MOVEM.L (SP)+,D2/D3/D4/A2 ; Same as PULS D2,D3,D4,A2 (44~) {Coded as 4CDF-041Ch}

pulls the register contents back out, restoring the **System Stack Pointer** to its original value. Any Address register can be used in place of **A7**. In general, the time taken for a multiple Push is $8 + 8n$~ and multiple Pull is $12 + 8n$~, where n is the number of registers involved. Thus to Push a full register complement takes 132 clock cycles (16.5 μs at 8 MHz) against 224 clock cycles and 32 bytes of program memory using ordinary MOVEs.

The MOVEM instruction uses a post-word to the op-code to indicate which registers are involved, as shown in Fig. 4.1. If less than the full complement is involved, then the order of storage in the stack is still that shown in the register list. There is a word-sized MOVEM which only transfers the lower register words. This saves stack space and time; however, on return all registers — both Data and Address — are filled with the *sign-extended long* version of the stored word.

Less usefully, a fixed address can be used as MOVEM's address mode instead of Pre-Decrement (registers to memory) or Post-Increment (memory to registers). In this case no pointer marks the bottom of the dump, and the same address is used for both directions.

EXCHANGE (EXG) swaps around the complete 32-bit contents of *any* two registers, Data or Address. SWAP acts only on Data registers, and exchanges the lower and upper words. This is useful, for example, when using the Division operation, which produces a 16-bit quotient in the lower part of a Data register and the remainder in the upper 16-bits. Using SWAP makes getting at the remainder easier (see Table 4.12). The 68020 MPU has a byte-sized SWAP which exchanges the lower two bytes. The 68000 can use a ROL.W #8,Dn to perform the same function (see Table 4.3).

The 68000 provides for Addition, Subtraction, Multiplication and Division operations together with some ancillary instructions. The elementary Addition and Subtraction operations are straightforward, with at least one of the operands being a Data register, for example:

Table 4.2: Arithmetic operations.

Operation	Mnemonic	X	N	Z	V	C	Description
Add							Add source to destination
to Data reg.	ADD.s3 ea,Dn	√	√	√	√	√	[Dn] <- [Dn] + [ea]
to memory	ADD.s3 Dn,ea	√	√	√	√	√	[ea] <- [ea] + [Dn]
to Address reg.	ADDA.s2 ea,An	•	•	•	•	•	[An] <- [An] + [ea]
quick	ADDQ.s3[1] #d_3,ea	√	√	√	√	√	[ea] <- [ea] + #d_3[2]
immediate	ADDI.s3 #kk,ea	√	√	√	√	√	[ea] <- [ea] + #kk
with extend	ADDX.s3 Dy,Dx	√	√	[3]	√	√	[Dx] <- [Dx] + [Dy] + X
	ADDX.s3 -(Ay),-(Ax)	√	√	[3]	√	√	[-(Ax)] <- [-(Ax)] + [-(Ay)] + X
Clear							Clears destination
	CLR.s3 ea[4]	•	0	0	1	0	[ea] <- 00
Divide							Generates quotient and remainder (%)
signed	DIVS ea,Dn	•	√	√	√	0	[Dn(15:0)] <-[Dn(31:0)]÷[ea(15:0)]
unsigned	DIVU ea,Dn	•	√	√	√	0	[Dn(31:16)]<-[Dn(31:0)]%[ea(15:0)]
Extend							Sign Extend Data register
word	EXT.W Dn	•	√	√	0	0	[Dn(15:0)] <- [SEX|[Dn(7:0)]]
long	EXT.L Dn	•	√	√	0	0	[Dn(31:0)] <- [SEX|[Dn(15:0)]]
Load Effective Address							Effective Address to Address reg.
	LEA ea,An	•	•	•	•	•	[An] <- ea
Multiply							
signed	MULS ea,Dn	•	√	√	0	0	[Dn(31:0)]<-[Dn(15:0)]×±[ea(15:0)]
unsigned	MULU ea,Dn	•	√	√	0	0	[Dn(31:0)]<-[Dn(15:0)]× [ea(15:0)]
Negate							Reverses 2's complement sign
data	NEG.s3 ea	√	√	√	√	√	[ea] <- 00 - [ea]
with extend	NEGX.s3 ea	√	√	[3]	√	√	[ea] <- 00 - [ea] - X
Push Effective Address							Effective Address into Stack
	PEA ea	•	•	•	•	•	[-SP] <- ea
Subtract							Subtract source from destination
from Data reg.	SUB.s3 ea,Dn	√	√	√	√	√	[Dn] <- [Dn] - [ea]
from memory	SUB.s3 Dn,ea	√	√	√	√	√	[ea] <- [ea] - [Dn]
from Addr. reg.	SUBA.s2 ea,An	•	•	•	•	•	[An] <- [An] - [ea]
quick	SUBQ.s3[1] #d_3,ea	√	√	√	√	√	[ea] <- [ea] - #d_3[2]
immediate	SUBI.s3 #kk,ea	√	√	√	√	√	[ea] <- [ea] - #kk
with extend	SUBX.s3 Dy,Dx	√	√	[3]	√	√	[Dx] <- [Dx] - [Dy] - X
	SUBX.s3 -(Ay),-(Ax)	√	√	[3]	√	√	[-(Ax)] <- [-(Ax)] - [-(Ay)] - X

Note 1: Only Long and Word with Address register destination. Also CCR unchanged.
Note 2: d_3 is a 3-bit number 1 to 8.
Note 3: Cleared for non-zero, otherwise unchanged.
Note 4: Not Address register.

Its Instruction Set 93

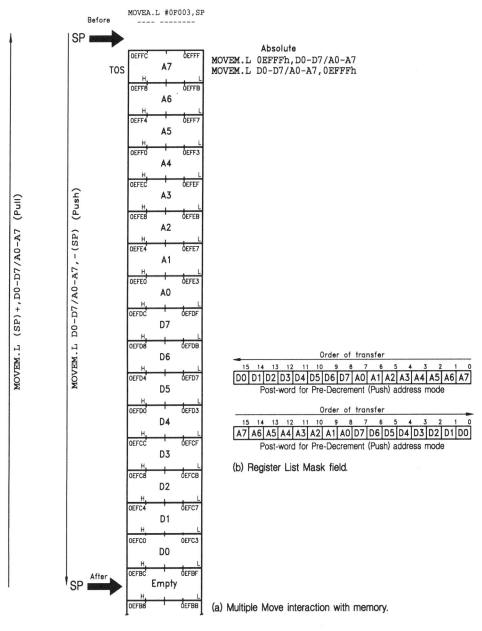

Figure 4.1: Multiple moves to and from memory.

```
ADD.B   D0,1234h  ; [1234]    <- [D0(7:0)] + [1234h].    Add <Source> to   <Destination>
SUB.W   1234h,D1  ; [D1(15:0)] <- [D1(15:0)] - [1234:5h]. Sub <Source> from <Destination>
ADD.L   D0,D1     ; [D1(31:0)] <- [D1(31:0)] + [D0(31:0)]. Add <Source> to  <Destination>
```

In all cases the result is stored at the destination. Notice that in subtraction the , can be read as from. When the destination is in memory, then it must of course be alterable memory, usually RAM. Amongst the instructions, only MOVE can have both operands in memory.

An Address register is not permitted as a destination, although legal as a source. Instead the special instructions ADDA and SUBA are used. As is usual, the **CCR** flags are not changed by any operation that alters an Address register, and only word and long-word sizes are permitted. Word results are always sign extended to a long-word.

The ADD IMMEDIATE QUICK and SUB IMMEDIATE QUICK instructions are used as a substitute for the missing Increment and Decrement operations. A constant between 1 and 8 can be added or subtracted from any Data or Address register or read/write memory location, for example:

```
ADDA.W  #1,A0      ; [A0(31:0)] <- [A0(31:0)] + 1.  Increment (12˜) {Coded as D0FC-0001h}
ADDQ.W  #1,A0      ; [A0(31:0)] <- [A0(31:0)] + 1.  Increment ( 8˜) {Coded as 5248h}
SUBQ.B  #1,1234h   ; [1234h]    <- [1234]   - 1.    Decrement (16˜) {Coded as 5338-1234h}
```

The constant is encoded as a 3-bit group in the op-code itself. As can be seen above, this halves the size of the instruction and therefore decreases execution time. If an Address register is targeted, the usual word or long-word sizes are permitted, with the latter being sign extended to the whole 32 bits. The **CCR** flags remain unaltered.

Notice that the last example above altered a memory location directly without using a Data register as an intermediary stop. The ADD IMMEDIATE and SUB IMMEDIATE instructions can be used where the data is greater than 8, for example:

```
SUBI.W  #500h,0C000h  ; [C000:1] <- [C000:1] - 500h
```

Where operands of greater than 32 bits are involved, then several sequential Adds or Subtracts may be used to form the multiple-precision sum or difference. In most processors the **Carry flag** provides the linkage between successive operations but, as noted on page 58, the **X flag** is used for this purpose in the 68000 family.

Figure 4.2 shows an example of a 96-bit addition made up of three 32-bit operations. The program for this is:

```
MOVEA.L #0C00Ch,A0   ; Point A0 to just before least significant long-word <Source>
MOVEA.L #0C10Ch,A1   ; and A1 to just before least significant long-word <Destination>
ADD.L   -(A0),-(A1)  ; Add LSLWs, sum in <Destination> LSLW
ADDX.L  -(A0),-(A1)  ; Add NSLWs, sum in <Destination> NSLW
ADDX.L  -(A0),-(A1)  ; Add MSLWs, sum in <Destination> MSLW
```

One main point to notice here is the use of the Pre-Increment Address Register Indirect address mode. As described in the next section, the Address register used to point to the operand (like an Index register) is automatically decremented by the appropriate number of bytes (by four here) before being used. With the arrangement of Fig. 4.2, the address will naturally creep towards the most significant bytes as we do each addition. This is the only memory targeted address mode that can be used by ADDX and SUBX to access data in memory. Alternatively both operands can lie in Data registers.

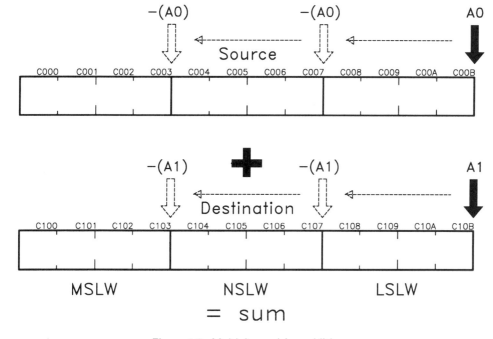

Figure 4.2: Multiple precision addition.

Wouldn't it be useful if you could tell whether the whole multiple sum or difference was zero? A normal Add or Subtract will set the Z flag if the result is zero otherwise it will clear it; thus the state of Z reflects the last addition/subtraction. However, ADDX/SUBX does not affect the Z flag when the result is zero, otherwise the flag is cleared. Thus setting the Z flag (and also clearing the X flag) and using all ADDX or SUBXs will give a final Z setting of 1 only if *all* outcomes in the sequence are zero. Use:

```
MOVE    #00000100b, CCR   ; Clears all flags, except Z = 1
```

to set up this condition.

An Address register cannot be zeroed using CLR; instead use a MOVEA #0,An or even SUBA An,An. NEGATE (NEG) is the normal 2's Complement operation (not on an Address register), but is rather unusually paired with a NEGATE WITH EXTEND (NEGX) instruction, which is used in a similar way to ADDX/SUBX for multiple-precision negations.

The use of LOAD EFFECTIVE ADDRESS (LEA) to move the result of a 6809 MPU's computed address into an Index register has been described in Sections 2.1 and 2.3. In the 68000 MPU, the destination is any Address register and the similar PUSH EFFECTIVE ADDRESS (PEA) inherently targets the System stack. We will discuss computed address modes in the next section, but some examples are:

```
LEA     8(SP),A0 ; [A0]   <- [SP] + 8, Point A0 to 8 bytes above SP
```

```
LEA  -200(PC),A1 ; [A1]   <- [PC] - 200, Point A1 to 200 bytes below PC
PEA  5(A0,D7.L)  ; [[-SP] <- [A0] + [D7(31:0)] + 5, Push into Stack the
                   contents of A0.L plus 32-bit contents of D7 plus 5
```

The middle example illustrates the use of LEA in position independent code (see Sections 2.2 and 4.2).

Signed and unsigned 16 × 16 multiplication is provided as a primitive. The Source can be anywhere in memory, a Data register or immediate data, whilst the destination must be a Data register, for example:

```
MULU 0C000h,D0   ; [D0(31:0)] <- [D0(15:0)] x [C000:1]
MULS #-7,D0      ; [D0(31:0)] <- [D0(15:0)] x -7
MULU D1,D2       ; [D2(31:0)] <- [D2(15:0)] x [D1(15:0)]
```

The Division instructions are more complex. These are designed to divide a 32-bit dividend by a 16-bit divisor, giving a 16-bit quotient in the lower word of the destination Data register and a 16-bit remainder in the upper word of the *same* register. The following code fragment shows how a dividend in **D0.L** is divided by 5000, with the quotient result placed in the the bottom of **D6** and the remainder in the bottom of **D7**:

```
DIVU #5000,D0 ; Divide the destination by the source
              ; [D0(15:0)]  <- [D0(31:0)] / 5000 (/ symbol is integer division)
              ; [D0(31:16)] <- [D0(31:0)] % 5000 (% symbol is integer remainder)
CLR.L  D6     ; Will hold the quotient
CLR.L  D7     ; Will hold the remainder
MOVE.W D0,D6  ; 16-bit quotient to D6.W
SWAP   D0     ; 16-bit remainder in lower D0
MOVE.W D0,D7  ; to D7
```

Preclearing **D6.L** and **D7.L** effectively promotes the word-moved unsigned quantities to 32 bits; it can be omitted if the upper 16 bits of these registers can be ignored. Alternatively, if DIVS is used, EXT can be utilized for a signed extension. Permitted operand address modes are the same as for MUL.

As only 16 bits are reserved for the quotient and the dividend is 32 bits, it is possible that overflow will occur. This is especially likely with a small divisor. In such cases the **V flag** will be set. If the source should be zero, then a trap will occur, as described in Section 6.2.

Four types of Shift operation are available, each in a right and left version, as shown in Table 4.3. Any Shift operation can be targeted to a *word* in read/write memory or a Data register. The former is limited to a single shift, for example:

```
LSR.W 0C000h ; Logic Shift Right the contents of C000:1 one place
```

Multiple shifts are possible if a Data register is targeted. Fixed shifts of 1 to 8 places are specified as a 3-bit code embedded in the op-code (like ADDQ). Thus:

```
LSR.L #4,D0  ; Shift all bits in D0 left 4 places
```

Table 4.3: Shifting instructions.

Operation	Mnemonic	X	N	Z	V	C	Description
Arithmetic Shift Right							Linear Shift Right keeping the sign
memory	ASR.W ea	b_0	√	√	1	b_0	
static Data reg.	ASR.s3 #d_3,Dn	b_0	√	√	1	b_0	
dynamic Data reg.	ASR.s3 Dx,Dy	b_0	√	√	1	b_0	
Logic Shift Right							Linear Shift Right
memory	LSR.W ea	b_0	√	√	0	b_0	
static Data reg.	LSR.s3 #d_3,Dn	b_0	√	√	0	b_0	
dynamic Data reg.	LSR.s3 Dx,Dy	b_0	√	√	0	b_0	
Arithmetic Shift Left							Linear Shift Left
memory	ASL.W ea	b_m	√	√	1	b_m	
static Data reg.	ASL.s3 #d_3,Dn	b_m	√	√	1	b_m	
dynamic Data reg.	ASL.s3 Dx,Dy	b_m	√	√	1	b_m	
Logic Shift Left [2]							Linear Shift Left
memory	LSL.W ea	b_m	√	√	0	b_m	
static Data reg.	LSL.s3 #d_3,Dn	b_m	√	√	0	b_m	
dynamic Data reg.	LSL.s3 Dx,Dy	b_m	√	√	0	b_m	
ROtate Right							Circular Shift Right
memory	ROR.W ea	•	√	√	0	b_0	
static Data reg.	ROR.s3 #d_3,Dn	•	√	√	0	b_0	
dynamic Data reg.	ROR.s3 Dx,Dy	•	√	√	0	b_0	
ROtate Left							Circular Shift Left
memory	ROL.W ea	•	√	√	0	b_m	
static Data reg.	ROL.s3 #d_3,Dn	•	√	√	0	b_m	
dynamic Data reg.	ROL.s3 Dx,Dy	•	√	√	0	b_m	
ROtate Right with eXtend							Circular Shift Right through X
memory	ROXR.W ea	b_0	√	√	0	b_0	
static Data reg.	ROXR.s3 #d_3,Dn	b_0	√	√	0	b_0	
dynamic Data reg.	ROXR.s3 Dx,Dy	b_0	√	√	0	b_0	
ROtate Left with eXtend							Circular Shift Left through X
memory	ROXL.W ea	b_m	√	√	0	b_m	
static Data reg.	ROXL.s3 #d_3,Dn	b_m	√	√	0	b_m	
dynamic Data reg.	ROXL.s3 Dx,Dy	b_m	√	√	0	b_m	

Note 1: Set IF most significant bit, b_m, changes, ELSE cleared.
Note 2: Identical with ASR except V flag cleared.

Alternatively, the number of shifts can be specified dynamically by the lower five
bits held in *another* Data register **Dx[4:0]**. For instance:

```
MOVEQ  #18,D7  ; [D7.L] <- 00000012h
.....  .....   ; Sometime later
LSR.L  D7,D0   ; Shift all bits in D0 left by [D7[4:0]], i.e. 18
```

As well as being able to specify a shift number larger than eight, this type of
specification has the advantage of variability, as it can be changed dynamically in
software as conditions warrant, for example in a loop.

The Logic Shift instructions simply shift in 0s from the left or right as appropriate, with the emerging bit being caught by flags **C** and **X**. ARITHMETIC SHIFT
LEFT and LOGIC SHIFT LEFT are the same, except that the V **flag** is set if the
MSbit changes. If the operand is a signed number, this would signal a sign change,
for instance $0,10011110 \rightarrow 1,0011100$. In the case of ARITHMETIC SHIFT RIGHT,
the sign bit propagates right; thus $1,1110100b$ (-12) becomes $1,1111010b$ (-6)
becomes $1,1111101b$ (-3) etc. and $0,0001100b$ $(+12)$ becomes $0,0000110b$ $(+6)$
becomes $0,0000011b$ $(+3)$ etc.

ROtate through the eXtend instructions (ROXL, ROXR) are similar to ADD with
eXtend, in that they can be used for multiple-precision operations. A ROtate
through eXtend takes in the **X** flag from any previous Shift and in turn saves its
ejected bit in **X**. As an example, a 48-bit number stored as three consecutive 16-bit
words in memory $\boxed{_{47} \quad M \quad _{32}|_{31} \quad M+2 \quad _{16}|_{15} \quad M+4 \quad _0}$ can be shifted
once right as follows[8]:

```
LSR    M     ;  0   →     ⇒  M   | b₃₂ → X
ROXR   M+2   ; b₃₂/X →    ⇒  M+2 | b₁₆ → X
ROXR   M+4   ; b₁₆/X →    ⇒  M+4 | b₀  → X
```

True circular ROtates are provided, where the shift is not through a flag (although the **C flag** still catches the emerging bit). This emerging bit is copied into
the other end of the operand word. Thus:

```
ROR.W  #8,D0  ; [D0(15:8)] <- [D0(7:0)], [D0(7:0)] <- [D0(15:8)]
```

moves the lower byte of **D0** up eight places and the next higher byte around to be
the new lower byte. This is the equivalent of SWAP.W D0 (only SWAP.L is available,
except in the 68020 MPU and up).

The three binary logic operations AND, OR, EXCLUSIVE-OR (EOR) and NOT
are provided, as shown in Table 4.4. The first two can bitwise operate on any Data
register or alterable memory location. EOR (rather inconsistently) can only use a
Data register as target. All three have an Immediate variant that can target an
alterable memory location directly or be used to change any bit or bits in the **CCR**
or **SR** (the latter only in the Supervisor state), for example:

```
ANDI.B #11111110b,CCR   ; Clear Carry flag, others unchanged
```

Table 4.4: Logic Instructions.

Operation	Mnemonic	Flags X	N	Z	V	C	Description
AND							Logic bitwise AND
to Data register	AND.s3 ea,Dn	•	√	√	0	0	[Dn] <- [Dn] · [ea]
to memory	AND.s3 Dn,ea	•	√	√	0	0	[ea] <- [ea] · [Dn]
immediate	ANDI.s3 #kk,ea[1]	•	√	√	0	0^2	[ea] <- [ea] · #kk
EOR							Logic bitwise EXclusive-OR
to Data register	EOR.s3 ea,Dn	•	√	√	0	0	[Dn] <- [Dn] ⊕ [ea]
immediate	EORI.s3 #kk,ea[1]	•	√	√	0	0^2	[ea] <- [ea] ⊕ #kk
NOT	NOT.s3 ea	•	√	√	0	0	[ea] <- $\overline{\text{[ea]}}$
OR							Logic bitwise OR
to Data register	OR.s3 ea,Dn	•	√	√	0	0	[Dn] <- [Dn] + [ea]
to memory	OR.s3 Dn,ea	•	√	√	0	0	[ea] <- [ea] + [Dn]
immediate	ORI.s3 #kk,ea[1]	•	√	√	0	0^2	[ea] <- [ea] + #kk

Note 1: Any alterable memory location, Data register, CCR or SR (privileged).
Note 2: With destination CCR or SR, all flags altered accordingly.

NOT is a single-operand instruction that inverts all 8, 16 or 32 bits in either a Data register or alterable memory. Some assemblers use COM (COMPLEMENT) as the mnemonic for this instruction.

Being able to get at individual bits of an operand directly is considered important for microcontrollers [9], but rather unusual in 16/32-bit MPUs. The 68000 MPU has four such instructions, listed in Table 4.5, which can clear, set or toggle any bit in a byte of alterable memory, or any of the 32 bits in a Data register. The bit number may be defined as a static immediate operand or dynamically held in another Data register (like the Shift instructions). All three instructions also affect the Z flag giving the state of the targeted bit *before* the operation.

The final instruction BTST does not alter the bit in question, but the Z flag still ends up reflecting its state; thus the code fragment:

```
LOOP:   BTST    #6,08080h ; How is the state of bit 6 in location 8080h?
        BEQ     LOOP      ; If it is still zero try again
```

circulates in a tight loop waiting for bit 6 of memory location $8080h$ to change to logic 1. This may be the Control register of a PIA, and thus effectively the program will be waiting for the active edge of handshake line CA2 (programmed as an input) to occur. Of course if that event never occurs, due to a hardware fault, then the system will hang up indefinitely. More about that later.

Strictly speaking BTST should be classified as a Data testing instruction, its purpose being not to change the operand but to sense its state, which is reflected in

Table 4.5: Bit-level instructions.

Operation		Mnemonic	X	N	Z	V	C	Description
Bit Test and Change								$Z = \bar{b}_n$. Toggle bit n
	dynamic	BCHG Dx,ea¹	•	•	$\bar{b}_n$	•	•	$b_{[Dx]} \leftarrow \bar{b}_{[Dx]}$
	static	BCHG #kk,ea¹	•	•	$\bar{b}_n$	•	•	$b_{\#kk} \leftarrow \bar{b}_{kk}$
Bit Test and Clear								$Z = \bar{b}_n$. Clear bit n
	dynamic	BCLR Dx,ea¹	•	•	$\bar{b}_n$	•	•	$b_{[Dx]} \leftarrow 0$
	static	BCLR #kk,ea¹	•	•	$\bar{b}_n$	•	•	$b_{\#kk} \leftarrow 0$
Bit Test and Set								$Z = \bar{b}_n$. Set bit n
	dynamic	BSET Dx,ea¹	•	•	$\bar{b}_n$	•	•	$b_{[Dx]} \leftarrow 1$
	static	BSET #kk,ea¹	•	•	$\bar{b}_n$	•	•	$b_{\#kk} \leftarrow 1$
Bit Test								$Z = \bar{b}_n$. Test bit n
	dynamic	BTST Dx,ea¹	•	•	$\bar{b}_n$	•	•	No change except in Z
	static	BTST #kk,ea¹	•	•	$\bar{b}_n$	•	•	No change except in Z

Note 1: Size is Byte if ea is out in memory, else Long if a Data register.

the **Z flag** to be used later by a Conditional Branch. The two other such instructions are COMPARE (CMP) and TEST (TST), as shown in Table 4.6. A COMPARE does a subtraction of the source operand from the destination operand (as does SUB), setting the flags accordingly but not putting the difference into the destination. A TEST FOR ZERO OR NEGATIVE is just a COMPARE with a zero source operand (i.e. TST D0 is the same as CMP #0,D0).

There are four varieties of COMPARE available. The 'plain vanilla' CMP can use any memory contents, immediate data, Data register or Address register as source to be compared with a Data register, for example:

```
CMP.W  #56,D0   ; Compare [D0(15:0)] with the number 56, [D0(15:0)]-56
CMP.B  123h,D1  ; Compare [D1(7:0)] with the contents of 123h, [D1(7:0)]-[123h]
CMP.L  A0,D2    ; Compare [D2(31:0)] with [A0(31:0)], [D2(31:0)]-[A0(31:0)]
```

Notice the comparison is *destination with source*, just as SUB is subtract source from destination. Some processor assemblers, such as for the PDP-11 minicomputer and 80x86 family MPUs, reverse the order.

CMPA is used with Address register destinations. Unlike other such targeted instructions (e.g. ADDA), the **CCR** flags are set normally, but with word-length source operands sign extended in the usual way to a long-word, for example:

```
CMPA.W #8000h,A0 ; [A0(31:0)] is compared with FFFF8000h (-32,768)
CMPA.L 1234h,A1  ; Compare [A1(31:0)] with [1234:5:6:7]
CMPA.L    D0,A2  ; Compare [A2(31:0)] with [D0(31:0)]
```

Table 4.6: Data testing instructions.

Operation	Mnemonic		Flags X N Z V C					Description
Compare								Non-destructive [destn] − [source]
Data reg. with	CMP.s3[1]	ea,Dx	•	√	√	√	√	[Dx] − [ea]
Addr. reg. with	CMPA.s2	ea,Ax	•	√	√	√	√	[Ax] − [ea]
Mem. with const.	CMPI.s3	#kk,ea	•	√	√	√	√	[ea] − #kk
Mem. with mem.	CMPM.s3	(Ay)+,(Ax)+	•	√	√	√	√	[[Ax]+] − [[Ay]+]
Test for Zero or Minus								Non-destructive [destination] − 0
	TST.s3 ea[2]		•	√	√	0	0	[ea]−00

Note 1: Only Word and Long if source is Address register.
Note 2: Only alterable memory and Data register, not Address register.

An immediate quantity can be compared to any alterable memory or Data register by using CMPI, for example:

```
CMPI.B  #64,1234h ; Compare [1234h] with 64
```

Memory can be directly compared to memory with a COMPARE MEMORY (CMPM). In this case only the Post-Increment address mode is available, as CMPM is primarily designed as a Block-Compare primitive. For instance, the following code fragment exits with the address+1 of the first pair of bytes which differ in two blocks of data or strings:

```
       MOVEA.L #BLOCK_1,A0  ; Point A0 to bottom of Block 1
       MOVEA.L #BLOCK_2,A1  ; Point A1 to bottom of Block 2
CLOOP: CMPM.B  (A0)+,(A1)+  ; Compare bytes and move each pointer on one
       BEQ     CLOOP        ; IF same THEN next
       .....   ..........   ; ELSE continue
```

The TEST primitive is represented by the TST instruction. This can check that the contents of any memory location or Data register is zero (sets **Z flag**) or negative (sets **N flag**), for example:

```
TST.B  1234h    ; Test contents of 1234h for zero or negative
TST.W  D0       ; Test lower 16 bits of D0 for zero or negative
```

The Block-Test code fragment above followed the Comparison operation by the Conditional Branch (BEQ). Branch instructions add an offset to the **Program Counter** if the condition is True ($Z = 1$ in the example) otherwise its state remains pointing to the following instruction. There is also an Unconditional Branch, BRA, which always adds the offset. Two sizes of Branches are available, Short (or byte)

Table 4.7: Instructions which affect the Program Counter.

Operation	Mnemonic	Description
Unconditional Program Transfer		Always goto
Branch to Label	BRA Offset[1]	Offset always added to PC, relative goto
Jump to Label	JMP ea	[PC] <- ea, absolute goto
Conditional Program Transfer		Goto IF condition is True
Branch to Label	Bcc[2] Offset	Offset added onto PC IF condition is met
Test, Decrement & Branch	DBcc[2] Dx,Offset	Repeat loop until any condition is met
		IF condition is True THEN exit loop
		ELSE
		[Dx(15:0)] <- [Dx(15:0)] - 1
		IF [Dx(15:0)] = -1 is True THEN exit loop
		ELSE
		[PC] <- [PC] + Offset (continue loop)
No Operation		Does nothing except increment PC by 2
	NOP	[PC] <- [PC] + 2, takes 4~

Note 1: Normally a label is specified here and the assembler works out the offset.
Note 2: The condition codes (cc) are:

			True on				True on
0000	T[3]	True always	Always	1000	VC	oVerflow Clear	$V = 0$
0001	F[3]	False always	Never	1001	VS	oVerflow Set	$V = 1$
0010	HI	HIgher than	$C+Z = 0$	1010	PL	PLus	$N = 0$
0011	LS	Lower or Same	$C+Z = 1$	1011	MI	MInus	$N = 1$
0100	CC	Carry Clear	$C = 0$	1100	GE	Greater or Equal	$N \oplus V = 0$
0101	CS	Carry Set	$C = 1$	1101	LT	Less Than	$N \oplus V = 1$
0110	NE	Not Equal	$Z = 0$	1110	GT	Greater Than	$\overline{N \oplus V} \cdot \overline{Z} = 1$
0111	EQ	EQual	$Z = 1$	1111	LE	Less or Equal	$\overline{N \oplus V} \cdot \overline{Z} = 0$

Note 3: Only for DBcc.

which carries an 8-bit signed offset as part of the op-code, and Word, where a 16-bit signed offset follows the op-code word. The 68020 allows a Long Branch.

There are 14 combinations of the C, Z, N and V flags which can be used as a test for a Conditional Branch. The X flag is reserved exclusively for multiple-precision arithmetic and does not take part in this exercise. With the exception of the somewhat useless BRANCH NEVER (BRN), all 6809 Conditional Branches listed in Table 2.6 are also available to the 68000 family. The mathematical significance of the various flag combinations are given on page 28 and will not be repeated here. In Table 4.7 these tests are listed as 4-bit code combinations (cc). All Branch op-codes start with 0110b followed by the cc code, followed on by the the 8-bit displacement if Short or all zeros if Word. In the latter case the 16-bit displacement follows the op-code. Thus the instruction BPL .06 (BRANCH IF PLUS six places on) is coded as 0110-1010-00000110b (6A06h).

In the 68000 family the cc tests can be used with other instructions, the most useful of which are the Decrement, Test and Branch loop operations. We have already used software loops, for example the Block-Compare routine on page 101. Essentially a loop is a mechanism in which a section of code can either be repeated a fixed number of times (the loop count) or exit when a certain condition or conditions are fulfilled, or both.

As an example of the latter situation, consider interfacing to a peripheral which sets bit 6 of an interface device's Control register (e.g. a 6821 PIA) when it has valid data it wishes to be read. This involves continually checking the state of bit 6 in a loop until it goes high; only then do we move on and read the data. But what happens if, say, due to hardware malfunction, this Data Ready signal is never sent? The software will then hang indefinitely. Perhaps it would be better to give up after a fixed number of times and go to an error routine if this sequence of events happens. To do this we would have to check the flag; if it is not set, then decrement the loop count, and if this hasn't fallen through zero (i.e. to −1) then repeat. Following the structure of Fig. 4.3 a possible coding is:

```
        MOVE.W    #n,D1        ; Set loop count n
LOOP:   BTST      #6,CONTROL   ; Test bit 6 of the Control register    16~
        BNE       EXIT         ; IF True THEN EXIT (cc=Not Equal Zero) 12~
        SUBQ.W    #1,D1        ; ELSE decrement loop count              4~
        BCC       LOOP         ; IF no Carry then [D1] is not -1       18~
EXIT:   CMP       #-1,D1       ; Exit with n = -1?
        BEQ       ERROR        ; IF True THEN error
        MOVE.B    PORT,D0      ; ELSE read data from port
        .....     .......      ; and continue
```

The alternative combines the decrement and two tests thus:

```
        MOVE.W    #n,D1        ; Set loop count n (max 65535)
LOOP:   BTST      #6,CONTROL   ; Test bit 6 of the Control register    16~
        DBNE      D1,LOOP      ; Decrement and repeat loop until True  18~
; Pass here either IF True that bit 6 is 1 OR True that n = -1
EXIT:   CMP       #-1,D1       ; Exit with n = -1?
        BEQ       ERROR        ; IF True THEN Error
        MOVE.B    PORT,D0      ; ELSE read data from port
        .....     .......      ; and continue
```

For applications where speed is important (not this example) reducing the time taken by the control mechanism is important, as this housekeeping overhead is executed on each pass through the loop body. In this case the Test and Control is 34~ as against 50~. Notice that BNE is shown with an execution time of 12~, whilst BCC is 18~. This is because Branches taken (i.e. True) for byte offsets take longer than Branches not taken (but the opposite for word offsets, 18~ and 20~!). Similarly DBcc has a variable execution time. As the number n used by DBcc is

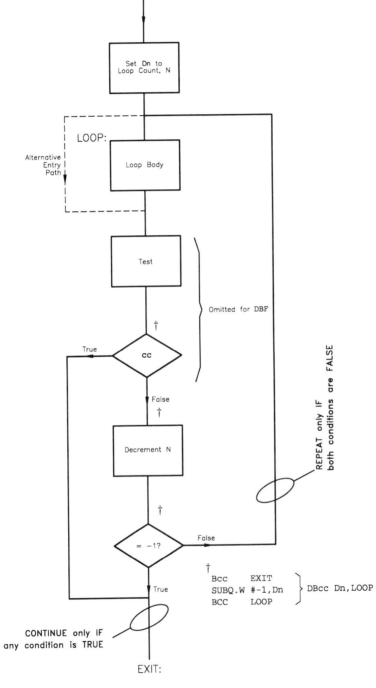

Figure 4.3: Using DBcc to implement a loop structure.

limited to 65,536, the ordinary Branch construction must be used where the default timeout parameter exceeds this number.

Some situations require the number of loop passes to be fixed. As the normal DBcc exits if either test is True, the variant DBF makes the first test always False, and so an exit only happens when the loop count reaches -1. The routine below, which is a fixed delay using an idle loop body, shows this:

```
DELAY:  MOVE.W  #n,D0     ; n is the delay parameter   16~
LOOP:   NOP               ; Do nothing and take         4~
        DBF     D0,LOOP   ; one less pass              18~
```

The total delay here is $16 + (n + 1) \times 22$ (+8 extra when DBF is True), a total of $46 + 22n$ clock cycles. Thus a 0.1 s delay requires $46 + 22n = 8 \times 10^5 \, \mu s$ at a clock rate of 8 MHz, giving $n = 36,363$. Remember that n has a maximum value of 65,536 for DBF.

Using the DBcc construct, the number of loop passes is $n + 1$, where n is the word-sized number preloaded into a Data register. It is possible initially to enter the loop directly into the control mechanism, as shown dashed in Fig. 4.3, in which case the number of passes is just n. In this case no passes through the loop body will occur if $n = 0$ or the test is True (WHILE-DO loop construct) whereas the former situation always includes one pass irrespective (DO-WHILE loop construct). An example of this is shown in Table 5.4.

The DBcc instruction can be confusing because it operates in the opposite sense to the analagous Bcc. Thus BEQ LOOP causes control to be passed to LOOP if the conditional test outcome is True (i.e. $Z = 0$). The similar DBEQ LOOP does *not* transfer control to LOOP if the outcome is True, that is the processor escapes from the loop. Using the terminology 'Decrement and Branch *until* True' as opposed to 'Branch *if* True' may help clarify the situation. Table 4.8 summarizes the complete 68000/8 instruction set. For each instruction, its operand size is given and allowable address modes are given. Finally, its effect on the five flags in the Code Condition register is tabulated.

Table 4.8: Summary of 68000 instructions (*continued next page*).

Instruction	Size	#	Dn	An	(An)	(An)+	-(An)	$\pm d_{16}$(An)	$\pm d_8$(An,Ri)	A.L	A.W	$\pm d_{16}$(PC)	$\pm d_8$(PC,Ri)	X	N	Z	V	C
ABCD Dx,Dy	B													√	U	√	U	√
ABCD -(Ax),-(Ay)	B													√	U	√	U	√
ADD [ea],Dx	BWL	*	*	*	*	*	*	*	*	*	*	*	*	√	√	√	√	√
ADD Dx,[ea]	BWL				*	*	*	*	*	*	*			√	√	√	√	√
ADDA [ea],Ax	WL	*	*	*	*	*	*	*	*	*	*	*	*	•	•	•	•	•
ADDI #K,[ea]	BWL		*		*	*	*	*	*	*	*			√	√	√	√	√
ADDQ #K_3,[ea]	BWL		*	*	*	*	*	*	*	*	*			√	√	√	√	√
ADDX Dx,Dy	BWL													√	√	√	√	√
ADDX -(Ax),-(Ay)	BWL													√	√	√	√	√
AND [ea],Dx	BWL	*	*		*	*	*	*	*	*	*	*	*	•	√	√	0	0
AND Dx,[ea]	BWL				*	*	*	*	*	*	*			•	√	√	0	0
ANDI #K,[ea]	BWL		*		*	*	*	*	*	*	*			•	√	√	0	0
ANDI #K,CCR	B													√	√	√	√	√
ANDI #K,SR^P	W													√	√	√	√	√
ASL/R Dx,Dy	BWL													√	√	√	√	√
ASL/R #d_3,Dx	BWL													√	√	√	√	√
ASL/R [ea]	BWL				*	*	*	*	*	*	*			√	√	√	√	√
Bcc [label]	BW													•	•	•	•	•
BRA [label]	BW													•	•	•	•	•
BSR [label]	BW													•	•	•	•	•
BCHG Dx,[ea]	BL		L		B	B	B	B	B	B	B			•	•	√	•	•
BCHG #K,[ea]	BL		L		B	B	B	B	B	B	B			•	•	√	•	•
BCLR Dx,[ea]	BL		L		B	B	B	B	B	B	B			•	•	√	•	•
BCLR #K,[ea]	BL		L		B	B	B	B	B	B	B			•	•	√	•	•
BSET Dx,[ea]	BL		L		B	B	B	B	B	B	B			•	•	√	•	•
BSET #K,[ea]	BL		L		B	B	B	B	B	B	B			•	•	√	•	•
BTST #K,[ea]	BL		L		B	B	B	B	B	B	B			•	•	√	•	•
BTST Dx,[ea]	BL		L		B	B	B	B	B	B	B			•	•	√	•	•
CHK [ea],Dx	W	*	*	*	*	*	*	*	*	*	*	*	*	•	√	U	U	U
CLR [ea]	BWL		*		*	*	*	*	*	*	*			•	0	1	0	0
CMP [ea],Dx	BWL	*	*	*	*	*	*	*	*	*	*	*	*	•	√	√	√	√
CMPA [ea],Dx	WL	*	*	*	*	*	*	*	*	*	*	*	*	•	√	√	√	√
CMPI #K,Dx	BWL		*		*	*	*	*	*	*	*			•	√	√	√	√
CMPM (Ax)+,(Ay)+	BWL													•	√	√	√	√
DBcc Dx,[label]	W													•	•	•	•	•
DIVS [ea],Dx	W	*	*		*	*	*	*	*	*	*	*	*	•	√	√	√	0
DIVU [ea],Dx	W	*	*		*	*	*	*	*	*	*	*	*	•	√	√	√	0
EOR Dx,[ea]	BWL		*		*	*	*	*	*	*	*			•	√	√	0	0
EORI #K,[ea]	BWL		*		*	*	*	*	*	*	*			•	√	√	0	0
EORI #K,CCR	W													√	√	√	√	√
EORI #K,SR^P	W													√	√	√	√	√
EXG Dx,Dy	L													•	•	•	•	•
EXT Dx	WL													•	√	√	0	0
ILLEGAL														•	•	•	•	•
JMP [ea]					*			*	*	*	*	*	*	•	•	•	•	•
JSR [ea]					*			*	*	*	*	*	*	•	•	•	•	•
LEA [ea],Ax	L				*			*	*	*	*	*	*	•	•	•	•	•
LINK Ax,#K														•	•	•	•	•
LSL/R Dx,Dy	BWL													√	√	√	0	√
LSL/R #d_3,Dx	BWL													√	√	√	0	√
LSL/R [ea]	BWL				*	*	*	*	*	*	*			√	√	√	0	√

√ : Flag operates in the normal manner.   • : Not affected.   U : Undefined.
P : Privileged.   * : Available. d_n : n-bit displacement.
#K_m : m-bit immediate number.   ± : Sign extended

Table 4.8: (*continued*) Summary of 68000 instructions.

Instruction	Size	#	Dn	An	(An)	(An)+	-(An)	±d_{16}(An)	±d_8(An,Ri)	A.L	A.W	±d_{16}(PC)	±d_8(PC,Ri)	X	N	Z	V	C
MOVE [ea],[ea]	BWL	*^{S}	*	*^{S}	*	*	*	*	*	*	*	*^{S}	*^{S}	•	√	√	0	0
MOVE [ea],CCR	W	*	*		*	*	*	*	*	*	*	*	*	√	√	√	√	√
MOVE SR,[ea]	W		*		*	*	*	*	*	*	*			•	•	•	•	•
MOVE [ea],SR^P	W	*	*		*	*	*	*	*	*	*	*	*	√	√	√	√	√
MOVE USP,Ax^P	L			*										•	•	•	•	•
MOVEA Ax,USP^P	L													•	•	•	•	•
MOVEA [ea],Ax	WL	*	*	*	*	*	*	*	*	*	*			•	•	•	•	•
MOVEM [ΣR],[ea]	WL				*		*	*	*	*	*			•	•	•	•	•
MOVEM [ea],[ΣR]	WL				*	*		*	*	*	*	*	*	•	•	•	•	•
MOVEP Dx,±d_{16}(Ay)	WL							*						•	•	•	•	•
MOVEP ±d_{16}(Ay),Dx	WL							*						•	•	•	•	•
MOVEQ #±K_8,Dn	L													•	√	√	0	0
MULS [ea],Dx	W	*	*		*	*	*	*	*	*	*	*	*	•	√	√	0	0
MULU [ea],Dx	W	*	*		*	*	*	*	*	*	*	*	*	•	√	√	0	0
NBCD [ea]	B		*		*	*	*	*	*	*	*			√	U	√	U	√
NEG [ea]	BWL		*		*	*	*	*	*	*	*			√	√	√	√	√
NEGX [ea]	BWL		*		*	*	*	*	*	*	*			√	√	√	√	√
NOP														•	•	•	•	•
NOT [ea]	BWL		*		*	*	*	*	*	*	*			•	√	√	0	0
OR [ea],Dx	BWL	*	*		*	*	*	*	*	*	*			•	√	√	0	0
OR Dx,[ea]	BWL				*	*	*	*	*	*	*			•	*	√	0	0
ORI #K,[ea]	BWL	*			*	*	*	*	*	*	*			•	*	√	0	0
ORI #K,CCR	B													√	√	√	√	√
ORI #K,SR^P	W													√	√	√	√	√
PEA [ea]	L				*			*	*	*	*	*	*	•	•	•	•	•
RESET^P														•	•	•	•	•
ROL/R Dx,Dy	BWL													•	√	√	0	√
ROL/R #d_3,Dx	BWL													•	√	√	0	√
ROL/R [ea]	BWL				*	*	*	*	*	*	*			•	√	√	0	√
ROXL/R Dx,Dy	BWL													√	√	√	0	√
ROXL/R #d_3,Dx	BWL													√	√	√	0	√
ROXL/R [ea]	BWL				*	*	*	*	*	*	*			√	√	√	0	√
RTE^P														√	√	√	√	√
RTR^P														√	√	√	√	√
RTS														•	•	•	•	•
SBCD Dx,Dy	B													√	U	√	U	√
SBCD -(Ax),-(Ay)	B													√	U	√	U	√
Scc [ea]	B		*		*	*	*	*	*	*	*			•	•	•	•	•
STOP^P														√	√	√	√	√
SUB [ea],Dx	BWL	*	*	*	*	*	*	*	*	*	*	*	*	√	√	√	√	√
SUB Dx,[ea]	BWL				*	*	*	*	*	*	*			√	√	√	√	√
SUBA [ea],Ax	WL	*	*	*	*	*	*	*	*	*	*	*	*	•	•	•	•	•
SUBI #K,[ea]	BWL		*		*	*	*	*	*	*	*			√	√	√	√	√
SUBQ #K_3,[ea]	BWL		*	*	*	*	*	*	*	*	*			√	√	√	√	√
SUBX Dx,Dy	BWL													√	√	√	√	√
SUBX -(Ax),-(Ay)	BWL													√	√	√	√	√
SWAP Dx	W													•	√	√	0	0
TAS [ea]	B		*		*	*	*	*	*	*	*			•	√	√	0	0
TRAP #K_4														•	•	•	•	•
TRAPV														•	•	•	•	•
TST [ea]	BWL		*		*	*	*	*	*	*	*			•	√	√	0	0
UNLK Ax														•	•	•	•	•

√ : Flag operates in the normal manner.   • : Not affected.   U : Undefined.
P : Privileged.   S : Source only.   * : Available.
d_n : n-bit displacement.   #K_m : m-bit immediate number.   ± : Sign extended

4.2 Address Modes

Except for the few inherent operations which do not require data, such as RETURN FROM SUBROUTINE (RTS), some part of the instruction must be used to specify where or how to calculate the whereabouts of the operand(s). Broadly there are three methods of specifying an **effective address** (ea):

1. Constant (fixed) data: Immediate. Here the data is part of the instruction and usually follows the op-code. Some instructions have quick varieties, such as ADDQ, which embed small immediate numbers (e.g. 1 to 8) in the op-code itself.

2. Fixed location: Absolute memory or Register direct. The fixed memory address follows the op-code, or a register is specified as part of the op-code.

3. Variable location: Address register or Program Counter register Indirect with optional fixed and variable offsets, where a register points to the operand. As such register contents can be changed in software at run time, the effective address is a variable.

The use of the more complex address modes of category 3 are important in high-level language where data is often allocated space relative to a Stack Pointer rather than in absolute addresses. Computed addresses are also useful in accessing data structures such as arrays and in producing position independent code, see page 40.

As an illustrated example, consider the problem of clearing an array of 1024 bytes located between E000 and E03Fh. Using only Absolute addressing, the routine would look something like this:

```
CLEAR_ARR:   CLR.B  0E000h      ; Clear ARRAY[0]
             CLR.B  0E001h      ; and ARRAY[1]
             CLR.B  0E002h      ; each CLR occupies 6 bytes
             CLR.B  0E003h      ; of program memory
             CLR.B  0E004h      ; and takes 20 clock cycles
             CLR.B  0E005h      ; Keep on going
                 ............
                 ............
             CLR.B  0E3FFh      ; Clear ARRAY[1023]. Phew!
```

This routine occupies 6144 bytes of program memory and takes 20,480 clock cycles (2560 μs at 8 MHz) to execute.

As we need to repeat the same operation 1024 times, clearly we have a prime candidate for using a loop construction, thus:

```
CLEAR_ARR: MOVEA.L #0E000h,A0  ; Point A0 to ARRAY[0]
           MOVE.W  #1023,D0    ; Set up loop count less 1 in D0.W
CLOOP:     CLR.B   (A0)+       ; While [D0.W] > -1
; Clear Array element pointed to by A0 and move pointer on one byte
           DBF     D0,CLOOP    ; Decrement loop count, exit on D0.W = -1
```

This routine occupies $6 + 4 + 2 + 4 = 16$ bytes of program memory and takes $4 + 4 + (12 \times 1024) + (10 \times 1023) + 14 = 22,540$ clock cycles ($2817.5\,\mu s$ at 8 MHz). In the first two instructions Immediate addressing is used to place constants. The loop body uses Address Register Indirect with Post-Increment addressing to walk through the array. **Address register_A0** holds the address of the array element, and, after that address has been put out on the bus, is automatically incremented. Although the execution time of this address mode is shorter than for Absolute, as the address does not have to be fetched after the op-code, this is more than made up for by the overhead of the loop control DBF instruction, which takes 10~ when the loop is re-entered and 14~ for the final exit. Thus the quid pro quo for the reduction of program memory by a factor of 38,400% is an increase in execution time of around 10%.

The rest of this section looks at the available address modes. Sizes are given for single-operand instructions, double-operands may require additional extension words.

Inherent

op-code

Inherent instructions make implicit reference to a register or registers. Thus RTS implies the use of the **SSP** and **PC** registers. The Branch instructions are sometimes listed under this category, implying the **PC** register; however, they can also be thought of as using a type of Program Counter with Displacement address mode.

Immediate, #kk

op-code		3 or ±8-bit (Quick)
op-code	constant	8/16-bit (.B or .W)
op-code	constant	32-bit (.L)

Here the operand is the data itself, not an address or pointer to an address. Generally the constant follows the op-code as one or two words. Three instructions have Quick-Immediate variants where the data is embedded in the op-code itself, MOVEQ reserves 8 bits for the signed constant (+127 to −128) and ADDQ/SUBQ can only be used for unsigned 3-bit constants 1 to 8 ($000b$ represents 8 here). The instruction variants ADDI/SUBI permit constants of any applicable size to be added or subtracted directly on alterable memory locations, rather than on Data registers. Some examples are:

```
ADD.L   #1,D0        ; [D0(31:0)]<-[D0(31:0)]+1    (16~) {Coded D0BC-0000-0001h}
ADDQ.L  #1,D0        ; [D0(31:0)]<-[D0(31:0)]+1    ( 8~) {Coded 5280h}
ADDQ.W  #1,0E000h    ; [E000:1]  <-[E000/1]  +1    (20~) {Coded 5279-0000-E000h}
ADDI.W  #56h,0E000h  ; [E000:1]  <-[E000/1]  +56h  (24~) {Coded 0679-0056-0000-E000}
```

Notice the difference in size and execution time between the top two examples, which do the same thing. Of course ADDQ is limited to operand sizes of up to only eight. The difference between ADDQ and ADDI for alterable memory destinations is not so great, but still significant.

Direct or Absolute modes

Three submodes are available which specify that the operand is in either a Data register, Address register or in absolute memory.

Data Register Direct, Dn

op-code

The vast majority of instructions use a Data register as the destination, source or both — as listed in Table 4.8. The op-code itself holds the register number(s) (see Fig. 4.4), so instructions using this address mode are short and also execute faster. Thus, where convenient, variables should be kept in a register. The first two examples under the Immediate heading also used Data Register Direct addressing as the destination; some other possibilities are:

```
ADD.L  D0,D1       ; [D1(31:0)] <- [D1(31:0)] + [D0(31:0)]  {Coded as D280h}
ADD.B  D1,0E000h; [E000]     <- [E000]     + [D1(31:0)]  {Coded as D339-0000-E000h}
```

Address Register Direct, An

op-code

Addresses stored in an Address register can point to data for most instructions, but only the special instructions ADDA, SUBA and MOVEA can also target and hence change these pointers. The ADDQ and SUBQ variants can also target any Address register in .W or .L sizes. They are useful to increment or decrement pointers. Some examples are:

```
ADD.L  A0,D0       ; [D0(31:0)] <- [D0(31:0)]+[A0(31:0)]   {Coded as D188h}
ADDA.W #8000h,A1 ; [A1(31:0)] <- [A1(31:0)]+FFFF8000h    {Coded as D2FC-8000h}
SUBQ.L #1,A1       ; [A1(31:0)] <- [A1(31:0)]-00000001h    {Coded as 5389h}
```

Note again that any operation changing Address register contents *always* acts on all 32 bits, and if word-sized (no byte size allowed), will be sign extended as shown in the second example above.

Memory Direct (or Absolute), M

op-code	address

op-code	address

±16-bit (Short)

32-bit (Long)

The absolute address itself directly follows the op-code in this mode. In the short-form version, only a 16-bit address is specified, and this is sign-extended in the usual manner before being sent out on to the address bus. The applicable range for this is *00007*FFF*h* to *00000000h* and *FFFFFFFFh* to *FFFF*8000*h*. Conceptualizing the memory map as a grand circle, this can be thought of as a range from +0 up to +32,767 and back to −32,768. The long form will of course specify any address directly, but occupies an extra word of program memory and thus takes an extra Read cycle (4˜) during the fetch phase. Two examples are:

```
MOVE.W 500h,D0  ; [D0(15:0)] <- [00000500:1] {Coded as 3038-0500h}
MOVE.W 9000h,D0 ; [D0(15:0)] <- [00009000:1] {Coded as 3039-0000-9000h}
```

Absolute addresses are by definition constant as part of the program (except with risqué self-modifying code) and as such are most useful for specifying data from I/O ports, which are fixed in the memory map by virtue of their hardware decoder.

Register Indirect Modes

The most flexible of the address modes; this group generates the effective address (ea) as a simple function of the contents of an Address register or the **Program Counter**. As the state of such a register is not constant, it may be changed at any time to reflect the current storage requirements of the program, and may be systematically advanced or retarded to deal with arrays or other data structures. The opening example of this section on page 108 demonstrated this flexibility.

Address Register Indirect, (An)

| op-code |

Here an Address register holds the location of the operand in memory, that is points to the operand. The term Indirect is used, as the register does not hold the data itself. Thus:

```
MOVEA.L #0E100h,A0 ; [A0(31:0)] <- #0000E100     {Coded as 207C-0000-E100h}
ADD.B   (A0),D0    ; Adds contents of E100h to D0 {Coded as D010h}
```

has the same affect as ADD.B 0E000h,D0, but of course once **A0** is set up, the shorter and faster indirect access can be used, and the target address dynamically altered by changing the contents of **A0**.

Address Register Indirect with Displacement, $\pm d_{16}$(An) or ($\pm d_{16}$,An)

| op-code | displacement |

Similarly to the previous mode, a 16-bit displacement is used to define a signed offset of between +32,767 to −32,768. As an example, if we assume that we have two arrays, one starting at E000*h* and the other at E200*h*, then, assuming **A0** has been pointed to E100*h* by the previous example, the sequence:

```
MOVE.B  -100h(A0),D0  ; Get ARRAY_1[0]              {Coded as 1028-FF00h}
ADD.B    100h(A0),D0  ; and add to it ARRAY_2[0]    {Coded as D028-0100}
```

puts the sum of the first two array elements in **D0.B**.

Of course the displacement is a fixed part of the program, but if necessary we can still change the base address in **A0**.

Address Register Indirect with Pre-decrement/Post-increment, -(An)/(An)+

> op-code

There are two modes here, both of which automatically modify the designated Address register, which points to the operand. The former decrements the effective address by one, two or four for a byte, word or long object respectively *before* the operation. In the latter case the Address register holds the ea, which, *after* the operation is complete, is incremented by the appropriate one, two or four.

We have already illustrated these modes in use, see Fig. 4.1 and the opening example on page 108, where we cleared an array. As a further example, which also uses the previous indirect modes, consider the problem of digitally low-pass filtering this same array. Taking the 1024 byte-array elements already stored between locations **E000** and **E3FF**h as samples in advancing time, originating from, say, an analog to digital converter, then the 3-point algorithm [9] is given as:

$$Y[n] = \frac{X[n]}{2} + \frac{X[n-1]}{4} + \frac{X[n-2]}{2}$$

where n is the sample number, $X[n]$ the existing nth array sample and $Y[n]$ the new filtered nth array element.

The following listing starts at the top of the X array and works its way down *overwriting* this with the new Y array:

```
       MOVEA.L  #0E400h,A0  ; Point A0 to one past X[1023]
LOOP:  MOVE.B   -(A0),D0    ; Decrement pointer and then get X[n]
       LSR.B    #2,D0       ; Divide by 4
       MOVE.B   -1(A0),D1   ; Get X[n-1] (A0 unchanged)
       LSR.B    #1,D1       ; Divide by 2
       ADD.B    D1,D0       ; Y[n] = X[n]/4 +X[n-1]/2
       MOVE.B   -2(A0),D1   ; Get X[n-2] (A0 unchanged)
       LSR.B    #2,D1       ; Divide by 4
       ADD.B    D1,D0       ; Y[n] = X[n]/4 + X[n-1]/2 + X[n-2]/4
       MOVE.B   D0,(A0)     ; Overwrite X[n] by Y[n]
       CMPA.L   #0E002h,A0  ; Check for end, cannot go lower than X[2]
       BNE      LOOP        ; IF not repeat with n to be decremented
       RTS                  ; Exit
```

Notice that **A0** points to $X[n]$ and it is automatically decremented on each pass through the loop. Notice also the edge effect in that $Y[0] = X[0]$ and $Y[1] = X[1]$.

Address Register Indirect with Index, $\pm d_8$(An,X.W) / $\pm d_8$(An,X.L) or ($\pm d_8$,An,X.W) / ($\pm d_8$,An,X.W)

| op-code | X reg./disp. |

This mode offsets the contents of a designated Address register with both a constant and a variable to give the effective address. The variable index can be the contents of any Address or Data register. Either the entire 32 bits (.L) or a sign-extended 16 bits (.W) can be used. The constant is a signed 8-bit byte. Thus we have:

$$\text{<ea>} = \pm d_8 + \text{X.L} + \text{An} \quad \text{or} \quad \pm d_8 + \text{SEX|X.W} + \text{An}$$

As an example consider a subroutine to convert a decimal 0–9, passed in **D0.B** to its 7-segment equivalent returned in the same place. The 7-segment equivalents are stored sequentially as a table (array) of 10 bytes following the subroutine. We assume the subroutine starts at $0600h$.

```
(600/605)                              MOVEA.L #TABLE_BOT,A0 ; Point A0 to table
(606/609 1030-0000)                    MOVE.B  0(A0,D0.W),D0 ; Get element [D0(15:0)]
(60A/60B 4E75)                         RTS                   ; and return

(60C/610 01-4F-12-06-4C) TABLE_BOT:.BYTE  1,4Fh,12h,6,4Ch; 7-segment code
(611/615 24-20-0F-00-0C)          .BYTE  24h,20h,0Fh,0,0Ch
```

If we assume that **D0.W** is $0004h$ on entry, then the first instruction puts the absolute address of the first table element ($060Ch$) into **A0**. The effective address calculated in the following instruction is 00 + [A0] + SEX|D0(15:0), in this case $00 + 0000060C + 00000004 = 00000610h$. The data in this byte is $4Ch$, and this is the value moved to **D0(7:0)** prior to return.

As can be seen from this example, this mode is useful for random access into an array, with the array number (or a multiple of, for word or long-word arrays) being in the Index register. It is instructive to compare this example with its equivalent 6809 code on page 40, which used an Accumulator to hold the variable offset and one of the Index registers to hold the base address.

Program Counter Indirect with Displacement, $\pm d_{16}$(PC) or ($\pm d_{16}$,PC)

| op-code | displacement |

This is similar to Address Register Indirect with Displacement but this time the Program Counter is the specified register. For example in:

```
MOVE.B   200h(PC),D0  ; [D0(7:0)] <- [[PC]+200h]
```

the data $200h$ bytes on from where the **PC** is (actually pointing to the next instruction) is placed in **D0.B**. Of course this is not an absolute address, as only the *distance* from the instruction is of interest. Like the relative Branch instructions, a label is normally used for the destination and the assembler evaluates the appropriate offset.

Program Counter Indirect modes are used to generate position independent code (PIC) as described on page 40. As an example, referring back to the 7-segment decoder just listed, we see in line 1 that the absolute address of the table base, 0000060Ch, is placed in Address register **A0.L**. If, say, the subroutine were to be relocated to start at 1780h, then the ROM would have to be reprogrammed to change the extension word of the MOVEA instruction from 060Ch to 178Ch, the rest of the code remaining the same. Here is a PIC version of the same subroutine:

```
(600/603 41FA-0006)         LEA        6(PC),A0     ; Point A0 to table
(604/607 1030-0000)         MOVE.B     0(A0,D0.W),D0 ; Get element [D0(7:0)]
(608/609 4E75)              RTS                     ; and return
(60A/13  ....) TABLE_BOT:.BYTE    etc.              ; 7-segment code
```

The only difference between the two programs is in line 1. Previously the *absolute* address of the table bottom was put into **A0**. In the PIC case, **A0** is loaded with the contents of **PC** plus 6, which is again the address of the bottom of the table, but is calculated at run time. If we were to relocate the subroutine to start at 1780h, nothing would change.

In practice, if the first line of the program were:

```
LEA   TABLE_BOT(PC),A0
```

the assembler would produce the same code (41FA-0006h), evaluating the *difference* between TABLE_BOT and the location of the following instruction, that is 6 bytes. The absolute value of TABLE_BOT is not used as the offset — as in the case of Branch instructions.

Note the use of LOAD EFFECTIVE ADDRESS to move the ea generated by any address mode (except Pre-Decrement and Post-Increment) into an Address register. Some other examples are:

```
LEA      20(A7),A7       ; Move Stack Pointer up 20 bytes
LEA 20(A0,D7.L),A1       ; Add A0.L to D7.L plus 20 and put into A1.L
```

LEA is long-word sized only, and must solely target an Address register.

Program Counter Indirect with Index, $\pm d_8(PC,X.W)$ / $\pm d_8(PC,X.L)$ or
$(\pm d_8,PC,X.W)$ / $(\pm d_8,PC,X.L)$

| op-code | X reg./disp. |

This is similar to Address Register Indirect with Index in that a constant offset plus a variable offset in either an Address or Data register is added to the **PC** to give an effective address. The assembler permits a label to be used as the constant, and will calculate the required difference. Using this mode the 7-segment program reduces to:

```
(600/603 103B-0002)         MOVE.B   TABLE_BOT(PC,D0.W),D0
(604/5   4E75)              RTS
(606/F   ....) TABLE_BOT:   .BYTE    etc.
```

Note the offset of 02 in the machine code generated by the first instruction.

The offset permissible for this mode is only $+127$ to -128, which represents a considerable limitation compared to the plain offset-mode with a range of $+32,767$ to $-32,768$ (both ranges have been extended for the 68020 MPU).

The twelve address modes covered there are summarized in Table 4.9. Except for the two Register Direct modes, additional time is needed to calculate the effective address. Some of this may be due to the necessity to fetch one or more extension words, and some due to the address arithmetic. As an example, the base time to CLEaR a memory byte is 8 clock cycles (4 to read the op-code and 4 to send out the zero on the data bus). Thus from the table, CLR.B (An) takes $8 + 4 = 12\tilde{\ }$, CLR.B 0E04567h takes $8 + 12 = 20\tilde{\ }$. Reference [4] gives timings for all instructions. The 68008 takes longer to generate eas for most operations due to its byte-sized Data bus.

Table 4.9: A summary of 68000 address modes.

Address mode	ea	Extra cycles 68000/8			Code
		Byte	Word	Long	Mode:Register
Dn	Dn	0/0	0/0	0/0	000:rrr[1]
An	An	0/0	0/0	0/0	001:rrr
(An)	[An]	4/4	4/8	8/16	010:rrr
(An)+	[An]+	4/4	4/8	8/16	011:rrr
−(An)	[−An]	6/6	6/10	10/18	100:rrr
$\pm d_{16}$(An)	[An+d_{16}]	8/12	8/16	12/24	101:rrr
$\pm d_8$(An,X)[2]	[An+X+d_8]	10/14	10/18	14/28	110:rrr
$\pm d_{16}$(PC)	[PC+d_{16}]	8/12	8/16	12/24	111:010
$\pm d_8$(PC,X)[2]	[PC+X+d_8]	10/14	10/18	14/26	111:011
abs.W	sex|<abs value>	8/12	8/16	12/24	111:000
abs.L	<abs value>	12/20	12/24	16/32	111:001
#immediate	...	4/8	4/8	8/16	111:100

Note 1: A 3-bit code indicating the target register for modes $000b$ to $110b$, otherwise a submode.

Note 2: The Index register, which can be any Data or Address register, is specified as a 4-bit code in the extension word, which also carries the 8-bit offset.

Not all address modes are legitimate in many situations. For example, an Immediate operand by definition cannot be specified as the destination ea. Also, but not so obviously, the two Program Counter Indirect modes are also illegal for a destination operand. This is because it is considered bad practice to modify program code, and in any case the area around the **PC** will frequently be in ROM and therefore cannot be altered. The group of address modes excluding PC Relative and Immediate are referred to as Alterable. Those also excluding Address Register Direct are categorized as Data Alterable. In general, except for special instructions such as `ADDX`, all address modes may be used as a source operand. The destination operand may be a Data register only, an Address register only or, in more comprehensive operations, such as `MOVE` and `ADD`, a Data Alterable mode may be specified. Except for `MOVE`, one of the operands must be a register. Table 4.8 summarizes the permitted address modes for each instruction.

Table 4.9 also lists a 6-bit code against each mode. This is the bit pattern used in the op-code to specify the address mode for both source (if present) and destination. Two examples are given in Fig. 4.4. Of course it is not necessary for the programmer to work out the binary code for an instruction, unless he or she suspects the assembler's integrity — I did once find an assembler which incorrectly coded one instruction–address mode combination. After all this is the main raison d'être for using an assembler.

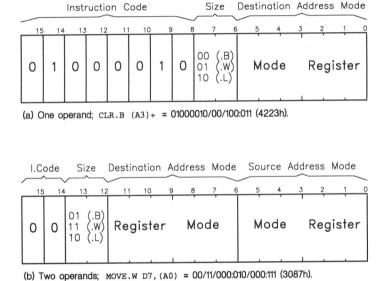

Figure 4.4: Two examples of machine coding.

4.3 Example Programs

The last few sections used program fragments to illustrate various instruction/address mode combinations. Here we finish our introduction to 68000 software by developing three programs of a slightly more elaborate nature. These will implement similar functions to those coded in 6809 assembly language in Section 2.3, and this will allow comparison between the software of the two processors.

As in Section 2.3 we are using the Real Time Systems XA8 cross-assembler, the syntax and format rules of which were discussed at that point. There are two minor differences which are relevant here. 6809 assembly language assigns the source operand to the operand field and destination to the instruction mnemonic, for instance:

```
LDB   1234h       ; [B] <- [1234h]        {[<Destination>] <- [<Source>]}
LDY   #0E000h     ; [Y] <- #E000h         {[<Destination>] <- <Source>}
```

In 68000 assembly language, the mnemonic does not contain any operand information, and any operands appear explicitly or implicitly in the operand field as <source>,<destination>, for example:

```
MOVE.B   1234h,D0   ; [D0(7:0)]  <- [1234h]    {[<Destination>] <- [<Source>]}
MOVEA.L #0E000h,A0  ; [A0(31:0)] <- #0000E000h {[<Destination>] <- <Source>}
```

However, the size of the operands are indicated in the mnemonic field by the extension .B, .W or .L as appropriate. *Both* operands are the same size.

One quirk peculiar to the XA8 cross assembler is the treatment of the MOVE MULTIPLE (MOVEM) instruction. The standard Motorola way of representing a range of registers is to use the - range operator, for example D0-D3 meaning D0/D1/D2/D3. Thus the two ways of indicating a Push of the registers **D0** to **D3** and **A0** on to the System stack are:

```
MOVEM.L      D0-D3/A0, -(A7)     ; Not used by XA8 assembler
MOVEM.L   D0/D1/D2/D3/A0, -(A7)  ; Applicable to all assemblers
```

The XA8 assembler unfortunately does not support the - range operator.

Each program module is written in the form of a complete subroutine, with data assumed present on entry in some place, usually a Data register, and terminated by a RETURN FROM SUBROUTINE (RTS) instruction. We will look at subroutines in some detail in Chapter 5.

Our first program generates the sum of all integers up to a maximum n of 65,535 (FFFFh). We assume that n is passed to the subroutine in the lower word of **D0**.

Table 4.10: Object code for sum of n integers program.

```
1                          .processor m68008
2  ; *******************************************************************
3  ; * FUNCTION : Sums all unsigned word numbers up to n (max 65,535) *
4  ; * ENTRY    : n is passed in Data register D0.W                   *
5  ; * EXIT     : Sum is returned in Data register D1.L               *
6  ; *******************************************************************
7  ;
8                          .psect _text         ; Direct code into text area
9  ; for (sum=0;n>=0;n--){
10 000400 02800000FFFF SUM_OF_N: and.l #0000FFFFh,d0 ; n promoted to long
11 000406 4281                   clr.l d1            ; Sum initialized to 00000000
12 000408 D280         SLOOP:    add.l d0,d1         ; sum = sum + n
13 00040A 51C8FFFC               dbf   d0,SLOOP      ; n--, REPEAT WHILE N>-1
14 00040E 4E75         S_EXIT:   rts
15                               .end
```

The maximum possible sum of 2,147,450,880 fits comfortably in the 32-bit **D1** for return. Compare this with the $n = 255$ limit in the 6809 equivalent on page 46 due to its smaller registers, although of course external memory could have been used for larger operands.

The algorithm used in the listing of Table 4.10 simply clears **Data register D1**, which will hold the 32-bit sum, and also the upper 16 bits of **D0**. This latter operation effectively promotes the word-sized parameter n, passed to the subroutine in **D0.W**, to long-sized. The equality is necessary for the addition of line 12, which adds the progressively decrementing n to the partial sum. The loop control DBF implements this decrementation using n both as the operand and the loop counter. When n drops below zero, the loop terminates and the final sum is in **D1.L** as specified.

The object code shown in Table 4.10 is the result of passing the source code file through the assembler and then the linker-loader, as described in Section 7.2. All 68000-based programs in this book assume ROM from $0400h$ up for the program sections designated _text and RAM from $E000h$ up for the _data sections. Only _text is needed in this case. The program is 16 bytes long and takes $54 + 14n$ clock cycles to execute (maximum 114,694.75 μs at 8 Mhz).

The alternative direct algorithm:

$$\text{sum} = n \times \frac{(n+1)}{2}$$

is shown coded in Table 4.11. This copies n into **D1.W**, adds one, multiplies to give the long $n \times (n+1)$ and then divides by two using a Shift Right once operation. Only 10 bytes in length, it takes 104 clock cycles to execute (13 μs at 8 MHz) irrespective of n. However, like its 6809 equivalent of Table 2.10, one value of n will give an erroneous zero answer. It is left to the reader to determine which, and

Table 4.11: A superior implementation.

```
1                        .processor m68008
2       ;********************************************************************
3       ; * FUNCTION : Sums all unsigned word numbers up to n (max 65,535) *
4       ; * ENTRY    : n is passed in Data register D0.W                    *
5       ; * EXIT     : Sum is returned in Data register D1.L                *
6       ; * EXIT     : No other registers disturbed                         *
7       ;********************************************************************
8       ;
9                        .psect   _text   ; Direct code into text area
10      ; sum = n*(n+1)/2
11 000400 3200 SUM_OF_N: move.w  d0,d1    ; Copy n into d1.w
12 000402 5241            addq.w  #1,d1    ; which becomes n+1
13 000404 C2C0            mulu    d0,d1    ; n*(n+1) now in d1.l
14 000406 E289            lsr.l   #1,d1    ; Divide to give n*(n+1)/2
15 000408 4E75 S_EXIT:   rts
16                        .end
```

to devise a means to avoid this problem.

Our second example involves converting a binary number to a string of ASCII-coded digits, terminated with $00h$ (ASCII NULL). In Fig. 2.4 we implemented this by evaluating the nth digit as the number of successful subtractions by 10^n, starting with the maximum n and moving down to zero. The values of 10^n were stored as a table of constants. We used this technique in preference to the usual algorithm of continually dividing by ten, with the remainders giving the digits, as the 6809 MPU has no Division operation. This is not the case for the 68000 family, and so this is the approach taken in the listing of Table 4.12. As an example:

$65536 \div 10 = 06553\,r\,6$ Fifth digit
$06553 \div 10 = 00655\,r\,3$ Fourth digit
$00655 \div 10 = 00065\,r\,5$ Third digit
$00065 \div 10 = 00006\,r\,5$ Second digit
$00006 \div 10 = 00000\,r\,6$ First digit

The conversion loop simply divides repetitively by ten the long binary number passed in **D0**, producing the 16-bit remainder in the top of **D0** and the 16-bit quotient at the bottom. SWAP (line 19) is used to reverse the order of these, and with the quotient safely at the top, the following Convert to ASCII and Move-byte operations leave this undisturbed (lines 20 to 22). Finally, clearing the remainder and swapping again restores the quotient as a 32-bit quantity ready for the next $32 \div 16$ bit DIVU.

Data register_D1.W is used with DBF to give 5 passes around the loop, and **A0** is used as a pointer to the next RAM byte for the string digits, in conjunction with the Post-Increment address mode. MULTIPLE MOVEs at the start and end of the subroutine Push and Pull all use registers into the System stack, and ensure that the internal state (except the **CCR**) is returned unaltered on completion.

Unlike the 6809 equivalent in Table 2.12, the binary number is not restricted to $FFFFh$ (65,535). As we have coded the algorithm for five digits, the upper limit is 99,999. Changing line 16 to MOVEQ #5,D1 (i.e. six digits) will increase this to 655,359 before overflow occurs. The reason the limit is not 999,999 is the 16-bit quotient produced by DIVU. The 68020 MPU has a 32×32 divide, giving a 32-bit quotient and remainder (e.g. DIVUL #10,D0:D1 puts the 32-bit quotient in **D0** and 32-bit remainder in **D1**). With the 68000's DIVU, one approach is initially to divide the binary number by 10,000, the quotient then holding the upper five digits and the remainder the lower five digits. Each half is then processed as shown. The limit thus is 4,294,967,295. Coding this is left as an exercise for the reader.

Our final example is the evaluation of the factorial of an integer n passed to the subroutine in the lower byte of **Data register D0**. $n!$ is returned as a long-word in **D1**. As we observed on page 51, this restricts n to no more than 12, and to signal a value outside this range, **D0.L** is used to return an error status, −1 for error and 0 for success.

As in Section 2.3, there are two techniques for tackling problems of this nature. The direct method uses the mathematical definition of factorial as the product of all integers up to and including n (with the exception of $0! = 1$), as shown in Fig. 2.5. Although the 6809 MPU has a multiplication instruction, its 8×8 field size meant that the necessary 32×8 products had to be evaluated as four separate operations together with the necessary shifting and addition. Furthermore the growing product had to be kept externally in four memory bytes, all of which led to the messy coding of Table 2.13.

Matters are somewhat improved in the 68000 with its 16×16 multiply and 32-bit Data registers. Implementing a 32×8 multiplication now involves the process:

Table 4.12: Binary to decimal string conversion.

```
1                          .processor m68000
2    ; **********************************************************************
3    ; * Converts binary code (max 99,999 decimal) to a string of five      *
4    ; * ASCII-coded characters, terminated by 00 (NULL)                    *
5    ; * EXAMPLE : 0000FFFF -> '6''5''5''3''5'NUL (36/35/35/33/35/00h)      *
6    ; * ENTRY   : Binary in D0.L                                           *
7    ; * EXIT    : Decimal string in 6 RAM bytes starting from DEC_STRG     *
8    ; * EXIT    : All register contents except CCR unchanged               *
9    ; **********************************************************************
10                          .list     +.text
11                          .psect    _text        ; Direct code into text area
12   ; Initialize data and pointer
13   000400 48E7C080 BIN_2_BCD: movem.l d0/d1/a0,-(sp); Save everything except CCR
14   000404 207C000E006       movea.l #DEC_STRG+6,a0; Point a0 to top of string
15   00040A 4220              clr.b   -(a0)        ; Put a null at this point
16   00040C 7204              moveq   #4,d1        ; Loop counter 5-1 = 4
17   ; Divide by 10 five times, the remainders giving the decimal digits
18   00040E 80FC000A BLOOP:   divu    #10,d0       ; Divide by ten
19   000412 4840              swap    d0           ; Remainder to lower word
20   000414 06000030          add.b   #'0',d0      ; converted to ASCII (add 30h)
21   000418 1100              move.b  d0,-(a0)     ; Move down one char & put it out
22   00041A 4240              clr.w   d0           ; Zero this remainder
23   00041C 4840              swap    d0           ; & get quotient back in word form
24   00041E 51C9FFEE          dbf     d1,BLOOP     ; Dec count & repeat unless -1
25   ;
26   000422 4CDF0103          movem.l (sp)+,d0/d1/a0; Return everything except CCR
27   000426 4E75              rts
28   ; **********************************************************************
29   ; This is the area of RAM where the number string is returned in order
30   ; TEN_THOU THOU HUNDS TENS UNITS NULL from DEC_STRG to DEC_STRG+5
31                            .psect   _data       ; Variable data space
32   00E000           DEC_STRG:.byte   [6]         ; Reserve six bytes for string
33                            .end
```

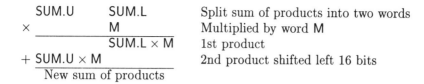

Firstly the 32-bit sum of products is split into two words each of which is multiplied by n (promoted to word size in line 12). The second product is shifted left 16 places and the two products added to give the new sum. Repeating this with M decrementing from n to 1 gives the loop algorithm of Table 4.13, lines 21–34.

Splitting up the sum of products, using a word MOVE from **D1** (holding the 32-bit sum) to **D2.W**, gives the 16-bit SUM.L. Moving all of **D1** to **D3** and then swapping words (SWAP D3) puts the 16-bit SUM.U in the lower word of **D3**. The two MULUs of lines 26 and 27 then give the two sub-products. The second of these is moved left 16 places by doing a SWAP and clearing the lower 16 bits. Finally,

Table 4.13: Mathematical evaluation of factorial n.

```
 1                               .processor  m68000
 2   ;****************************************************************
 3   ; * EXAMPLE : n = 12; n! = 479,001,600                           *
 4   ; * ENTRY   : n in lower byte of d0; maximum value 12            *
 5   ; * EXIT    : n! in 32-bit d1                                    *
 6   ; * EXIT    : d0.1 = -1 (FFFFFFFFh) if error (n>12) ELSE 0       *
 7   ;****************************************************************
 8                               .define ERROR = -1
 9   ; Initialize
10                               .psect   _text
11 000400 48E73000 FACTORIAL:    movem.l  d2/d3,-(a7)  ; Save these registers on Stack
12 000404 024000FF               and.w    #00FFh,d0    ; n extended to 16 bits
13   ; Error conditions
14 000408 0C00                   cmp.b    #12,d0       ; IF n>12 THEN error condition
15 00040C 6304                   bls      CONTINUE     ; ELSE continue
16 00040E 70FF                   moveq    #ERROR,d0    ; FFFFFFFFh in d0 signals error
17 000410 6022                   bra      ERR_EXIT     ; and exit with it
18   ;
19 000412 7201      CONTINUE:    moveq    #1,d1        ; Initialize sum to 00000001
20   ; N<=1?
21 000414 0C000001  OUTER_LOOP:cmp.b   #1,d0        ; IF n<=1 then answer is in d1
22 000418 6318                   bls      FEXIT
23 00041A 3401      MUL_LOOP:    move.w   d1,d2        ; Lower word of sum to d2
24 00041C 3601                   move.w   d1,d3
25 00041E 4843                   swap     d3           ; Upper word to d3
26 000420 C4C0                   mulu     d0,d2        ; First product (n*sum.l) in d2
27 000422 C6C0                   mulu     d0,d3        ; Second product (n*sum.u) in d3
28 000424 4843                   swap     d3           ; Move it to the upper word
29 000426 02430000               and.w    #0,d3        ; Zeroing the lower word
30 00042A 2202                   move.l   d2,d1        ; Begin to build the new sum
31 00042C D283                   add.l    d3,d1        ; Sum of products
32   ; n=n-1
33 00042E 5300                   subq.b   #1,d0
34 000430 60E2                   bra      OUTER_LOOP
35 000432 4280      FEXIT:       clr.l    d0           ; Zero indicates no error
36 000434 4CDF000C  ERR_EXIT:    movem.l  (a7)+,d2/d3  ; Get used registers from Stack
37 000438 4E75                   rts
38                               .end
```

they are summed in **D1** (lines 30 and 31) to give the grand total. Decrementing n (line 33) completes the loop.

Once again this example is easier to implement with the 68020 MPU, which has a 32×32-bit multiply MULU.L. This would avoid the need to split the multiplicand in two and later combine the two sub-products.

On entry to the loop, n is tested for 1 or 0, and if True the subroutine is exited with **D0.L** cleared. The alternative exit if $n > 12$ (lines 14 and 15) puts FFFFFFFFh (-1) in **D0.L** to signal error and bypasses the clearing operation.

Table 4.14: Factorial using a look-up table.

```
1                              .processor m68000
2    ;******************************************************************
3    ; * EXAMPLE : n = 12; n! = 479,001,600                             *
4    ; * ENTRY   : n in lower byte of d0; maximum value 12              *
5    ; * EXIT    : n! in 32-bit d1                                      *
6    ; * EXIT    : d0.1 = -1 (FFFFFFFFh) if error (n>12) ELSE 0         *
7    ;******************************************************************
8                              .define ERROR = -1
9                              .list   +.text
10   ; Initialize
11                             .psect  _text
12 000400 48E70080 FACTORIAL:  movem.l a0,-(a7)  ; Save a0.L on Stack
13 000404 024000FF             and.w   #00ffh,d0 ; n extended to 16 bits
14 000408 207C00000426         movea.l #TABLE,a0 ; Point a0 to bottom of table
15   ; Error conditions
16 00040E 0C00000C             cmp.b   #12,d0    ; IF n>12 THEN an error condition
17 000412 6304                 bls     CONTINUE  ; ELSE continue
18 000414 70FF                 moveq   #ERROR,d0 ; Put FFFFFFFFh in d0 signals error
19 000416 6008                 bra     ERR_EXIT  ; and exit with it
20   ;
21 000418 E508     CONTINUE:   lsl.b   #2,d0     ; Multiply n by 4 as table is 4-wide
22 00041A 22300000             move.l  0(a0,d0.w),d1 ;Get long-wrd at [a0]+[d0] to D1
23 00041E 4280     FEXIT:      clr.l   d0        ; Zero indicates no error
24 000420 4CDF0100 ERR_EXIT:   movem.l (a7)+,a0  ; Retrieve a0.1 from Stack
25 000424 4E75                 rts
26   ;******************************************************************
27   ; Now the table of factorials which is in the text (ROM) area
28 000426          TABLE:      .double 1, 1, 2, 6, 24, 120, 720, 5040, 40320,
                                       362880, 3628800, 39916800, 479001600
          00000001
          00000002
          00000006
          00000018
          00000078
          000002D0
          000013B0
          00009D80
          00058980
          00375F00
          02611500
          1C8CFC00
29                             .end
```

Where no simple mathematical algorithm exists to specify a function, using a table of outcomes is the only approach, for example the 7-segment decoder of page 113. Although this is not the case here, there are only 13 successful outcomes to the subroutine, and the use of a look-up table is an attractive proposition.

Using this approach, the resulting coding of Table 4.14 shows the active portion of the program (i.e. excluding error checking and reporting, which is the same as the previous listing) to be only lines 21 and 22. The first multiplies n by four to match the size of the table entries. This is then used as the Index register (**D0.W**) to point into the table, with **A0** holding the base address $0426h$ (TABLE). For example, if $n = 4$ then [**D0(15:0)**] becomes $10h$ (4×4) and MOVE.L 0(A0,D0.W),D1 effectively moves the 4 bytes starting at $0 + [A0] + [D0(15:0)] = 0 + 0426h + 10h = 0436h$ to **D1.L**. The contents of 0436:7:8:$9h$ are 24 (00 00 00 $18h$), as required for $n!$.

References

[1] Starnes, T.W.; Powerful Instructions and Flexible Registers of the 68000 Make Programming Easy, *Electronic Design*, **28**, no. 9, April 1980, pp. 171–176.

[2] Wakerly, J.F.; *Microcomputer Architecture and Programming: The 68000 Family*, Wiley, 1989, Section 8.4.1.

[3] Motorola; *M68000 16/32-bit Microprocessor Programmer's Reference Manual*, 5th ed., 1986.

[4] Leventhal, L.A.; *68000 Assembly Language Programming*, McGraw-Hill, 2nd ed., 1986.

[5] Leventhal, L.A. and Cordes, F.; *Assembly Language Subroutines for the 68000*, McGraw-Hill, 1989.

[6] Kelly-Bootle, S. and Fowler, B.; *68000, 68010, 68020 Primer*, H.W. Sams, 1985.

[7] Van de Goor, A.J.; *Computer Architecture and Design*, Addison-Wesley, 1989, Section 4.1.2.

[8] Cahill, S.J.; *The Single-Chip Microcomputer*, Prentice-Hall, 1987, Section 1.2.

[9] Cahill, S.J.; *The Single-Chip Microcomputer*, Prentice-Hall, 1987, Section 9.1.

CHAPTER 5

Subroutines, Procedures and Functions

A **subroutine** may be defined as a self-standing sequence of instructions which may be called from anywhere and, having been run, will return control whence it was called. Thus, for example, the code for the calculation of $\sin(x)$ may be stored offside the main program. To exercise the function:

```
y = sin(x);
```

the program must jump out to the code, carrying with it the value of x. After execution, the outcome y will be found at some prearranged location.

Subroutines are primarily used to reduce the size of the overall code, since they may be successively called from many points outside, including other subroutines, and even from inside itself (when they are known as **recursive**)! For example, the calculation of sine may be needed at five different parts of the program, but if it is coded as a subroutine, only one implementation is necessary. Furthermore, subroutines can be nested, with one subroutine calling another. For example, a call to a cosine subroutine will invariably have recourse to the use of the sine function.

Sets of useful subroutines are often organized in a **library**. These libraries are scanned at link time (see Section 7.2) and the relevant entries referred to in the user's program, extracted and added to the final code. To be used in this manner, each subroutine must be documented with well-defined parameter-passing protocols. Libraries may be built up by the user or be available as a commercial package. High-level languages usually come with several such packages.

Aside from saving space, subroutines are the vehicle normally used to implement modular programming [1]. A structured approach to hardware design decomposes the system into functional modules, for example oscillator, gate, counter, decoder, display. Each module has a relatively simple function and may be designed, implemented and tested as a separate entity, with the appropriate stimuli. This may not produce the smallest, most efficient circuit, but it is likely that the product will come to fruition earlier and be more maintainable due to its testability.

The software module is analogous to its hardware cousin as it too can be inserted into its motherboard (the main program), takes one or more signals (parameters, e.g. x) and has an outcome (return values, e.g. $\sin(x)$). A software module, invariably in the form of a subroutine, is normally self-standing with its own area of code (usually in ROM) and data storage in RAM. Good programming techniques are used to enforce a single entry and exit point and a minimum of interaction with data areas used by other modules.

The expression **function** is commonly used in high-level languages to describe a callable module. In Pascal the name **procedure** is reserved for the special case of a function that returns no value, that is a **void function**. In common with assembly language, Fortran uses the name subroutine. Irrespective of the name used, assembly-level subroutines are normally used to implement these high-level modules. Thus an understanding of the structure of subroutines is the key to comprehending the operation of these important aspects of high-level languages. This is the objective of this chapter.

5.1 The Call-Return Mechanism

In essence, getting to a subroutine involves nothing more than placing the address of its opening instruction in the **Program Counter** (**PC**), that is doing a Jump or Branch. Thus, if we take as an example a subroutine which evokes a delay of 0.1 s (i.e. does nothing for 100 ms) and starts at E100h, then JMP 0E100h will transfer control. In practice the programmer will probably not know the absolute address of the subroutine, especially if it is hidden in a library. However, a subroutine entry point is normally identified with a label, and the assembler or linker will evaluate the appropriate address, for example JMP DELAY (see Table 5.2).

The problem lies not in getting there, but returning afterwards. As can be seen from Fig. 5.1, the jumping-off point may be from anywhere in the main program or indeed from another subroutine — the latter process is known as **nesting**. Thus the microprocessor (MPU) needs to remember the value of its **PC** (which is already pointing to the instruction following the Jump or Branch after its fetch) before its contents are overwritten.

One possibility is to move the contents of the **PC** to a designated memory location or Address register, for example LEAX 0,PC (LOAD EFFECTIVE ADDRESS 0 + **PC** to the 6809's **X** register) or LEA 0(PC),A0 (LOAD EFFECTIVE ADDRESS 0 + **PC** into the 68000's **A0** register). Then the subroutine can be terminated by moving this pre-saved jumping-off address back to the **PC** (JMP 0,X or JMP (A0)).

This approach breaks down when a subroutine wishes to call another, for the secondary subroutine will overwrite the return address of the primary. To get around this problem, the jumping-off address could be pushed down into a stack, rather than using a fixed register or memory location. As each subroutine is called, this Stack Pointer is moved down automatically by the appropriate number of bytes. Returning inwards simply involves the mirror operation of pulling up out of the stack back into the **PC**. The Stack Pointer moves up accordingly. This last-in first-out sequence, necessary for nesting, exactly describes the structure supported by the System stack/Stack Pointer.

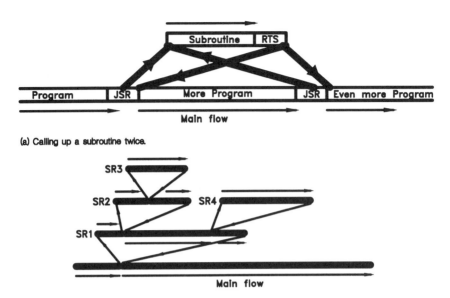

(a) Calling up a subroutine twice.

(b) Nested subroutine with Main calling SR1, which in turn calls SR2, which in turn calls SR3. SR1 also calls SR3.

Figure 5.1: Subroutine calling.

Using this technique gives us:

$$\left.\begin{array}{l}\text{PSHS PC}\\ \text{JMP DELAY}\end{array}\right\} = \text{JSR DELAY} \dots\dots\dots\dots\dots\dots\dots\dots\dots\text{PULS PC = RTS}$$

for the 6809 MPU, and

$$\left.\begin{array}{l}\text{PEA 0(PC)}\\ \text{JMP DELAY}\end{array}\right\} = \text{JSR DELAY} \dots\dots\dots\dots\dots\dots\dots\dots\dots\text{JMP (SP)+ = RTS}$$

for the 68000 MPU.

Notice how we simulated a Pull operation for the 68000 MPU, which does not have an explicit Pull instruction. The Post-Increment Indirect address mode operation on **A7** (the System Stack Pointer) causes the **SSP** to move up (4 bytes) after the data (the return address) has been extracted. By definition, the Jump operation puts this extracted address in the Program Counter.

Calling and returning from a subroutine is a sufficiently frequent operation to warrant the specific Call and Return instructions of Table 5.1. These have exactly the same outcome as the generalized approach shown above. JUMP TO SUBROUTINE (JSR) and its relative BRANCH TO SUBROUTINE (BSR) push the return address on to the System stack before going off. The BSR variants follow the same rules as ordinary Branches (see Sections 2.2 and 4.2) and of course generate position-independent code (PIC). RETURN FROM SUBROUTINE (RTS) pulls the return address back from the System stack. 80x86 microprocessors use the mnemonics CALL and RET for the same purpose.

Table 5.1: Subroutine instructions.

Operation	Mnemonic	Description
Call		Transfer to subroutine
Jump to subroutine	JSR ea	Push PC onto Stack, PC <- <ea>
Branch to subroutine[1]		
short	BSR offset$_8$	Push PC onto Stack, PC <- PC + sex|offset$_8$
long	LBSR offset$_{16}$	Push PC onto Stack, PC <- PC + offset$_{16}$
Return		Transfer back to caller
from subroutine	RTS	Pull original PC back from Stack

(a) 6809 instructions.

Operation	Mnemonic	Description
Call		Transfer to subroutine
Jump to subroutine	JSR ea	Push PC onto Stack, PC <- <ea>
Branch to subroutine[1]		
short	BSR offset$_8$	Push PC onto Stack, PC <- PC + sex|offset$_8$
long	LBSR offset$_{16}$	Push PC onto Stack, PC <- PC + sex|offset$_{16}$
Return		Transfer back to caller
from subroutine	RTS	Pull original PC back from Stack
and restore CCR	RTR	Pull original CCR back from Stack
		Pull original PC back from Stack
Frame		Maintain a frame for local variables
Make	LINK An,#kk$_{16}$	An into Stack (save old Frame Pointer, An)
		An <- SP (Point An to Top Of Frame, TOF),
		SP <- SP + sex|kk$_{16}$ (SP to Bottom of Frame)
Close	UNLNK An	SP <- An (move SP back to TOF),
		Pull An (Get old Frame Pointer from Stack)

(b) 68000 instructions.

Note 1: Available in signed 8-bit (+127, −128) and 16-bit offset (+32,767, −32,768) varieties. Most assemblers can chose the appropriate versions automatically. The 68020 upwards have a full 32-bit offset Branch capability.

From Fig. 5.2 we see that the action of JSR/BSR and RTS on the System stack is the same for both 6809 and 68000 MPUs, except the latter requires four bytes. As is usual for Motorola MPUs, the lower byte is located in the higher address (i.e. the lower byte of the address is pushed out first). The 68000's **SSP** must always point to an *even address*, and this will be enforced even if a single byte is pushed out.

128 *Subroutines, Procedures and Functions*

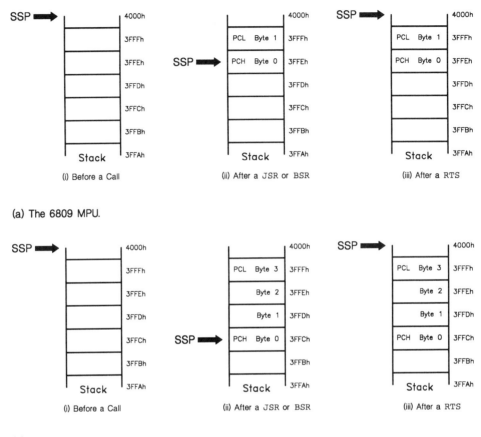

Figure 5.2: Saving the return address on the Stack. The SSP assumed a priori set to 4000h.

As an example, consider a subroutine to give a 0.1s delay. This is easily implemented by loading a constant into a register and decrementing to zero. Coding for the 6809 and 68000 processors is shown in Table 5.2. Other than the terminating RTS, the programs are perfectly normal routines. Strictly, in calculating their delay, the time to get to the subroutine should be considered, and this can differ according to how far away the subroutine is from the caller and which Call instruction and/or address mode is used. This also illustrates that there is a time overhead in using a subroutine, and where speed is of the essence, in-line code should be used.

Notice that in both cases illustrated in Table 5.2, one of the registers (**X** or **A0**) will be returned in an altered state, the same being true of the **Code Condition register** (**CCR**). Provided that such changes are well documented, this will frequently be of little consequence. However, it is often preferable to make subroutines **transparent** in that all registers, or perhaps a subset, remain unaltered. This can be accomplished by pushing all relevant registers into a stack at the beginning of the

Table 5.2: A simple subroutine giving a fixed delay of 100 ms when called.

```
1                           .processor m6809
2  ; **********************************************************************
3  ; * This subroutine does nothing and takes 0.1s to do it               *
4  ; * ENTRY : Non                                                        *
5  ; * EXIT  : X Address register = 0000, CCR destroyed                   *
6  ; **********************************************************************
7                           .define  N =12500-(5+3/8)
8  E000 8E30D3  DELAY: ldx   #N    ; Delay factor, 3~
9  E003 301F    DLOOP: leax  -1,x  ; Decrement     , Nx5~
10 E005 26FC           bne   DLOOP ; to zero       , Nx3~
11 E007 39             rts         ;               , 5~
12                     .end
```

(a) 6809 code; 1 MHz clock, $\sim = 1\,\mu s$.

```
1                           .processor m68000
2  ; **********************************************************************
3  ; * This subroutine does nothing and takes 0.1s to do it               *
4  ; * ENTRY : Non                                                        *
5  ; * EXIT  : D0.W Data register = 0000, CCR destroyed                   *
6  ; **********************************************************************
7                           .define  N = (200000-8-14)/14
8  000400 303C37CC DELAY: move.w #N,d0  ; Delay factor, 8~
9  000404 5340     DLOOP: subq.w #1,d0  ; Decrement    , Nx4~
10 000406 66FC            bne    DLOOP  ; to zero      , Nx10/8~ (taken/not)
11 000408 4E75            rts           ;              , 16~
12                        .end
```

(b) 68000 code: 8 MHz clock, $\sim = 0.5\,\mu s$.

subroutine and pulling them out again just before the final exit RTS. This is easy in the 6809 MPU, as any combination of registers, including the **CCR**, can be Pushed or Pulled with a single instruction, see Table 5.3(a). There is a slight problem with the 68000 MPU. The MOVEM instruction used for Pushing and Pulling only acts on Address and Data registers. There is a MOVE SR,-(SP) instruction which copies the whole **Status register**, of which the **CCR** is the lower byte. The opposite Pull operation is supported, that is MOVE (SP)+,CCR! Although the latter only pulls out a byte, the **SSP** moves up two bytes. This is necessary to obey the rule that the **SSP** always points to an even address, and thus preserves the integrity of the System stack. Interestingly the 68010 and higher family members have gained the missing MOVE CCR,<ea> instruction, which matches the MOVE <ea>,CCR instruction.

From Table 5.1(b), we see that the 68000 family has a second Return instruction, RTR (RETURN AND RESTORE CCR). This is used as an equivalent to the sequence:

MOVE   (SP)+,CCR
RTS

and assumes that the **CCR** has been saved out onto the System stack at the beginning of the subroutine before any other stack-based operations have altered the **SSP**. Notice from Table 5.3(b) that the **CCR** is saved first (line 8), before the Data register is Pushed. The Pull sequence at the end of the subroutine is then in the reverse order. Failure to observe this can lead to spectacular crashes! The

130 *Subroutines, Procedures and Functions*

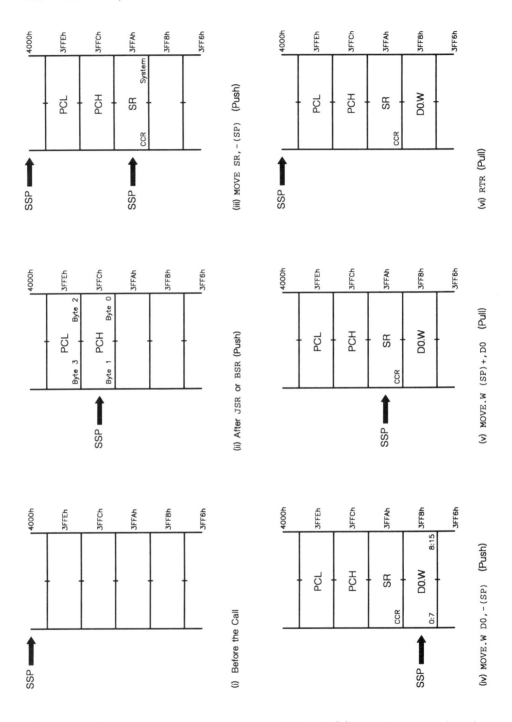

Figure 5.3: The stack when executing the code of Table 5.3(b), viewed as word-oriented.

Table 5.3: Transparent 100 ms delay subroutine.

```
1                           .processor m6809
2     ;************************************************************
3     ; * This subroutine does nothing and takes 0.1 s to do it    *
4     ; * ENTRY : None                                             *
5     ; * EXIT  : No change                                        *
6     ;************************************************************
7                           .define N =12500-(5+3+7+7/8)
8  E000  3411   DELAY: pshs   x,cc    ; Save Address reg and CCR , 7~
9  E002  8E30D2        ldx    #N      ; Initial delay factor     , 3~
10 E005  301F   DLOOP: leax   -1,x    ; Decrement                , Nx5~
11 E007  26FC          bne    DLOOP   ; to zero                  , Nx3~
12 E009  3511          puls   x,cc    ; Get registers back       , 7~
13 E00B  39            rts            ;                          , 5~
14                     .end
```

(a) 6809 code. Note lines 12 and 13 could be replaced by puls x,cc,pc.

```
1                           .processor m68000
2     ;************************************************************
3     ; * This subroutine does nothing and takes 0.1 s to do it    *
4     ; * ENTRY : None                                             *
5     ; * EXIT  : No change                                        *
6     ;************************************************************
7                           .define N = (200000-8-18-14-8-8)/14
8  000400 40E7 DELAY: move   sr,-(sp)   ; Save CCR (in SR)       , 14~
9  000402 3F00        move.w d0,-(sp)   ; and Data reg d0(15:0)  , 8~
10 000404 303C37CC    move.w #N,d0      ; Initial delay factor   , 8~
11 000408 5340 DLOOP: subq.w #1,d0      ; Decrement              , Nx4~
12 00040A 66FC        bne    DLOOP      ; to zero                , Nx10/8~(taken/not)
13 00040C 301F        move.w (sp)+,d0   ; Retrieve old d0(15:0)  , 8~
14 00040E 4E77        rtr               ; Retrieve CCR then RTS  , 20~
15                    .end
```

(b) 68000 code. Note rtr is equivalent to 6809 code puls cc,pc.

equivalent instruction PULS CCR,PC is sometimes used to terminate a 6809 subroutine (see Table 8.3).

Apart from its convenience, transparency is necessary to support the **recursive** use of a subroutine. A subroutine is recursive if it calls itself. Clearly register variables used in the subroutine will be wiped out when used again by the next recursion. Similarly, static memory locations cannot be used to store variables for a subroutine which is to be recursive, but variables can be saved in a stack, as shown in the next section, where they are known as **automatic variables**.

5.2 Passing Parameters

The simple fixed-delay subroutine used as the example in the previous section is unusual, in that no information was passed from the caller and none returned. Another example of a **double-void subroutine** would be a function actuating an external relay, where the very act of calling is sufficient. The actuation is sometimes referred to as a side effect.

Table 5.4: Using a register to pass the delay parameter. The call-up sequence shown above passed a constant (ten) to the subroutine.

```
1                               .processor m68000
2   ;************************************************************************
3   ; * This subroutine does nothing and takes Zx0.1s to do it              *
4   ; * EXAMPLE : Z = 10; delay = 1 second                                  *
5   ; * ENTRY   : Z passed in lower 16 bits of D1                           *
6   ; * EXIT    : D1(15:0) = FFFF, D2(15:0) = 0000, CCR destroyed           *
7   ;************************************************************************
8                               .define N = (200000-8-10)/14
9   000400 6008       DELAY:    bra     LOOPTEST    ; Check Z = 0            , 10~
10  000402 303C37CC   OUTERLOOP: move.w #N,d0       ; 100ms delay factor     , 8~
11  000406 5340       INNERLOOP: subq.w #1,d0       ; Decrement              , Nx4~
12  000408 66FC                 bne     INNERLOOP   ; to zero                , Nx10/8
13  00040A 51C9FFF6   LOOPTEST: dbf     d1,OUTERLOOP ; One less 100ms click, 10/14~
14  00040E 4E75                 rts                 ;                        , 16~
15                              .end
```

Consider the situation where the total delay is to be an integer (0 to 65,535) multiple of 0.1 seconds, depicted as DELAY(Z), where Z is the aforementioned integer passed to the subroutine by the caller. In Table 5.4 it is assumed that the caller has set up the **D1.W register** accordingly. Thus to invoke a 1 s delay, the call would be something like this:

```
MOVE.W  #10,D1    ; Ten ticks = 1 second
BSR     DELAY     ; Go to it!
```

The coding itself uses an inner loop (lines 11 and 12) identical to that in Tables 5.2 and 5.3, with DBF being employed in conjunction with **D1.W** (i.e. Z) to count the number of passes through this inner core (i.e. 0.1 s ticks). This DBF Decrement and Test is exercised immediately the subroutine is entered, to ensure a speedy exit should Z be zero. The delay due to line 9 only happens once, and can be thought of together with the caller's JSR/BSR as a constant error independent of Z. No data is returned from this void subroutine.

If the delay parameter is a variable, for example data read from an analog to digital converter, and stored somewhere in memory at MEM_Z, then:

```
MOVE.W  MEM_Z,D1  ; Copy the delay variable to D1
BSR     DELAY     ; to pass to subroutine
```

will do the necessary. Note that the parameter passed is a *copy* of the variable (still in MEM_Z), not the variable itself. Thus when **D1.W** is decremented in the subroutine, Z will not be altered, just its clone. Passing copied parameters is known as **call by value** [2]. We will look at ways of directly affecting variables through a subroutine later.

Using registers to pass parameters is convenient, fast and efficient. Furthermore, with some modification, it is suitable for recursion (subroutines that call

Table 5.5: Using a static memory location to pass the delay parameter.

```
1                            .processor m68000
2    ;************************************************************
3    ; * This subroutine does nothing and takes Zx0.1 s to do it   *
4    ; * EXAMPLE : Z = 10; delay = 1 s                             *
5    ; * ENTRY   : Z passed in memory location 6000/6001h          *
6    ; * EXIT    : D1(15:0) = FFFF, D2(15:0) = 0000, CCR destroyed *
7    ;************************************************************
8                            .define   N = (200000-8-10)/14
9  000400 32386000 DELAY:     move.w   6000h,d1       ; Get delay parameter, 12~
10 000404 6008               bra      LOOPTEST       ; Check Z = 0         , 10~
11 000406 303C37CC OUTERLOOP: move.w   #N,d0          ; 100 ms delay factor, 8~
12 00040A 5340     INNERLOOP: subq.w   #1,d0          ; Decrement           , Nx4~
13 00040C 66FC               bne      INNERLOOP      ; to zero             , Nx10/8
14 00040E 51C9FFF6 LOOPTEST:  dbf      d1,OUTERLOOP   ; 1 less 100 ms click, 10/14~
15 000412 4E75               rts                     ;                     , 16~
16                            .end
```

themselves), supports re-entrant code (subroutines which can be interrupted and then called again by the service routine, see Section 6.1) and is position independent. Its main problem is lack of generality, as the complement, range and type of registers available vary considerably between devices. Thus the 6502 MPU has two 8-bit Address registers and one 8-bit Data register, the 8086 with four 16-bit Data registers and three 16-bit Address registers, while the 68000 has eight 32-bit registers each of both types. This is especially a problem with high-level language compilers, which attempt to be portable between processors.

Another technique, used especially with MPUs having a small complement of registers, is to use assigned memory locations as a common area between caller and subroutine. Where the location is fixed, this is known as **static allocation**. In Table 5.5, a single memory word is used to pass the static variable Z, with the caller copying the delay parameter thus:

```
MOVE.W   MEM_Z,6000h    ; Copy the delay variable from memory
BSR      DELAY          ; to pass to the subroutine via 6000h
```

If MEM_Z was actually the common memory location, then this copy would not need to be made, but care would have to be taken not to alter the variable itself (rather than the copy).

The use of common static memory has the advantage of being able to pass large numbers of parameters and structures such as arrays. However, as these locations are by definition fixed, such subroutines cannot be recursive or re-entrant. Also, unless different static locations are used for each subroutine, nesting can lead to unfortunate side effects as one subroutine inadvertently alters another subroutine's variables. This makes debugging difficult, as routines other than the one being tested may interact in unpredictable ways. Such common areas can be used to hold **global variables**, which are known throughout all linked program modules.

Table 5.6: Using the stack to pass the delay parameter.

```
 1                          .processor m68000
 2  ; **********************************************************************
 3  ; * This subroutine does nothing and takes Zx0.1 s to do it            *
 4  ; * EXAMPLE : Z = 10; delay = 1 s                                      *
 5  ; * ENTRY   : Z passed in Stack at SP+4/SP+5                           *
 6  ; * EXIT    : D1(15:0) = FFFF, D2(15:0) = 0000, CCR destroyed          *
 7  ; **********************************************************************
 8                          .define  N = (200000-8-10)/14
 9  000400 322F0004 DELAY:   move.w  4(sp),d1      ; Get delay parameter, 12~
10  000404 6008              bra     LOOPTEST      ; Check Z = 0         , 10~
11  000406 303C37CC OUTERLOOP: move.w #N,d0        ; 100 ms delay factor, 8~
12  00040A 5340     INNERLOOP: subq.w #1,d0        ; Decrement           , Nx4~
13  00040C 66FC              bne     INNERLOOP     ; to zero             , Nx10/8
14  00040E 51C9FFF6 LOOPTEST: dbf    d1,OUTERLOOP  ; 1 less 100 ms click, 10/14~
15  000412 4E75              rts                   ;                     , 16~
16                           .end
```

Many of these problems can be overcome by using a stack to pass variables back and forth, or preferably putting them there in the first place [3, 4]. This situation is depicted in the listing of Table 5.6 and Fig. 5.4. Now to call up DELAY, a copy of the delay variable Z is pushed onto the System stack before calling the subroutine. On return the System Stack Pointer must be moved back up again to balance this Push and be returned to its original position. Using LEA 2(SP),SP is an alternative to ADDQ #2,SP (or ADDA +2,SP), and can be used for operands up to 32,767. The 8086 MPU family has a convenient RET #n instruction which is equivalent to LEA +n(SP),SP after a RTS. Similarly, the 68010 and up has a RTD #n equivalent (RETURN AND DEALLOCATE PARAMETERS) where n is a 16-bit immediate parameter sign-extended to 32 bits.

```
MOVE.W   MEM_Z,-(SP)   ; Copy delay variable to the System stack
BSR      DELAY         ; to pass to the subroutine
LEA      +2(SP),SP     ; Clean up stack after return
```

Comparing Tables 5.6 and 5.5, we see that the only change is of Address mode in line 9. From Fig. 5.4, we see that Z lies 4:5 bytes up from where the SSP points to on arrival. Its effective address is thus 4(SP).

Passing parameters using dynamic allocation permits nesting, recursion and re-entrancy as the SSP automatically moves down for each call and up again on each return. Essentially such variables are **local** (sometimes called **automatic**) and are known only to their own subroutine. The technique is general to all processors supporting a stack, and is used by block-structured high-level languages such as Algol, Pascal and C [5]. It is also possible to return values on a stack in a similar manner.

All our examples so far have involved copying the value of a variable to pass to the subroutine. The actual variable itself is somewhere out in read/write memory

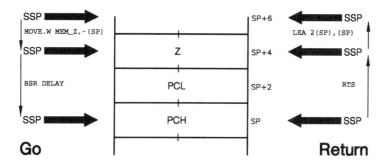

Figure 5.4: The Stack corresponding to Table 5.6.

and is not altered by processes in the subroutine. It is possible to use a subroutine to affect a variable directly by passing the *address* of that variable. This is known as **call by reference** [2]. Now that the subroutine knows where the variable lives, it can be modified. Passing addresses is also useful in pointing out to a subroutine where a large data structure, such as an array, is stored without having to send all its elements over. Only a **pointer** to the first element and its length need be passed.

A rather more sophisticated example of a program making use of a stack to pass both a copy of a variable and pointers is given in Table 5.7. The program specification is to make a copy of a block of data from one area of memory to another area of read/write memory. Parameters passed are pointers to the start of the source and destination blocks, and the length of the original block (assumed to be not greater than 64 kbytes). A successful copy is signalled by returning the code -1 (FFFFh) in **D0.W**. If any copy action is unsuccessful, then the subroutine is exited with **D0.W** holding the block length less the number of successfully transferred bytes. The caller can then subsequently calculate LENGTH$-$D0.W to give the number of bytes actually transferred. Other than the error status return, all other registers are to be unaltered. A typical application of such a subroutine would be to copy a table of initialized variables stored in ROM by a compiler to RAM where they can be modified later (see Table 10.12). Initial values cannot be stored in RAM, as such memory is volatile. Usually the compiler will generate the necessary constants, such as block start addresses and length, at link time.

The core of the program is contained in lines 21 to 26 of Table 5.7. Each byte is moved directly from memory to memory using Address registers to point to the two locations. A comparison tests for a successful copy as well as advancing the pointer. The DBNE loop control exits if it is true that the two bytes are not equal (i.e. unsuccessful) otherwise decrements the count in **D0.W** (originally set to LENGTH $-$ 1 in line 19) and repeats. The residue in **D0.W** will be FFFFh if each copied byte is verified, otherwise its exit state reflects the number of loop passes taken.

The System stack, as seen in Fig. 5.5, is used for three purposes. Firstly the three parameters are pushed out prior to the call, in a sequence such as:

136 *Subroutines, Procedures and Functions*

```
MOVE.W    #LENGTH,    -(SP)      ; Word length parameter pushed    (2 bytes)
MOVE.L    #RAM_START, -(SP)      ; Pointer to start of RAM pushed  (4 bytes)
MOVE.L    #ROM_START, -(SP)      ; Pointer to start of ROM pushed  (4 bytes)
BSR       BLOCK_COPY             ; Go to it
LEA       10(SP),SP              ; After return, clean up Stack
```

Then the actual call places the **PC** on the System stack automatically. Finally, as the subroutine is to be transparent, the System stack is used to save any used registers, apart from **D0**.

The code shown in Table 5.7 uses offsets from the **SSP** to obtain the three parameters, for example MOVE.W 22(SP),D0. This can cause problems, since in the body of many subroutines, the **SSP** is used to Push and Pull temporary results of evaluation into and out of the System stack. In particular local variables (that is variables used only by the subroutine and forgotten about after return) are also frequently kept on this stack. All this means that the parameter offsets from the

Table 5.7: Making a copy of a block of data of arbitrary length.

```
 1                       .processor m68000
 2  ; *************************************************************************
 3  ; * Copies a block of data from one area (e.g. ROM) to another (e.g. RAM)*
 4  ; * ENTRY : Constant LENGTH passed in SP+22/23 (up to 65,535)            *
 5  ; * ENTRY : Constant address RAM_START passed in SP+18/19/20/21          *
 6  ; * ENTRY : Constant address ROM_START passed in SP+14/15/16/17          *
 7  ; * EXIT  : Block of data from ROM_START to ROM_START+(LENGTH-1)         *
 8  ; * EXIT  : copied to RAM_START to RAM_START+(LENGTH-1)                  *
 9  ; * EXIT  : D0.W = FFFFh if copy successful                              *
10  ; * EXIT  : else LENGTH-D0.W is number of successful bytes transferred   *
11  ; *************************************************************************
12  ;
13 000400 42E7    BLOCK_COPY: move    CCR,-(SP)         ; Save CCR
14 000402 48E740C0            movem.l A0/A1/D1,-(SP)   ; Save used registers
15  ; Get length parameter and check for zero
16 000406 302F0016            move.w  22(SP),D0         ; Get LENGTH out from stack
17 00040A 4A40                tst.w   D0                ; Is it zero?
18 00040C 6714                beq     EXIT              ; IF yes THEN exit
19 00040E 5340                subq.w  #1,D0             ; ELSE redress DBNE'S n+1 loop
20  ; Now do the move loop
21 000410 206F0012            movea.l 18(SP),A0         ; Point A0 to RAM
22 000414 226F000E            movea.l 14(SP),A1         ; Point A1 to ROM
23 000418 1219    CLOOP:      move.b  (A1)+,D1          ; Move byte ROM to D1
24 00041A 1081                move.b  D1,(A0)           ; and hence up to RAM
25 00041C B218                cmp.b   (A0)+,D1          ; Did it get there ok?
26 00041E 56C8FFF8            dbne    D0,CLOOP          ; IF so THEN dec. and repeat
27  ; Pass here IF LENGTH is zero, OR error occurs OR copy is finished
28 000422 48DF0302    EXIT:   movem.l (SP)+,A0/A1/D1    ; Restore registers
29 000426 4E77                rtr                       ; and CCR before return
30                           .end
```

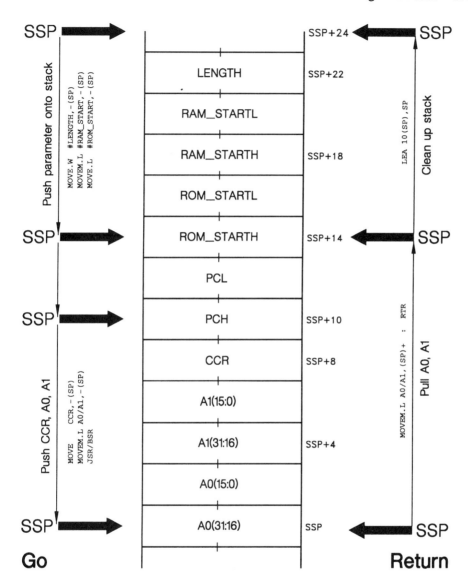

Figure 5.5: The Stack used for the BLOCK_COPY subroutine.

SSP will be in a constant state of flux. To get around this problem another Address register is frequently pointed to the top of the System stack at the beginning of the subroutine and this remains as a fixed point of reference for the duration of the subroutine, irrespective of what is happening to the **SSP**. This is known as the **Frame Pointer** (FP), with the space used on the System stack after entry being the **Frame**.

Our final example is used to illustrate the concept of a Frame. Consider a subroutine where an analog signal must be sampled as rapidly as possible for a

138 *Subroutines, Procedures and Functions*

variable number of times, using an 8-bit analog to digital converter, after which the resulting array is to be processed in some manner. Typical processes are filtering, averaging and peak detection. To keep our program as simple as possible, we will assume that we wish to return the simple sum of not more than 256 of these samples. To comply with the injunction that sampling should be as quick as possible, it will be necessary to allocate space to store temporarily up to 256 bytes. After this burst of sampling, the process can be carried out on the array now in situ in this RAM buffer.

Our first implementation is based on the 6809 MPU, as an example of a processor without any specific Frame-handling instructions. The System stack reflecting the coding of Table 5.8 is shown in Fig. 5.6. The variable i representing the number of samples to be taken is pushed on to this stack in the normal way prior to the subroutine call. The subroutine itself commences by saving the contents of the User

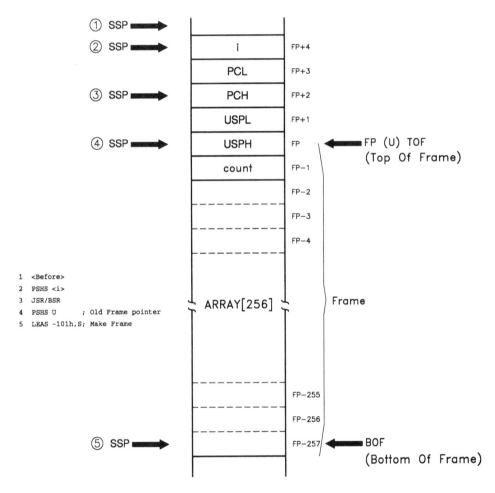

Figure 5.6: The 6809 System stack organized by the array averaging subroutine.

Stack Pointer (USP) on the System stack. The USP is to point to the Top Of Frame (TOF) and is thus to be the Frame Pointer. Transferring the contents of the System Stack Pointer (SSP) to the USP effectively points the Frame Pointer to the TOF, and then the SSP is moved down 257 bytes, one to hold the temporary (local) variable holding the count and 256 for the array (lines 11–13). At this point, the SSP points to the bottom of the frame (BOF) but, as all references in Table 5.8 use the Frame Pointer (e.g. line 21, DEC -1,U), it can be used subsequently for other purposes.

After the body of the subroutine, the Frame is closed by copying the Frame Pointer to the SSP — that is moving it up to the TOF — and pulling out the old Frame Pointer, before RTS (lines 34 and 35). Of course, after return the System stack will need to be cleaned up to compensate for passing i.

Table 5.8: Using a frame to acquire temporary data; 6809 code.

```
1                         .processor m6809
2   ; **********************************************************************
3   ; * Burst acquires up to 256 analog samples and returns the sum        *
4   ; * ENTRY : i is the number of samples on the stack                    *
5   ; * EXIT  : The sum of i 8-bit samples in Accumulator D                *
6   ; * EXIT  : X, CCR altered. U is used as the Frame Pointer             *
7   ; **********************************************************************
8   ;
9                         .define A_D = 6000h ; Where the A/D converter lives
10  ; First make the frame
11  E000 3440    ARRAY_AV: pshs  u         ; Save current USP on stack, old FP
12  E002 1F43             tfr   s,u        ; Point Frame Pointer to TOF
13  E004 32E9FEFF         leas  -101h,s    ; Open frame of 257 bytes, SP to BOF
14  ; Initialize to acquire data
15  E008 E644             ldb   4,u        ; Copy i into frame
16  E00A E75F             stb   -1,u       ; to initialize count (= i)
17  E00C 305F             leax  -1,u       ; X to just above ARRAY[0] (array ptr)
18  ; Burst sample
19  E00E F66000  GET_LOOP: ldb  A_D        ; Get data
20  E011 A782             sta   ,-x        ; Put it in frame, decrement pointer
21  E013 6A5F             dec   -1,u       ; count = count - 1
22  E015 26F7             bne   GET_LOOP   ; and repeat
23  ; Initialize to sum data
24  E017 E644             ldb   4,u        ; Copy i back into frame again
25  E019 E75F             stb   -1,u       ; to initialize count (= i)
26  E01B 305F             leax  -1,u       ; Point X to just above ARRAY[0] again
27  E01D 4F               clra             ; Clear sum (Acc.D)
28  E01E 5F               clrb
29  ; Now do the summation
30  E01F E382    ADD_LOOP: addd ,-x        ; Add byte to sum, decrement pointer
31  E021 6A5F             dec   -1,u       ; count = count - 1
32  E023 26FA             bne   ADD_LOOP   ; and repeat
33  ; Close frame
34  E025 1F34             tfr   u,s        ; Move SP back up
35  E027 3540             puls  u          ; and get back old frame pointer, USP
36  E029 39               rts              ; and return
37                        .end
```

The core program in lines 15–32 is unremarkable. The Frame Pointer is copied into the **X Index register** to permit the use of the Pre-Decrement Index Address mode in stepping through the array, yet leaving the Frame Pointer untouched (lines 17 and 20). The passed parameter i is copied into the Frame to initialize the loop counter in both instances (lines 16 and 25). It would be more efficient to use an Accumulator as a loop counter, but the 6809 MPU does not have enough registers to make the use of such register variables a feasible proposition. One quirk exhibited by this implementation is the need to pass i = 0 to sample 256 times, as a byte can only represent up to 255.

The 68000 System stack of Fig. 5.7, reflecting the code in Table 5.9, is very similar to its 6809 counterpart. This time i is passed as a word to preserve the evenness of the **System Stack Pointer** (a byte sized Pre-Decrement/Increment MOVEM via **A7**, i.e. Push and Pull, always results in a word being transferred to/from the

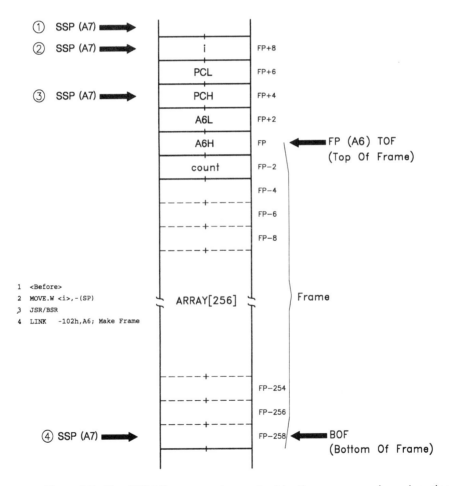

Figure 5.7: The 68000 System stack organized by the array-averaging subroutine.

Table 5.9: Using a Frame to acquire temporary data; 68000 code.

```
 1                              .processor m68000
 2    ; ************************************************************************
 3    ; * Burst acquires up to 256 analog samples and returns the sum         *
 4    ; * ENTRY : i is the number of samples on the stack                     *
 5    ; * EXIT  : The sum of i 8-bit samples in Data register D7              *
 6    ; * EXIT  : A0, D0.W and CCR altered. A6 is used as the Frame Pointer   *
 7    ; ************************************************************************
 8    ;
 9                              .define A_D = 6000h ; Where the A/D converter lives
10    ; First make the frame
11 000400 4E560102 ARRAY_AV: link    a6,#102h    ; Make 258-byte frame, A6 as FP
12    ; Initialize to acquire data
13 000404 3D6E0008FFFE         move.w 8(a6),-2(a6); Copy i into frame
14 00040A 41EEFFFE             lea    -2(a6),a0   ; Point A0 to just above ARRAY[0]
15    ; Burst sample
16 00040E 11386000 GET_LOOP: move.b A_D,-(a0)    ; Get data into frame & dec pntr
17 000412 536EFFFE             subq   #1,-2(a6)   ; count = count - 1
18 000416 66F6                 bne    GET_LOOP    ; and repeat
19    ; Initialize to sum data
20 000418 3D6E0008FFFE         move.w 8(a6),-2(a6); Copy i back into frame again
21 00041E 41EEFFFE             lea    -2(a6),a0   ; A6 to just above ARRAY[0] anew
22 000422 4247                 clr.w  d7          ; Clear sum (D7)
23 000424 4240                 clr.w  d0          ; Use D0 to extend byte to word
24    ; Now do the summation
25 000426 1020     ADD_LOOP: move.b -(a0),d0    ; EXtend ARRAY[n] to word, ptr--
26 000428 DE40                 add.w  d0,d7       ; Add to word sum
27 00042A 536EFFFE             subq   #1,-2(a6)   ; count = count - 1
28 00042E 66F6                 bne    ADD_LOOP    ; and repeat
29    ; Close frame
30 000430 4E5E                 unlk   a6          ; SSP back up and restore old FP
31 000432 4E75                 rts                ; and return
32                             .end
```

System stack, the upper byte of which is null).

The coding shown in Table 5.9 is designed to reflect the 6809 equivalent, rather than using the more efficient features of the 68000, such as DBF. The LINK A6,#102h instruction in line 11 replaces the three equivalent 6809 instructions in lines 11–13 of Table 5.8. The old Frame Pointer (A6 in this example, but any Address register except A7 could be used) is firstly saved in the System stack. Then it is overwritten by the SSP to become the new Frame Pointer to TOF. Finally, the SSP is moved down to open the 102-byte Frame. The opposite UNLINK (UNLK) instruction of line 30 undoes these three actions also in one go. Table 5.1(b) lists the behavior of this pair of instructions. Note that LINK An,#kk is a word operation, with kk being sign extended to a 32-bit constant and then added to SSP. Effectively this limits the frame size to 32,768 bytes. With relatively little modification, the code given below could deal with sampled arrays of this size. The 68020 MPU has a long LINK variant.

The core of the program is straightforward, with the only problem lying in lines 25 and 26. Here a byte sample is to be added to a word sum. As both source and destination operands must be the same size, the byte variable is promoted to word size by moving into previously cleared **D0.W**. This is then added to **D7.W**. In stepping an Address register through the array, **A0** fulfils the same role as the **X Index register** in the 6809 equivalent, leaving the Frame Pointer **A6** untouched (lines 16 and 25).

The 68000 family are blessed with a generous complement of registers. It would thus be more efficient to use a Data register to hold the loop counter rather than operate directly in memory. The C high-level language allows the programmer to declare local (known as Auto) variables as Register variables. The compiler will then make an attempt to lodge such variables in a register.

The last two examples have returned their single parameter in a Data register. High-level languages such as Pascal and C permit only one return variable, which is defined as the value of the function. Thus expressions in C such as:

```
if (block_copy(rom_start, ram_start, length) = -1)
    {do this, as no error has occurred;}
else
    {do that, on an error situation;}
```

are possible, where function block_copy() (see Table 5.7) is called up (with the passed parameters indicated in brackets) and its value compared to −1. Its 'value' is in fact the returned value.

In C and Pascal, larger numbers of variables can be altered by passing pointers (as in this example) or by declaring variables as global. Global variables are stored in fixed RAM locations, and are thus accessible to any function.

The System stack itself may be used to pass back multiple variables. In such cases, room is normally left on this stack, just below the pass-to variables, before moving control to the subroutine. On return, the **SSP** will then point to the returned parameters, which can be extracted before the stack is cleaned up.

References

[1] Yourdon, E.; *Techniques of Program Structure and Design*, Prentice-Hall, 1975, Section 3.4.

[2] Goor, A.J. van de; *Computer Architecture and Design*, Addison-Wesley, 1989, Section 8.3.

[3] Wakerly, J.K.; *Microcomputer Architecture and Programming: The 68000 Family*, Wiley, 1989, Section 9.3.6.

[4] Maurer, W.D.; Subroutine Parameters, *BYTE*, 4, no. 7, July 1979, pp. 226–230.

[5] Wakerly, J.K.; *Microcomputer Architecture and Programming: The 68000 Family*, Wiley, 1989, Section 9.2.

CHAPTER 6

Interrupts plus Traps equals Exceptions

A microprocessor used as a controller spends much of its time detecting and measuring events happening in the outside world. These external events happen in their own time and are in no way synchronized to the MPU's internal processes. A simple example of this is shown in Fig. 6.1, where we wish to measure the time in 1 ms 'ticks' between each cycle of an electrocardiograph (ECG or EKG) signal (heart wave). One possibility would be to use hardware to count 1 kHz oscillations and to detect the fiducial point [1]; indeed this hardware could itself be a MPU-based circuit. When this reference point (the signal peak in the diagram) occurs, the master microprocessor must be alerted to the fact. A response must be made within 1 ms of the event, as the counter continues incrementing.

One approach would be to use the peak detector's output to set a **flag** (latch). This latch output is buffered to the data bus, and can be accessed at some address. Thus the MPU could regularly read the flag at intervals of no less than 1 ms, and get the counter data only when the flag was set. Resetting the latch at this point prepares for the next event. However, in this example this will typically only happen around once per second, a 0.1% hit rate! This **polling** approach is fine if there are only a few events being measured and the background processing task is not too onerous. However, in this instance we may also be measuring blood pressure, temperature etc. for a whole ward of patients. In that case the MPU will spend most of its time polling, leaving little time for processing.

To circumvent this problem, all MPUs have at least one input labelled Interrupt. When its Interrupt line is tugged (usually by going low or by a low-going edge) the MPU will temporarily suspend its operation and go to an **interrupt service routine (ISR)**. This is just a subroutine entered via an external (hardware) signal. At the end of this routine, control is passed back to the background program. However, interrupts as seen from the MPU happen at random, so care must be taken that the machine state has not been disrupted when control does return. Furthermore, when several devices can request an interrupt, some means must be found to determine the source of the service request, and prioritize when more than one peripheral requires attention.

All this refers to hardware-generated interrupts. Most MPUs can generate in-

144 Interrupts plus Traps equals Exceptions

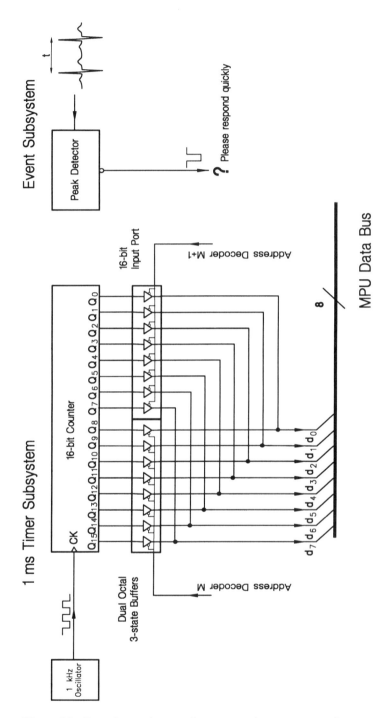

Figure 6.1: Detecting and measuring an asynchronous external event.

terrupts when some exceptional condition occurs internally, for example using a zero divisor for the DIVU and DIVS 68000 instructions. Allied to these **traps** are explicit instructions which can cause the processor to act in much the same way as a hardware interrupt. These are sometimes known as **software interrupts**. A generic term for all hardware and software interrupts is an **exception** (for exceptional circumstances).

Processors handle exceptions in differing ways. In this chapter we will look at the general concepts involved in interrupt handling, and how the 6809 and 68000 processors implement exceptions.

6.1 Hardware Initiated Interrupts

Although the minutia of the response to an interrupt request varies considerably from processor to processor, the following phases can usually be identified:

1. Finishing the current instruction.

2. Ignoring the request if the appropriate mask (if any) is set.

3. Saving at least the state of the **PC** and **CCR** registers.

4. Entering the appropriate service routine.

5. Identifying the source of the interrupt (if not done in phase 4).

6. Executing the defined task.

7. Restoring the processor state and returning to the point in the program where control was first transferred.

Interrupts are by definition asynchronous to system operation. Their apparent randomness means that the system response to such events must ensure that the interrupted program (the **background program**) is oblivious to the fact that the processor has 'gone away for a while' to service an external request. In some ways this is akin to transparency in subroutines (see page 128) but is more difficult to implement due to the erratic nature of the action.

At the very least, transparency to interrupts demands that the state of the MPU must be saved before going to the interrupt service routine, and restored on exit. This implies that instructions be treated as indivisible, as saving the MPU state part of the way through an instruction is difficult and to my knowledge is not implemented by any current MPU. Thus, although an interrupt request signal may be internally latched by the MPU at any time, usually on a clock edge, it will not be examined until the end of the current instruction execution.

As a consequence of this, care must be taken when dealing with data objects greater than the natural size of the processor. As an example, consider incrementing a 4-byte variable N in 6809 code. Assuming that this is stored in mem to mem+3 we have:

```
1   LDD    mem+2    ; Add one to lower word
2   ADDD   #1       ; stored in mem+2:mem+3
3   STD    mem+2    ; Lower word now incremented
4   LDD    mem      ; Add carry to upper word stored in mem:mem+1
5   ADCB   #0       ; one byte at a time
6   ADCA   #0       ; as there is no Double Add with Carry
7   STD    mem      ; N++ at last!
```

This is simple enough. But consider that N = FFFF FFFFh. If an interrupt strikes in-between lines 2 and 7, and the interrupt service routine uses N, then the value it will see is FFFF 0000h rather than 0000 0000h. Although problems like this can be avoided at assembly level, they are difficult to overcome when using high-level languages, as the machine-level code produced by the compiler is not directly under the control of the programmer. This is particularly true as high-level instructions are not entities as seen by an interrupt. In general do not share data between interrupt service routines and other code, see Section 10.2. However, avoiding the use of global variables is easier said than done.

Most interrupts can be inhibited during 'sensitive moments', such as described above, by setting the appropriate mask in the Code Condition register. Specifically the 6809 MPU supports three interrupt lines. These are labelled in Fig. 6.2(a) as $\overline{\text{IRQ}}$ (for $\overline{\text{Interrupt\_ReQuest}}$), $\overline{\text{FIRQ}}$ (for $\overline{\text{Fast\_Interrupt\_ReQuest}}$) and $\overline{\text{NMI}}$ (for $\overline{\text{Non\_Maskable\_Interrupt}}$). The former two are inhibited by mask bits **I** and **F** respectively. These are automatically set when the MPU is Reset, so that peripheral interface devices and relevant variables can be allocated their initial state before dealing with an interrupt. The ANDCC instruction can be used at any point in the program to clear either or both mask bits, for example ANDCC #10101111b enables both $\overline{\text{IRQ}}$ and $\overline{\text{FIRQ}}$ lines. Conversely the ORCC instruction can be used to inhibit, for example ORCC #01000000b disables $\overline{\text{FIRQ}}$.

The 6809 has one non-maskable interrupt line. This cannot be locked out, and as such must be used with caution. Unlike $\overline{\text{IRQ}}$ and $\overline{\text{FIRQ}}$ which are activated by a low voltage *level* ⎯⎯⎯ at the appropriate pin, $\overline{\text{NMI}}$ is triggered by a low-going voltage ⎤⎽ that is *edge triggered*. This voltage may stay low after the event, and will not cause another interrupt until the signal goes high and then low again. In the event of one type of interrupt being interrupted by another, the $\overline{\text{NMI}}$ will have top priority, that is $\overline{\text{NMI}}$ can interrupt an IRQ or FIRQ service routine, or even itself. $\overline{\text{IRQ}}$ has the lowest priority, and can be interrupted by a $\overline{\text{FIRQ}}$, as well as $\overline{\text{NMI}}$. As we shall see, the interrupt handling mechanism requires the use of the System stack. After the 6809 is Reset a $\overline{\text{NMI}}$ interrupt event is latched, but not acted upon, until the first load into the **System Stack Pointer**, which it is assumed sets up the System stack, for instance LDS #0400h.

The interrupt structure of the 68000 MPU as shown in Fig. 6.2(b) is somewhat more complex. Here too there are three interrupt lines, and in a minimum system these can be used to give three different responses. However, the processor is actually designed to differentiate between seven different interrupt requests, which it interprets from the 3-bit pattern on the **Interrupt Priority Level** $\overline{\text{IPL2}}\,\overline{\text{IPL1}}\,\overline{\text{IPL0}}$.

Hardware Initiated Interrupts 147

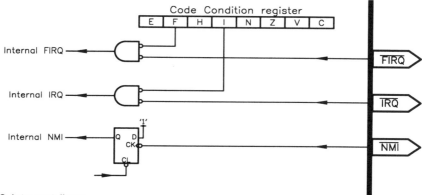

(a) 6809 Interrupt lines.

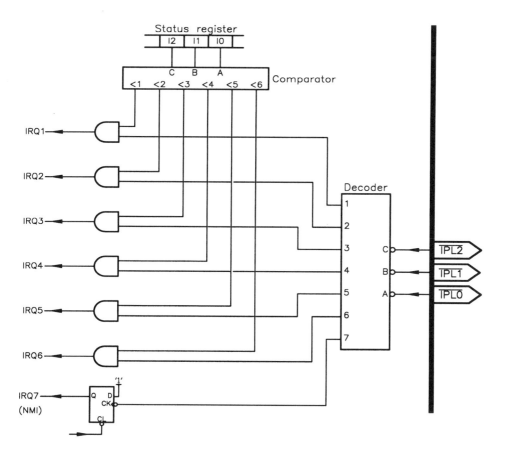

(b) 68000 Interrupt lines.

Figure 6.2: Interrupt logic for the 6809 and 68000 processors.

Thus 100b (active low 011b) is considered a level 3 interrupt request. A level 0 request ($\overline{\text{IPL2}}\,\overline{\text{IPL1}}\,\overline{\text{IPL0}}$ = 111) is ignored (no interrupt), whilst level 7 is non-maskable, and like the 6809's $\overline{\text{NMI}}$ equivalent, is edge triggered, an edge here being defined as a transition from a lower level.

The mask structure also echoes the level-oriented interrupt request. Three **mask bits** in the **Status register** (see Fig. 3.1) set the level above which a request is honored. Thus if **I2 I1 I0** is set at 100 (e.g. ANDI #11111 100 11111111b,SR) then any request from level 5 to 7 will result in the relevant internal IRQ line being activated. On Reset the three interrupt mask bits are set to 111, locking out all except level 7, the non-maskable interrupt.

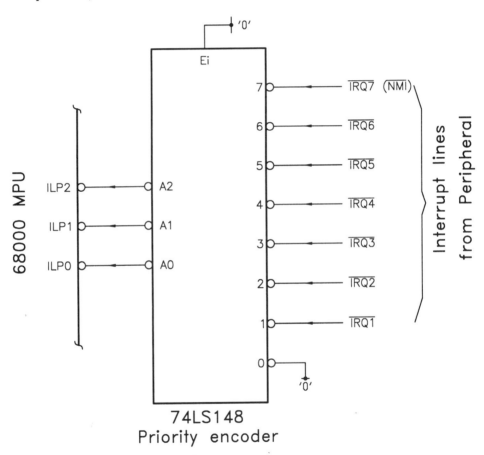

Figure 6.3: Using a priority encoder to compress 7 lines to 3-line code.

Interrupt request lines from three peripheral interfaces may be directly connected to $\overline{\text{IPL2}}\,\overline{\text{IPL1}}\,\overline{\text{IPL0}}$, having level 1, 2 or 4 priorities. Up to seven interrupt sources can be handled using external circuitry to encode these lines to 3-bit binary. The most common approach shown in Fig. 6.3 uses a 74LS148 priority encoder [2]. This has eight active-low inputs and three active-low outputs. The 74LS148 gives a 3-bit coded equivalent of the *highest* active input line. Thus if devices 6 and 1 simultaneously request service ($10111101b$), then the output will be 6 ($001b$, active-low). Once device 6 has been serviced and its interrupt request line lifted, the 74LS148's output will change to $110b$ (active low 1), and device 1 will then be eligible for service (if not masked out by **I2 I1 I0**). Similar considerations apply to the 68008 MPU, although as we can see from Fig. 3.2 $\overline{\text{IPL0}}$ and $\overline{\text{IPL2}}$ are internally connected, effectively allowing only levels 2 (*101*), 5 (*010*) and 7 (*000*) to be acceded to. The higher the level of interrupt request, the greater is its priority. Thus if a level 5 interrupt is in progress, it can only be interrupted by a level 6 or 7 request.

Once the MPU accepts an interrupt, it must change from executing the **background** program, and move to the appropriate interrupt service routine or **foreground** program. This is similar to switching to a subroutine, but the change-over is dictated by an apparently random call from outside. As this can happen anywhere in the background program, the state of all the MPU's registers (its context) used in the background program must be saved before the change-over. On return these are restored, leaving the state of the MPU unchanged. Making the interrupt process invisible in this manner allows the MPU apparently to execute more than one task in parallel. **Multitasking** in this manner is of course a serial process, and carries the overhead of the time to switch context between background and foreground [3].

There are two approaches to **context switching**. At the very least the Program Counter and Code Condition register/Status register must be saved. The former, so that control can be passed back to the background program at the point of the break, as in the case of a subroutine call. The latter, because the **CCR** will be altered by any but the most trivial interrupt service routine. Any additional registers altered by the service routine can be saved by Pushing and Pulling via a stack, in the manner shown in Table 5.3. Some early microprocessors, such as the 6800, save all internal registers automatically on the System stack when an interrupt response is initiated and return them at the end. This entire-state context switching is convenient, but in processors with a significant complement of registers, the resulting time overhead can have a noticeable impact on system response. This is not justified where only a few registers are actually used in the service routine. Early processors have few registers and/or stack-oriented instructions (the 6800 has one Address register, two Data registers and cannot directly Push or Pull the former), and thus an automatic whole-state context switch is efficient. Both types of context switching use the System stack to save the register states.

The 6809 MPU has both partial and full context switching. The **IRQ** and **NMI** responses automatically cause *all* registers to be Pushed on to the System stack, in the order shown in Fig. 6.4(b). The **FIRQ** response saves only the **PC** and **CCR**, leaving the rest up to the programmer (see Table 6.1(b)). The **E** flag in the **CCR** is set after the Push if the Entire state has been saved. It is used by the RETURN FROM INTERRUPT instruction, which terminates all 6809 interrupt service routines. RTI reverses the context switch and restores the MPU to its original state.

150 Interrupts plus Traps equals Exceptions

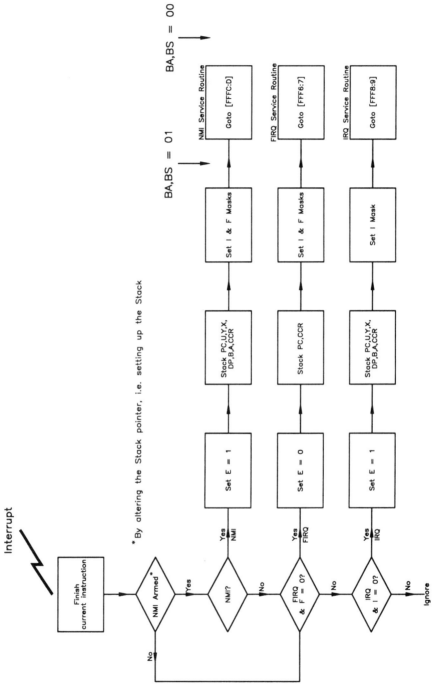

Figure 6.4: How the 6809 responds to an interrupt request (*continued next page*).

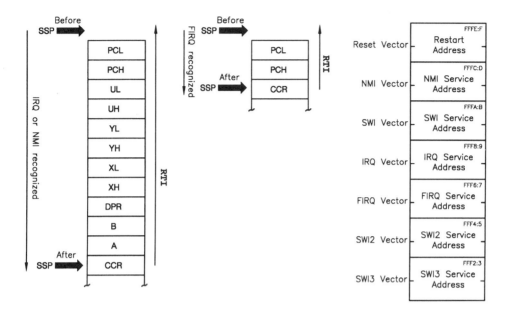

(b) Context switching, saving the state on the System stack.   (c) Vector table.

Figure 6.4: (*continued*) How the 6809 responds to an interrupt request.

The FIRQ response automatically sets the I and F mask bits in the **CCR** before entering its service routine, in order to ensure that it cannot be further interrupted by any other than the non-maskable interrupt. Only the I mask is set in the IRQ response. Consequently an IRQ service routine can be interrupted by a FIRQ response as well as a NMI. Of course when the old value of the **CCR** is returned, these changes vanish.

When a 68000 processor recognizes a level-n request, it saves the **SR** in a temporary internal register. Then the three interrupt mask bits are updated to level n, permitting only interrupts at a higher priority level to be further recognized during the level-n service routine. Also the **T flag** is cleared, to prevent Trace interrupts (see page 167), and the **S flag** is set. The latter means that the processor switches into the Supervisor state (if not already there). Thus when the **PC** and **SR** are saved, as shown in Fig. 6.5(b), the **Supervisor Stack Pointer** (**SSP**) and not the **User Stack Pointer** (**USP**) is used to delineate the context stack. The **SR** saved in this manner is the original copied into the internal register and not the modified version. Thus the interrupt service termination RETURN FROM EXEMPTION (RTE) (equivalent to the 6809's RTI) will move the processor back to the User state, if this was the interrupted state, as well as restoring the mask bits to their original value.

With everything put away on the System stack, the processor is ready to go to the start of the appropriate service routine. The simplest approach to this is to have the entry addresses stored in predetermined locations. The 6809 MPU

152 *Interrupts plus Traps equals Exceptions*

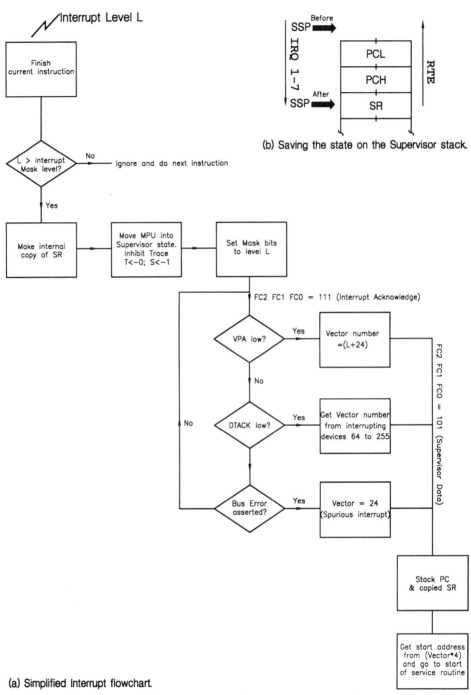

(a) Simplified Interrupt flowchart.

Figure 6.5: How the 68000 responds to an interrupt request (*continued next page*).

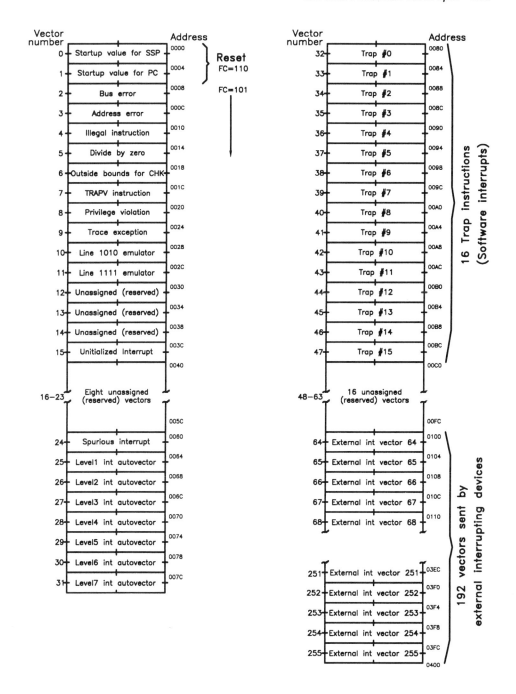

(c) Exception table (unassigned may be used for 68020+ processors).

Figure 6.5: (*continued*) How the 68000 responds to an interrupt request.

reserves 14 bytes at the *top* of its memory space to hold the seven start addresses of its three hardware, three software and one Reset interrupt, as shown in Fig. 6.4(c). For example, when the MPU responds to an IRQ request, it will find the start of the IRQ service routine in FFF8:9h. Normally this vector table is in ROM, and this is a necessity for the Reset vector in FFFE:Fh, as the address for the main routine must be present at power up (cold start). In systems where no actual memory exists at these locations, the Address decoder must be designed to enable physical memory when these addresses are output by the MPU. If necessary, clever address decoding can be used to place locations FFF2 – FFFDh in RAM where they may be dynamically altered by the program, although this is rare.

As an example, consider an extension to the system shown in Fig. 6.1. An external 16-bit counter records 1 ms ticks, whilst a detector circuit records signal peaks. An array of 256 peak to peak times in milliseconds is to be displayed on an oscilloscope. Two digital to analog converters are to be used to drive the X and Y oscilloscope plates — see Fig. 11.3. The background program is to scan this array sending its analog equivalent to the Y plates at the same time as the X plates drive is being incremented from 0 to 255 (0 to full-scale analog). This occurs as a continuous loop, giving a flicker-free display. Whenever a peak is detected, the processor is to switch from its background display task to updating the array with the latest period. When the array is full (256 peaks), the process is to be repeated, over-writing the oldest values. Provided that this foreground task is accomplished quickly, this switch back and forth will not be noticed on the display.

We need not concern ourselves with the details of the Address decoder nor the interfacing digital to analog converters here, but we must consider the problem of driving the MPU's interrupt input from the peak detector. Taking the 6809 MPU for our first solution, we will use the $\overline{\text{FIRQ}}$ input to keep the response time short. Now $\overline{\text{FIRQ}}$ (and $\overline{\text{IRQ}}$) are active as long as their *level* is low. We have not specified the duration of the peak detector's active output, but in this situation it is likely to be anything up to 250 ms, to avoid multiple triggering due to noise around the peak. Thus if $\overline{\text{FIRQ}}$ is still low after the return to the background program, then another interrupt response will be immediately set in train. In this case the whole 256-word array will probably be updated in one go!

As shown in Fig. 6.6, interposing a D flip flop solves the problem. As the flip flop is edge-triggered, its D input is only clocked in on the falling *edge* (in this case). This **interrupt flag** is thus 'lowered'. After the processor vectors to the service routine, the act of reading the counter also activates the flip flop's $\overline{\text{Preset}}$ input, which sets it to logic 1 (raises the flag). Thus on return, the interrupt line is no longer active, irrespective of the indeterminate length of the source request. Edge-triggered interrupts, such as $\overline{\text{NMI}}$, can be directly driven without using an external flag.

Peripheral devices designed specifically to interface to a MPU normally incorporate such flags as part of a Status or Control register. For example the 6821 PIA of Fig. 1.9 uses bits 6 and 7 of each **Control register** for this purpose [4]. Reading the appropriate Data register clears these flags automatically.

The 6809 code implementing our specification is shown in Table 6.1. This

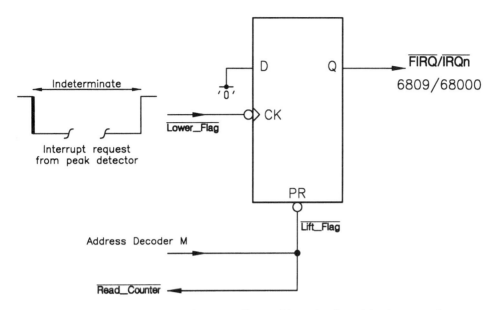

Figure 6.6: Using an external interrupt flag to drive a level-sensitive interrupt line.

comprises three separate source modules:

1. The background module DISPLAY which extracts the 256 array values using them to drive the oscilloscope Y plates as it ramps up the X plates. This module runs continually except when interrupted by the foreground module.

2. The foreground module UPDATE is entered only when an external event occurs. It reads the counter, evaluates the time since the last event, inserts the outcome in the array and moves the array index on one.

3. The VECTOR module simply sets up the Interrupt and Reset vectors. The actual values are put into memory at load time, that is when the EPROM is programmed (or program downloaded into RAM in a Microprocessor Development System). It does not execute as such at run time, it is simply in situ in a supporting role to the two previous modules.

Each of these three modules are separately assembled and subsequently linked together to give the listing of Table 6.1. We will discuss this linkage process in the following chapter, here it is sufficient to note that the assembler reserves 256 words in its Data program space .psect _data (line 17 of Table 6.1(a)), the start address of which is called ARRAY. This name is made known to the other separately assembled modules through the linker by declaring it .public in line 16. The foreground module needs to use this address, its value at assembly time being unknown, and it gets round this problem by declaring ARRAY as .external in line 20 of Table 6.1(b). This directive is really saying to the assembler 'hold your fire, the actual address will be supplied at a later date via the linker'. Of course this is an array of words, as the counter is 16-bits wide. In a similar manner, the address of both run-time modules are made known to module VECTOR by declaring their start address .public. They are consequently declared .external in line 9 of Table 6.1(c).

In the background module, the ramp count x (the Scan pointer) is also used as array index (i.e. at position x display ARRAY[x]). However, as each array element is a double byte, x is first promoted to 16 bits (line 24) and then multiplied by two (lines 25 and 26). The resulting value in Accumulator_D is then used as the offset to the X Index register — which is pointing to ARRAY[0] — to abstract ARRAY[x]. The same mechanism is used in the update interrupt module to convert the Update

Table 6.1: 6809 code displaying heart rate on an oscilloscope (*continued next page*).

```
1                              .processor m6809
2     ;*********************************************************************
3     ; * Background program which scans array of word data (ECG periods)  *
4     ; * Sends out to oscilloscope Y plates in sequence                   *
5     ; * At same time incrementing X plates                               *
6     ; * so that ARRAY[0] is seen at the left of screen                   *
7     ; * and ARRAY[255] at the right of screen                            *
8     ; * ENTRY : None                                                     *
9     ; * EXIT  : Endless loop                                             *
10    ;*********************************************************************
11    ;
12                        .define DAC_X=6000h, ; 8-bit X-axis D/A converter
13                                DAC_Y=6001h  ; 12-bit Y-axis D/A converter
14    ;
15                        .psect _data    ; Data space
16                        .public ARRAY   ; Make the array global
17 0000          ARRAY:   .word  [256]    ; Reserve 256 words for the array
18 0200          X_COORD: .byte  [1]      ; and a byte for the X co-ordinate
19    ;
20                        .psect _text    ; Program space
21                        .public DISPLAY ; Make program known to the linker
22 E000 10CE0800 DISPLAY: lds  #0800h     ; Define Top Of Stack
23 E004 F60200   DLOOP:   ldb  X_COORD    ; Get X co-ordinate
24 E007 4F                clra            ; Expand to word size
25 E008 58                lslb            ; Multiply by two
26 E009 49                rola            ; to give array index in Acc.D
27 E00A 8E0000            ldx  #ARRAY     ; Point to ARRAY[0]
28 E00D 308B              leax d,x        ; now to ARRAY[X]
29 E00F F60200            ldb  X_COORD    ; Get back X co-ordinate
30 E012 F76000            stb  DAC_X      ; Send it out to X plates
31 E015 EC84              ldd  0,x        ; Get ARRAY[X] word
32 E017 FD6001            std  DAC_Y      ; and send it to the Y plates
33 E01A 7C0200            inc  X_COORD    ; Go one on in X direction
34 E01D 20E5              bra  DLOOP      ; and show next sample
35                        .end
```

(a) The background array-display module.

Table 6.1: (*continued*) 6809 code displaying heart rate on an oscilloscope.

```
 1                          .processor m6809
 2   ; **************************************************************************
 3   ; * Interrupt service routine to update one array element                   *
 4   ; * with the latest ECG period, as signalled by the peak detector           *
 5   ; * ENTRY : Via a FIRQ interrupt                                            *
 6   ; * ENTRY : Location of ARRAY[0] is globally known through the linker       *
 7   ; * EXIT  : ARRAY[i] updated, where i is a local index                      *
 8   ; * EXIT  : MPU state unchanged                                             *
 9   ; **************************************************************************
10   ;
11                          .define   COUNTER =9000h, ; The 16-bit period Counter
12                                    INT_FLAG=9800h  ; The external Interrupt flag
13   ;
14                          .psect    _data   ; Data space
15 0201        UPDATE_I:    .byte     [1]     ; Space for the array update index
16 0202        LAST_TIME:   .word     [1]     ; and for the last counter reading
17   ;
18                          .psect    _text   ; Program space
19                          .public   UPDATE; Make routine known to the linker
20                          .external ARRAY   ; Get ARRAY from another module
21 E01F 3436   UPDATE:      pshs  a,b,x,y     ; For FIRQ save used registers
22 E021 7F9800              clr   INT_FLAG    ; Reset external Interrupt flag
23 E024 FC9000              ldd   COUNTER     ; and get the count from outside
24 E027 1F02                tfr   d,y         ; Put in Y register for safekeeping
25 E029 B30202              subd  LAST_TIME   ; Sub frm last cnt gives new period
26 E02C 10BF0202            sty   LAST_TIME   ; and update last counter reading
27 E030 1F02                tfr   d,y         ; Y now holds the new period
28 E032 F60201              ldb   UPDATE_I    ; Get the update array index
29 E035 4F                  clra              ; Expand to word
30 E036 58                  lslb              ; Multiply by 2 to cope with
31 E037 49                  rola              ; the word nature of ARRAY[]
32 E038 8E0000              ldx   #ARRAY      ; Point to ARRAY[0]
33 E03B 10AF8B              sty   d,x         ; Put new value (in Y) in ARRAY[I]
34 E03E 7C0201              inc   UPDATE_I    ; Move update marker on one
35 E041 3536                puls  a,b,x,y     ; Return machine state
36 E043 3B                  rti
37                          .end
```

(b) The foreground interrupt service routine updating the array.

```
 1                          .processor m6809
 2   ; **************************************************************
 3   ; * Sets up Interrupt and Reset vector at top of ROM           *
 4   ; * using globally known labels through the linker             *
 5   ; **************************************************************
 6   ;
 7                          .psect    _text
 8                          .public   VECTOR          ; Make this routine known globally
 9                          .external UPDATE,DISPLAY; These will be got thru the linker
10 E7F6 E01F   VECTOR:.word            UPDATE         ; Addr of the FIRQ service routine
11 E7F8                    .word      [3]             ; Skip IRQ, SWI, NMI not used here
12 E7FE E000   RESET: .word           DISPLAY         ; Go to DISPLAY routine on Reset
13                          .end
```

(c) The Vector table.

pointer i to the array pointer i (lines 29–31). Both the Scan and Update pointers conveniently wrap around from 255 to zero after incrementing. Numbers other than 255 would need to be actively zeroed.

Notice how in module VECTOR (lines 10–12) the start addresses for the Reset and FIRQ service routines are located in their appropriate place. In practice other vector addresses, not used in this fragment, would be defined here.

The 6809 MPU can only deal directly with three interrupt requests from separate sources. Some applications require many more than can be handled in this way. Wiring these n request lines through open-collector gates is a convenient way of channelling n lines to one interrupt line, see left side of Fig. 6.7. Normally the n service request lines are high and the open-collector gates are off, letting $\overline{IRQ}$ rise through the pull-up resistor to $+V$. If one or more request lines go low then $\overline{IRQ}$ goes low. MPU-compatible peripheral interface devices, such as the 6821 and 68230 PIAs, have integral open-collector buffers at their interrupt output lines.

Given that the MPU has gone to the service routine, how is it to distinguish between the various possible sources? A simple procedure is to examine each interrupt flag in turn, until the source is found. Where MPU-compatible peripherals are used, this is accomplished by examining the relevant bits in the appropriate peripheral Control/Status register.

Polling in this manner is rather slow but does have the advantage of simplicity, and a priority scheme of arbitrary complexity can be implemented in software.

There are many schemes which speed up the process of distinguishing between interrupting peripherals [5], one of which is shown in Fig. 6.7. Here, four events (e.g. peak detectors) trigger interrupt flags in the manner of Fig. 6.6. These four service requests are combined together with open-collector buffers to drive the MPU's $\overline{IRQ}$ line. The state of these four lines can be read at any time through 3-state buffers at address Vector. Assuming that unconnected data lines read as logic 0, we have:

Request	Vector	
0	00000100	(4)
1	00001000	(8)
2	00010000	(16)
3	00100000	(32)

If more than one request is simultaneously received, intermediate vector values will be generated. The appropriate software filtering routine can then separate and prioritize the requests, or a priority encoder can be used as a hardware solution. The MPU can then go to the appropriate routine.

As an extension to this scheme, the vector buffers could be enabled whenever the addresses FFF9:Ah are detected on the address bus with the Status bits BA BS = 01, rather than the ROM. Thus the address of the program start is directly generated as a response to the interrupt, but appears to originate at the appropriate vector address. In this situation it would be better to read the Service Request lines through a priority encoder to remove ambiguities caused by more

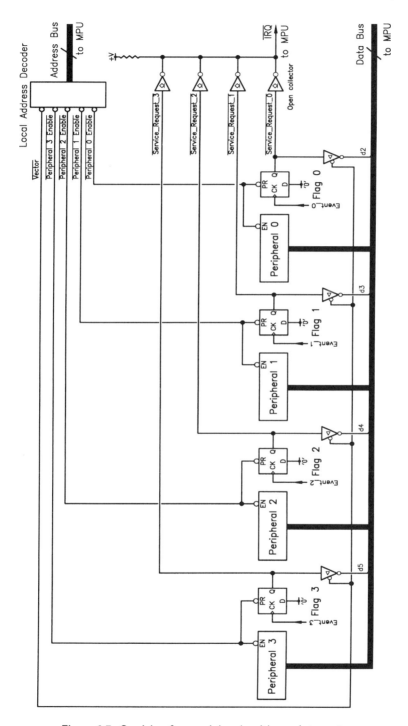

Figure 6.7: Servicing four peripherals with one interrupt.

than one peripheral requesting service at the same time [6]. Direct vectoring by device is the fastest technique available, but is expensive in hardware.

The 68000 family also makes use of a Vector table to service its various exceptional events, but in a rather more flexible manner. The lowest 256 long-words of memory, 000000–0003FFh, hold addresses potentially pointing to the beginning of 255 service routines as shown in Fig. 6.5. Of these, the bottom two long-words are reserved for the critical Reset vector thus:

SSP	0000–0003h
PC	0004–0007h

} Double long-word Reset vector

When the 68000 MPU is Reset, the initial setting of the Supervisor Stack Pointer (not the User Stack Pointer) is fetched from long-word 0 (000000–000003h), followed by the start value of the Program Counter in long-word 1 (000004–000007h). This dual vector *must* be in ROM to ensure a successful cold start (i.e. from power up), as must be the equivalent 6809 Reset vector at the top of memory. The remaining 254 vectors are normally also located in ROM, but clever address decoding can be used to overlay these vectors in RAM. This latter procedure allows the software dynamically to relocate exception service routines. The external decoder can distinguish between vectors 0 and 1, and 2 to 255 from the state of the Function Code status pins, which are 110b for the former (Supervisor Program) and 101b for the latter (Supervisor Data) — see page 70. As the Supervisor Stack Pointer is set up after the MPU leaves its Reset start-up, interrupts can be immediately serviced. The Interrupt Mask bits in the Status register are set to 111b, locking out all but level 7 interrupts (i.e. non-maskable).

When a 68000 MPU receives an interrupt request of a higher priority than its mask setting, it commences an Interrupt Acknowledge read cycle [7]. The level is echoed on Address lines $a_1 a_2 a_3$, with all other address lines going high. The Function Code lines FC2 FC1 FC0 are set to 111 and a normal Read cycle is implemented. Depending on external hardware, two things can happen. If the interrupting device wishes to use the fixed internal autovector table it responds by bringing $\overline{\text{VPA}}$ low during this Read cycle. More sophisticated peripheral interface devices specifically designed for the 68000 MPU can respond by putting a Vector number on its data bus and activating $\overline{\text{DTACK}}$ in the normal asynchronous way (see Fig. 3.6). The MPU multiplies this number by four (shift left twice) giving the address of the user interrupt vector somewhere in the table.

Referring to Fig. 6.8, we see that in both cases a 3 to 8-line decoder generates one of seven Interrupt Acknowledge signals IACKn from the 3-bit level address. This decoder is only active when the Function Code is 111, that is Interrupt Acknowledge. The rest of the address lines are logic 1 and the general address decoding must ensure that nothing else responds to this situation. The level, and hence which IACK line is active, is determined by the connection of the peripheral's service request to a 74LS148 Priority encoder, as described in Fig. 6.3.

First we look at a dumb interface, such as shown in Figs 6.1 and 6.6, which cannot generate its own Vector number. In Fig. 6.8 the level-1 request itself and

Hardware Initiated Interrupts 161

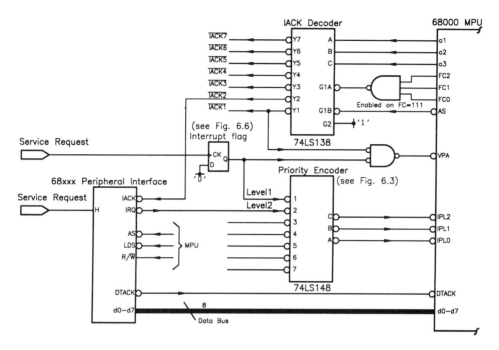

Figure 6.8: External interrupt hardware for the 68000 MPU.

acknowledgement (IACK1) are ANDed to drive $\overline{\text{VPA}}$ low. The MPU will go automatically to vector 25 (000064 − 7h) for its level-1 service routine. As previously described, the Interrupt flag must be lifted at this time.

Smart interfaces, such as the 68230 PI/T, are interfaced to the MPU in the normal way, see Fig. 3.13. Their IACK input is driven by the appropriate IACK decoder line, and the vector number put on the data bus during the concurrent Read cycle. This vector number is programmed into the appropriate interface register (the **Port Interrupt Vector register** in the 68230 [8]) during the setup routine. If we wanted to vector via address 000100h, the programmed-in vector number would be 40h (000100 ÷ 4h). Vector numbers 0 – 63 should not be used, although there is nothing physically to prevent this. Should a 68xxx peripheral interface not have its vector register set up when an interrupt occurs, a default vector 15 will be sent to indicate Uninitialized Interrupt.

The software for our example is given in Table 6.2. It matches the listing of Table 6.1 for the 6809 MPU, and the comments made there apply equally. Notice that the Interrupt Service routine UPDATE is terminated by RTE, the 68000 equivalent for RTI. I have assumed that a level-1 autovector is being used as a pointer to the service routine. A simple change of operand in line 12 of Table 6.2(c) would move the start address to any other appropriate vector number.

Vector 24 is described in Fig. 6.5(c) as a Spurious Interrupt. This startup address will be used if external circuitry asserts the $\overline{\text{Bus\_Error}}$ ($\overline{\text{BERR}}$) pin during an Interrupt Acknowledge Read cycle. The hardware designer may wish to do this

162 Interrupts plus Traps equals Exceptions

when $\overline{\text{DTACK}}$ (or $\overline{\text{VPA}}$) is not activated within a fixed time after the start of this cycle; to indicate a hardware problem. Such circuitry is frequently implemented as a retriggerable monostable which 'collapses' if not clocked frequently enough. Such a watch-dog timer can of course be used to indicate trouble 'out there' during a normal (i.e. not Interrupt Acknowledge) cycle. In such cases the MPU returns to the Supervisor state and enters the Bus Error exception service routine pointed to by Vector 2. Should the $\overline{\text{BERR}}$ signal persist when the status is being pushed out to the Supervisor stack on entry to the service routine, a catastrophic situation is assumed to have occurred. Such a Double-Bus fault causes the MPU to stop, with both $\overline{\text{Halt}}$ and $\overline{\text{Reset}}$ going low. This response will occur in general where a problem occurs when an exception (including a Reset) tries to Push out its registers, for example when the **Supervisor Stack Pointer** is odd.

Table 6.2: 68000 code displaying heart rate on an oscilloscope (*continued next page*).

```
 1                        .processor m68000
 2  ;************************************************************
 3  ; * Background program which scans array of word data (ECG points)  *
 4  ; * Sends out to oscilloscope Y plates in sequence                  *
 5  ; * At same time incrementing X plates                              *
 6  ; * so that ARRAY[0] is seen at the left of screen                  *
 7  ; * and ARRAY[255] at the right of screen                           *
 8  ; * ENTRY : None                                                    *
 9  ; * EXIT  : Endless loop                                            *
10  ;************************************************************
11  ;
12                        .define  DAC_X=6000h,; 8-bit X-axis D/A converter
13                                 DAC_Y=6001h ; 12-bit Y -axis D/A converter
14  ;
15                        .psect   _data         ; Data space
16                        .public  ARRAY         ; Make the array global
17 00E000         ARRAY:   .word    [256]         ; Reserve 256 words for the array
18 00E200         X_COORD: .byte    [1]           ; and a byte for the X co-ordinate
19  ;
20                        .psect   _text         ; Program space
21                        .public  DISPLAY       ; This program known to the linker
22 000400 4240   DISPLAY: clr.w    d0            ; Get X co-ordinate byte
23 000402 1039   DLOOP:   move.b   X_COORD,d0    ; expanded to word
       0000E200
24 000408 E348            lsl.w    #1,d0         ; x2 to give array index in D0.W
25 00040A 207C0000E000    movea.l  #ARRAY,a0     ; Point A0 to ARRAY[0]
26 000410 31F000006001    move.w   0(a0,d0.w),DAC_Y ; Get ARRAY[x] to Y plates
27 000416 31F90000E2006000 move.w  X_COORD,DAC_X    ; Send X coord to X plates
28 00041E 52390000E200    addq.b   #1,X_COORD    ; Go one on in X direction
29 000424 60DC            bra      DLOOP         ; and show next sample
30                        .end
```

(a) The background array-display module.

Table 6.2: (*continued*) 68000 code displaying heart rate on an oscilloscope.

```
1                              .processor m68000
2     ;************************************************************************
3     ;* Interrupt service routine to update one array element              *
4     ;* with the latest ECG period, as signalled by the peak detector      *
5     ;* ENTRY : Via a Level1 interrupt                                     *
6     ;* ENTRY : Location of ARRAY[0] is globally known through the linker *
7     ;* EXIT  : ARRAY[i] updated, where i is a local index                 *
8     ;* EXIT  : MPU state unchanged                                        *
9     ;************************************************************************
10    ;
11                         .define COUNTER=9000h, ; The 16-bit period Counter
12                                 INT_FLAG=9800h ; The external Interrupt flag
13    ;
14                         .psect    _data        ; Data space
15 00E201      UPDATE_I: .byte      [1]          ; Space for the array update index
16 00E202      LAST_TIME:.word      [1]          ; and for the last counter reading
17    ;
18                         .psect    _text        ; Program space
19                         .public   UPDATE       ; This routine known to the linker
20                         .external ARRAY        ; Get ARRAY from another module
21 000426 48E7C080 UPDATE: movem.l d0/d1/a0,-(sp); Save used registers
22 00042A 427900009800       clr   INT_FLAG     ; Reset external Interrupt flag
23 000430 303900009000       move.w COUNTER,d0  ; & get the count from the counter
24 000436 3200               move.w d0,d1       ; Put in D0.W for safekeeping
25 000438 92790000E202       sub.w  LAST_TIME,d1; Sub from last cnt for new period
26 00043E 33C00000E202       move.w d0,LAST_TIME; and update last counter reading
27 000444 4240               clr.w  d0          ; Prepare to get update array index
28 000446 30390000E201       move.w UPDATE_I,d0; expanded to word size
29 00044C E348               lsl.w  #1,d0       ; x2 to cope with word ARRAY[]
30 00044E 207C0000E000       movea.l #ARRAY,a0  ; Point A0.L to ARRAY[0]
31 000454 31810000           move.w d1,0(a0,d0.w); New value (D1.W) to ARRAY[I]
32 000458 52790000E201       addq.w #1,UPDATE_I; Move update marker on one
33 00045E 4CDF0103           movem.l (sp)+,d0/d1/a0; Return machine state
34 000462 4E73               rte
35                           .end
```

(b) The foreground interrupt service routine updating the array.

```
1                              .processor m68000
2     ; ************************************************************************
3     ; * Sets up interrupt and reset vectors at bottom of ROM               *
4     ; * using globally known labels through the linker                     *
5     ; ************************************************************************
6     ;
7                         .psect    _text
8                         .public   VECTOR       ; Make this routine known globally
9                         .external UPDATE,DISPLAY ; These will be got through the linker
10 000000 0000F000 VECTOR: SSP:.double 0F000h   ; Initial value of the System Stack
11 000004 00000400 PC:    .double   DISPLAY    ; Go to DISPLAY routine on Reset
12 000008                 .double   [23]       ; Other vectors not used here
13 000064 00000426 LEVEL1:.double   UPDATE     ; Addr of Level-1 IRQ serv routine
14                        .end
```

(c) The Vector table.

164   *Interrupts plus Traps equals Exceptions*

Another possibility is to assert $\overline{\text{BERR}}$ and $\overline{\text{Halt}}$ simultaneously. Then the failed bus cycle will be rerun, with the hope that a spurious failure occurred (perhaps due to noise) and that the situation can be redeemed [9].

6.2 Interrupts in Software

Interrupts occur when something outside requests assistance. The MPU responds by saving all or part of its internal state on the System stack and going to a service routine via a table of addresses. It is also possible to initiate a similar response by internal software means, either deliberately or via a dubious event, such as having a zero division for the DIVU/DIVS instruction. Software initiated exceptions are commonly known as Software Interrupts or **Traps**. In this section we will briefly consider these operations and other instructions associated with Exceptional operations.

Interrupt service routines are normal subroutines but terminated with an instruction or instructions that restore the state saved when responding to the interrupt. This is true of both hardware and software-initiated responses. In the 6809 processor the state returned by RETURN FROM INTERRUPT (RTI) depends on the setting of the E flag, either all registers if **E** is zero otherwise only the **PC** and **CCR**. The equivalent 68000 RETURN FROM EXCEPTION (RTE) always returns the **PC** and **SR** only. The same is true for the 8086 MPU family, where the instruction is INTERRUPT RETURN (IRET).

Most MPUs have at least one instruction which halts the processor until an interrupt (or reset) occurs. From Table 6.3(a), we see that the 6809 processor has two related instructions categorized as such. CLEAR AND WAIT allows the programmer to clear the F or I mask if desired prior to stopping. Thus CWAI #10111111b clears **F** and stops the processor after saving the entire machine state in the System stack (**E** set). If at some time in the future a $\overline{\text{NMI}}$ or $\overline{\text{FIRQ}}$ request is sent, the MPU will immediately go to the appropriate service routine. An $\overline{\text{IRQ}}$ will have no effect in this example, as it is masked out. Notice that unusually a $\overline{\text{FIRQ}}$ will enter its service routine with the entire machine state (context) saved.

The SYNCHRONIZE instruction is similar, although any **CCR** flags will have to be set by a preceding instruction. However, this time if the interrupt occurs but is masked out, then the processor will simply move on to the following instruction. If the interrupt is not masked out, and lasts for three clock cycles or more, then it will be answered in the normal way. Tri-state buses go high impedance during SYNC, allowing an external device to access memory directly [10].

The 68000's STOP instruction is comparable with the 6809's CWAI, but the immediate word operand is the new state of the **Status register**, rather than being ANDed with it. For example, STOP #001000 011 00000000b will halt the processor until an interrupt of level greater than 3 occurs. The MPU then responds in the normal way. STOP is privileged and thus can only be used in the Supervisor state. The machine context is not switched prior to the request. The equivalent HALT (HLT) for the 8086 family does not carry an immediate operand, but otherwise operates

Table 6.3: Exception related instructions.

Operation		Mnemonic	Description
ReTurn			Switch back to background
	from Interrupt	RTI	Pulls context back from System stack
Synchronize			Halt until interrupt
	Clear and WAIt	CWAI #kk$_8$	Clear CCR bits ([CCR] <- [CCR]·kk), save entire state and wait for interrupt
	SYNChronize	SYNC	Stop until interrupt occurs, THEN: continue if masked out, ELSE go to interrupt service routine
Trap			Software-initiated interrupt-like sequence
	SoftWare Interrupt	SWI	Save entire state and vector via FFFA:Bh and mask out I and F Hardware interrupts
	SoftWare Interrupt 2	SWI2	As above but vector FFF4:5h and no masking
	SoftWare Interrupt 3	SWI3	As above but vector FFF2:3h and no masking

(a) Relating to the 6809.

ReTurn			Switch back to background
	from Exception[1]	RTE	Pulls context back from Supervisor stack
Synchronize			Halt until interrupt
	STOP[1]	STOP #kk$_{16}$	[SR] <- #kk and wait for interrupt
Trap			Software-initiated interrupt-like sequence
	CHecK Bounds	CHK <ea>,Dn	IF 0 > Dn·W > ea THEN exception via vector 6
	ILLEGAL Instruction	ILLEGAL	Exception via vector 4
	TRAP	TRAP #kk$_4$	Sixteen software interrupts via vector 32 + #kk
	TRAP on oVerflow	TRAPV	IF V = 1 THEN exception via vector 7

(b) Relating to the 68000.

Note 1: Privileged instructions.

in the same manner.

The 6809 MPU has three instructions which explicitly initiate Software interrupt operations. SWI causes the entire state to be saved, sets the I and F masks to lock out all but $\overline{\text{NMI}}$ interrupts, and then vectors to the start of its service routine via FFFA:Bh. Instructions SWI2 and SWI3 are similar but using vectors FFF2:3h and FFF4:5h respectively to hold their start address, and not locking out the Hardware interrupts.

The 68000 MPU has 17 Software interrupts, known as TRAPs. Sixteen of these, TRAP #0 to TRAP #15 are unconditional and TRAPV is only implemented if the oVerflow flag is set at execution time. Looking at Fig. 6.5(c), we see that

TRAP #0 vectors via location 000080–3h up to TRAP #15 at 0000BC–Fh, Exception vectors 32 to 47. TRAPV has its service address located at 00001C–00001Fh. Like all other Exceptions, Traps execute in the Supervisor state.

Although what a Software interrupt/Trap does is clear enough, the reason for using one is not entirely evident. Consider an environment where an applications program is being written for a specific computer system. This system will have various means of communicating to the world, using typically a keyboard, VDU, serial and parallel ports, interrupts and various disk drives. Knowing the characteristics of all these input/output (I/O) devices, the programmer can write a suite of subroutines known as **device handlers**. Once this has been done, data can be transferred by calling up the appropriate handler. However, a change of environment to a different computer will likely require a complete rewrite of these handlers.

This approach is frequently adopted by the designers of embedded microprocessor systems, where the hardware infrastructure is usually highly individualistic. Some standardization is possible for mass-produced computing machines, such as engineering workstations and personal computers. These normally come with an **operating system**, which can be thought of as a shell around the applications software shielding the programmer from the hardware. Typical operating systems are UNIX [11] and MSDOS [12]. These systems are mainly disk-based loaded into RAM, but work in tandem with a Basic Input Output System (BIOS), usually located in ROM. The applications programmer can then call up the appropriate subroutine in the BIOS, to communicate with a peripheral. The BIOS ROM will vary with different machines, but in such a way as to hide the hardware details from the operating system. The use of an operating system leads to the concept of system-independent (portable) software.

Using a Trap call to communicate, rather than a subroutine, has the advantage that the address of the procedure need not be explicitly known, as the vector table will be in the BIOS. Hiding explicit details of the BIOS is important for portability. Thus, as an example, INT #25 in a MSDOS environment [12] will enable a Read from a magnetic disk (INT is the 8086 family mnemonic for TRAP). Parameters such as track, sector and drive are placed in registers prior to the Trap. In 68000-family based systems, the operating system normally resides in the Supervisor state, completely separated from the application program in the User state memory space.

The 68000 MPU has two additional explicit software interrupt instructions. The instruction ILLEGAL (op-code 4AFCh) causes a transfer via vector address 000010–13h and the CHECK REGISTER (CHK) instruction vectors via 000018–Bh if the lower word of the designated Data register is below zero or above the stated limit.

There are also a number of implicit traps, triggered by some internal event. These are:

Address Error, 00000C–0Fh

Entered when a word or long-word access to an odd address is attempted.

Illegal Instruction, 000010–13h

Entered if an illegal op-code is encountered, but see line A and line F Exceptions below.

Divide by Zero, 000014–17h

Entered when the divisor for DIVU/DIVS is zero.

Privilege Violation, 000020–23h

Entered when there is an attempt to execute a privileged instruction (e.g. STOP) while in the User state.

Trace, 000024–27h

Entered after each instruction if the T flag is set in the Status register. Used during debugging to monitor the state of the processor if the appropriate Trace Service routine is in situ [13].

Line A Op-Code, 000028–2Bh

Entered when the upper 4 bits of the op-code are 1010b. These op-codes are unused, but this facility provides the means for emulating unimplemented instructions in software.

Line F Op-Code, 00002C–2Fh

Entered when the upper 4 bits of the op-code are 1111b. Used as above (the 68020 MPU uses these codes for co-processor instructions, and therefore service routines are often used to simulate these missing instructions in software). All other unimplemented instructions vector via the Illegal instruction vector address above.

References

[1] Cahill, S.J. and McClure, G.; A Microcomputer-Based Heart-Rate Variability Monitor, *IEEE Trans. Biomed. Engng.*, **BME-30**, no. 2, Feb. 1983, pp. 87–92.

[2] Cahill, S.J.; *Digital and Microprocessor Engineering*, Ellis Horwood/Prentice-Hall, 2nd. ed., 1993, Section 3.2.1.

[3] Lawrence, P.D. and Mauch, K.; *Real-Time Microcomputer Design*, McGraw-Hill, 1987, Section 16.3.

[4] Cahill, S.J.; *The Single-Chip Microcomputer*, Prentice-Hall, 1987, Chapter 3.

[5] Leventhal, L.A.; *Introduction to Microprocessors*, Prentice-Hall, 1978, Chapter 9.

[6] Motorola Application Note AN866; *Vectoring by Device using Interrupt Sync Acknowledge with the MC6809/MC6809E*. Reprinted in *MCU/MPU Applications Manual*, **2**, 1984.

[7] Motorola Application Note AN1012; *A Discussion of Interrupts for the MC68000*, 1988.

[8] Miller, M.A.; *The 68000 Microprocessor*, Merrill Publishing, 1988, Section 8.8 – 8.11.

[9] Clements, A.; *Microprocessor Systems Design*, PWS, 2nd ed., 1992, Section 6.5.

[10] Motorola Application Note AN865; *The MC6809/MC6809E SYNC Instruction*, Reprinted in *MCU/MPU Applications Manual*, **2**, 1984.

[11] McIlroy, M.D.; UNIX Time-Sharing System, *The Bell System Technical Journal*, **57**, no. 6, part 2, 1978, pp. 1899 – 1904.

[12] Simrin, S.; *MS-DOS Bible*, H.W. Sams, 3rd ed., 1989.

[13] Leventhal, L.A.; *68000 Assembly Language Programming*, McGraw-Hill, 2nd ed., 1986, Chapter 19.

Part 2

C

The only reality as seen from a central processing unit, be it mainframe, mini or microprocessor, is in the patterns of binary states in memory. This is generally far removed from the human description of the task which is to be controlled by the processor hardware. In going from the problem specification to executable binary installed in memory involves many steps, both conceptual and in software. Many translation processes must occur on the way (see Fig. 7.1). Furthermore, testing, debugging and commissioning the system require additional skills and aids.

In Part 2 we look at these steps in some detail, how they interact and their limitations. In particular we will investigate the use of the high-level language C as a buffer between the problem-oriented human thought process and the machine-oriented assembly-level languages. Many of the concepts introduced here apply to other high-level languages, such as Pascal and Forth, but C is a small language which is widely available, especially in a cross form, popular, flexible and can run on inexpensive development systems. I can do no better than quote from the originators of the language:[1]

> C is a general-purpose programming language featuring economy of expression, modern control flow and data structure capabilities, and a rich set of operators and data types.
>
> C is not a 'very high-level' language nor a big one and is not specialized to any particular area of application. Its generality and an absence of restrictions make it more convenient and effective for many tasks than supposedly more powerful languages. C has been used for a wide variety of programs, including the UNIX operating system, the C compiler itself, and essentially all UNIX applications software. The language is sufficiently expressive and efficient to have completely displaced assembly language programming on UNIX.

---

[1]Ritchie, D.M. *et al.*; The C Programming Language, *The Bell System Technical Journal*, **57**, no. 6, part 2, July–August 1978, pp. 1991–2019.

CHAPTER 7

Source to Executable Code

Consider the fragment of code below. To a 68000-family MPU this makes perfect sense. Indeed a series of binary bits, typically represented by nominal 0 V and 5 V potentials stored in memory, is the only code that a MPU or any other type of computer, can understand. To the software engineer, interpreting programs in this pure **machine code** is virtually impossible. Writing code in this form is torturous, involving at the very least working out each op-code by hand, together with bits representing source, destination and any applicable data; evaluating relative offsets; and keeping tally of where data is stored.

```
0001000000111000     0001001000110100
0101110000000000
0001111000000000     0001001000110101
```

Even with a program written in such a form, some means must be found of putting or loading the code to its final place in memory. Very early computers did not use electronic memory at all, the code being configured by wire links. Using switches to set up each memory address and its corresponding data, in effect a kind of direct memory access, was still used up to the 1960s to enter a short startup program. This program was known as a bootstrap, as once in and executed, a paper tape reader could be controlled. Programs could then be read in from this source, that is the computer was able to pick itself up by its own bootstraps. A modern version of this is the resident BIOS in a PC, which allows the MPU to read in the operating system from magnetic disk after switch on, hence the term 'to boot up'.

Using the computer to aid in translating code from more user-friendly (human) forms to machine code and loading this into memory began in the late 1940s. At the very least it permitted the use of higher order number bases such as octal and hexadecimal. Using the latter, our code fragment becomes:

```
1038   1234
5C00
1E00   1235
```

A hexadecimal loader will translate this to binary and put the code in designated addresses. Hexadecimal coding has little to commend it, except that the number of keystrokes is reduced (but there are more keys!) and it is slightly easier to spot certain types of errors. Nevertheless, this technique was extensively used in the early 1970s for microprocessor software generation and is often still used in education as a first introduction to programming simple MPUs.

At the very least a symbolic translator or **assembler** is required for serious programming. This allows the use of mnemonics for the instructions and internal registers with names for constants, variables and addresses. We now have:

```
       .DEFINE   CONSTANT = 6
       MOVE.B    NUM1,D0         ; Get the number NUM1
       ADDQ.B    #CONSTANT,D0    ; Add the constant to it
       MOVE.B    D0,NUM2         ; is now the number NUM2
       .ORG      1234h           ; This is the data area
NUM1:  .BYTE     [1]             ; NUM1 lives at 1234h
NUM2:  .BYTE     [1]             ; and NUM2 at 1235h
```

Giving names to addresses and constants is especially valuable for long programs. Together with the use of comments, this makes code written in assembly level easier to maintain. Furthermore, programs can be written as separate modules with symbols defined in only one module and a **linker** program used to put them together with their actual values. This assembly of modules into one program gave the name assembly-level to this type of language [1]. Of course assemblers/linkers and their ancillary programs are rather more complex than simple hexadecimal loaders. Thus they demand more of the computer running them, especially in the area of memory and backup store. Because of this, their use in small MPU-based projects was limited until the early 1980s, when powerful personal computers (made possible by MPUs) appeared. Prior to this, either mainframe and minicomputers or target-specific microprocessor development systems (MDSs) were required. Any of these solutions were expensive.

Assembly-level language is machine-oriented, in that there is generally a one-to-one correspondence to the machine instructions. As such, code written at this level bears little relationship to the problem being implemented. The use of a high-level language permits a description of the problem in an algorithmically-oriented language. In C, our code fragment becomes:

```
#define    CONSTANT = 6
unsigned char NUM1,NUM2;     /* Define NUM1 and NUM2 as unsigned bytes */
{NUM1 = NUM2 + CONSTANT;}    /* The process                            */
```

Now we no longer need to keep track of exactly where NUM1 and NUM2 have to be stored. Also we have a large repertoire of mathematical and string functions, which do not have a one-to-one machine level counterpart. Notice that our program did not indicate which processor's machine code would eventually be produced, the target might well be a Z80 rather than a 68000 (see Table 10.15).

Of course there are problems in using high-level languages, especially when the target is an embedded MPU-based system. In general the further away the level is from the machine code, the more isolated the programmer is from the raw

172   *Source to Executable Code*

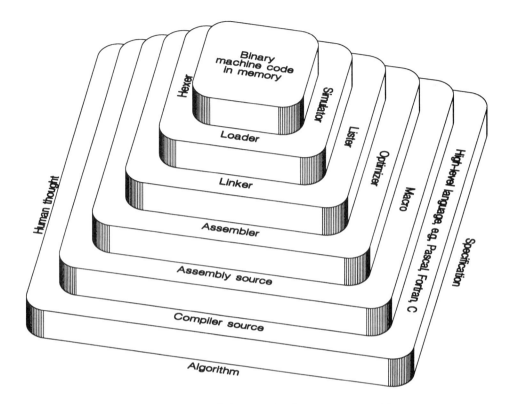

Figure 7.1: Onion skin view of the steps leading to an executable program.

hardware. A compiler also demands much more of its supporting computer, and for this reason only recently became popular as a tool in this type of design.

Many high-level languages compile their syntax into assembler-level source code which is then translated and linked in the same manner as 'hand written' assembly code. Thus in this chapter we will be looking at assemblers, linkers, loaders and their associated programs as well as compilation and related processes.

7.1 The Assembly Process

We have used assemblers at some length in Part 1 of this text, to present a more palatable interface to the reader of the (binary) software aspects of two microprocessors. Without going into any detail, we have seen that a Symbolic Assembler program (or assembler for short) allows us to use predefined symbols for the instructions and various processor registers, and to define names for constants, variables and memory locations. They take the drudgery out of calculating relative offsets and converting number bases. Comments, which are ignored at translation time, make maintenance easier than raw code. The use of a convenient editor allows alterations to be easily made to the source code, which can then be quickly retranslated with the updated symbolic and offset values [1, 2].

In faithfully reflecting the underlying structure of the hardware, assembler code can produce the smallest and quickest code of any of the symbolic languages. Even though it is furthest away from the problem algorithm, these advantages frequently mean that assembly-level routines are linked in with high-level code, or even used entirely to implement problems, especially when real-time operation is required.

Assemblers are one of a class of translator programs and are available from a wide range of originators for most target processors. Although some attempt has been made to standardize syntax [3, 4], normally each package has its own rules. Generally the MPU manufacturer's recommended mnemonics are adhered to reasonably closely. **Directives**, which are pseudo operators used to pass information to the assembler program, do differ considerably. Details of the layout and syntax for the assemblers used in Part 1 are given in Section 2.3 and will not be repeated here. Differences in other assemblers used later in this text are pointed out where they occur.

No matter which language is being used, the programmer must prepare the **source** form of the code in the appropriate format and syntax. This preparation involves the use of an editor program or word processor. The actual one used is irrelevant, provided that the text is stored in a form which can be read by the translator, usually plain ASCII. Most operating systems come with a basic editor, for example MSDOS's EDLIN and UNIX's ED. More sophisticated packages, such as Wordperfect, are usually favored for larger projects. Table 7.1 shows a slightly modified source form of the sum-of-integers program first presented in Table 4.10 (actually entered using EDLIN). This document, which is normally stored on magnetic disk, is the file presented to the assembler for translation. Conventionally the file name is postfixed .S, .SRC or .ASM for assembly source, thus the file printed in Table 7.1 was called list7_1.s.

Assemblers can be broadly classified as **absolute** or **relocatable**, according to the type of code they produce. The former normally generates a file with the machine code and its absolute location ready to be loaded into memory. This machine code file is a finished entity, to which no further alterations need be or should be made before loading. The output of a relocatable assembler is not yet complete, as it usually does not contain information regarding the eventual location of the machine code in memory. Furthermore, symbols may be used in the source code which are not defined at this juncture and which are assumed to be in modules coming from elsewhere. It will be the job of a Linker program to satisfy these unrequited references and to define code addresses.

Absolute assemblers tend to be simpler to use, as the path between source and machine code is more direct, as can be seen in Fig. 7.2(a). Despite their simplicity they are rarely used in major projects due to their lack of flexibility.

As a demonstration, consider the source code listed in Table 7.1. This is virtually identical to the source of Table 4.10, but with the directive .ORG replacing .PSECT. As this source is to be processed by an absolute assembler, the programmer must specify the start address or origin (ORG) of each section of code or data. The .ORG directive may be used as many times as required to locate the various sectors,

174 Source to Executable Code

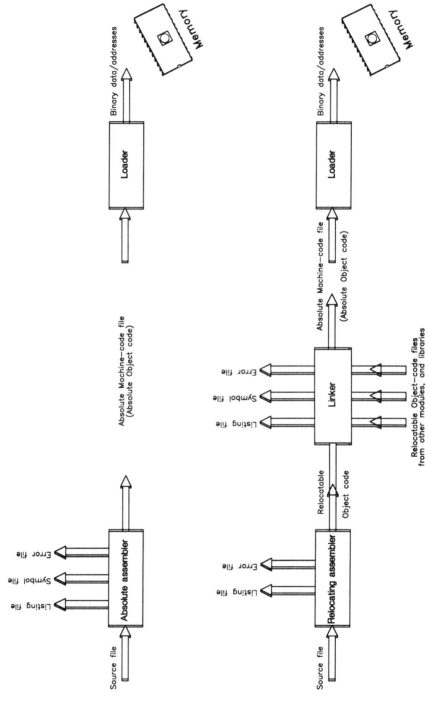

Figure 7.2: Assembly-level machine code translation.

Table 7.1: Source code for the absolute assembler.

```
        .processor m68008
; ******************************************************************
; * FUNCTION : Sums all unsigned word numbers up to n (max 65,535) *
; * ENTRY    : n is passed in Data register D0.W                    *
; * EXIT     : Sum is returned in Data register D1.L                *
; ******************************************************************
;
            .define LONG_MASK = 0000FFFFh ; Used to promote word to long
            .org     0400h                ; Program starts at 0400h
; for (sum=0;n>=0;n--){
SUM_OF_INT: and.l   #LONG_MASK,d0         ; n promoted to long
            clr.l   d1                    ; Sum initialized to 00000000
SLOOP:      add.l   d0,d1                 ; sum = sum + n
            dbf     d0,SLOOP              ; n--, n>-1? IF yes THEN repeat
SEXIT:      rts
            .end
```

thus if necessary each subroutine may be located at a specific start address.

In translating this source code input, the absolute assembler produces four kinds of output. Should there be a problem with the syntax of the source, an **error file** will be produced, giving the line in which it occurred and usually a short description. Sometimes a syntax error in one line can lead to problems in several other places. Table 7.2 is an example of such a file, it was generated by replacing the instruction AND in line 11 of our source by the illegal mnemonic **ANP** and the referenced label SLOOP in line 13 by LOOP, that is DBF D0,LOOP. The source file is referred to as a:list7_2.s.

If all goes well, zero errors will be produced. This does not of course guarantee that the program will work, only that there are no syntax errors! In this situation a **listing file** will be generated, as illustrated in Table 7.3. This shows the original source code together with addresses and the translated code. Other information may be provided as well. In this case a cross-reference table shows where names, other than reserved mnemonics and directives, are first defined and where they are

Table 7.2: A typical error file.

```
x68030 (1):

a:list7_2.s 11:   unknown op-code anp
a:list7_2.s 14:   LOOP not defined in file or include
a:list7_2.s 11:   anp not defined in file or include
a:list7_2.s 11:   system error <> #

a:list7_2.s: 4 errors detected
```

Table 7.3: Listing file produced from the source code in Table 7.1.

```
 1                         .processor m68008
 2   ;*********************************************************************
 3   ; * FUNCTION : Sums all unsigned word numbers up to n (max 65,535) *
 4   ; * ENTRY    : n is passed in Data register D0.W                   *
 5   ; * EXIT     : Sum is returned in Data register D1.L                *
 6   ;*********************************************************************
 7   ;
 8                      .define LONG_MASK = 0000FFFFh   ; Used to promote word to long
 9                           .org    0400h              ; Program starts at 0400h
10   ; for (sum=0;n>=0;n--){
11 000400 0280      SUM_OF_INT: and.l #LONG_MASK,d0     ; n promoted to long
          0000FFFF
12 000406 4281                  clr.l d1               ; Sum initialized to 00000000
13 000408 D280      SLOOP: add.l d0,d1                 ; sum = sum + n
14 00040A 51C8FFFC         dbf    d0,SLOOP            ; n--, n>-1? IF yes THEN repeat
15 00040E 4E75      SEXIT: rts
16                          .end

     SYMBOL   DEFIN  REFERENCES

   LONG_MASK  -----     8    11
       SEXIT    15
       SLOOP    13     14
  SUM_OF_INT    11
          d0  -----    11    13    14
          d1  -----    12    13
      m68008  -----     1
```

referred to. This can be useful when maintaining large programs. Listing files of this nature are for documentation only and have no executable function.

Symbol files list all symbols which occur in the program, giving name, location and sometimes other information. In Table 7.4 three labels are implicitly identified, SUM_OF_INT is located at $0400h$ (the 0x prefix is the hexadecimal indicator used in C), SLOOP at $0408h$, and SEXIT at $040Eh$. The suffix t indicates text (i.e. program section). The label LONG_MASK is explicitly valued and is suffixed a for absolute. See Table 7.11(a) for a more complex example.

Table 7.4: Symbol file produced from the absolute source of Table 7.1.

```
0x0000ffffa   LONG_MASK
0x0000040et   SEXIT
0x00000408t   SLOOP
0x00000400t   SUM_OF_INT
```

Symbol files are commonly used by simulator (see Section 15.2) and in-circuit emulator software (see Section 15.3) to replace addresses by their symbolic equivalents, to aid in the debug process. They are also useful as a documentation aid.

The most important output from an absolute assembler is the **machine-code file**. This is absolute **object code**, giving addresses and their contents, ready to be loaded into memory and run. In the microprocessor world there are several formats of machine code files which have been adopted as de facto standards. Although these have been developed by specific manufacturers, notably Intel, Motorola, Texas Instruments and Tektronix, in the main they can be used interchangeably by any processor. The type to be generated by an assembler can often be specified, and of course must be compatible with a format that can be accepted by the loader program.

Table 7.5 shows machine code files (often called hex files) produced by our example to several formats. The most common of these is the **Intel** hexadecimal object format, originally designed for the 8080 MPU. Each code record line comprises an initial colon marker followed by the number of code bytes. The record is terminated by a checksum byte, defined as the 2's complement of the modulo-256 (8-bit) sum of all the preceding bytes (two hexadecimal digits). As a check, the loader program sums all bytes plus the checksum for each downloaded line and accepts the accuracy of the data if the result is zero. There is a $\frac{255}{256}$ chance that a corrupted record will not pass this trial [5]. The last line should have a record type 01.

Expanding the (single) code record of Table 7.5(a) gives:

:	Start of line
10	Number of code bytes (16)
0400	The address of the first byte
00	Record type (code)
02800000FFFF4281D28051C8FFFC4E75	Code
80	Checksum (2's complement)

Originally developed for the 6800 MPU, the **Motorola S1/S9** object format is similar, with a starting marker of S followed by 1 for a code record and 9 for a termination line. This is succeeded by a count byte, which indicates the number of bytes trailing the S1 or S9 field (including itself), a 4-byte address field and then the code bytes. The checksum field is the 1's complement of the modulo-256 sum of all bytes following S1 or S9. The loader should sum each line including the checksum to FFh if the line has been correctly received. Using this format, we have from Table 7.5(b):

S1	Start of code line (S field)
13	Number of bytes after S field (19)
0400	The address of the first byte
02800000FFFF4281D28051C8FFFC4E75	Code
7C	Checksum (1's complement)

Table 7.5: Some common absolute object file formats.

```
:10 0400 00 02800000FFFF4281D24051C8FFFC4E75 80
:00 0000 01 FF
```
(a) Intel format.

```
S1 13 0400 02800000FFFF4281D24051C8FFFC4E75 7C
S9 03 0000 FC
```
(b) Motorola S1/S9 format.

```
S2 14 0FC400 02800000FFFF4281D24051C8FFFC4E75 AC
S8 04 0FC400 28
```
(c) Motorola S2/S8 format.

```
:02 0000 02 FC40 C0
:10 0000 00 02800000FFFF4281D28051C8FFFC4E75 84
:00 C400 01 3B
```
(d) Extended Intel format.

Neither of these object formats can handle addresses of more than 2-byte size. The Motorola S2/S8 format, developed for the 68000 MPU, is an extension to the S1/S9 format, but with a 3-byte address field. The S3/S7 format is used for 32-bit processors, which require 4-byte addresses. Table 7.5(c) shows the hex file for our example but originated at 0FC400h. The extended Intel hexadecimal equivalent is rather more complex as it was designed to cope with the segmented address space of the 80x86 family. This uses an extended address record (type 02) if the load address is over FFFFh. The data field here holds a 4-digit address which is shifted left four times by the loader (giving here F0000h) before being added to a subsequent 01 type data records' start addresses (here C400h) to give a 5-digit load address (i.e. 0FC400h).

The actual mechanism of the translation process used by the assembler is of little importance to us here. Most assemblers are described as **2-pass**, as historically all but the simplest read the source code, which was frequently on paper tape, twice through from beginning to end. During the first pass a location counter keeps track of where each instruction is to be placed in memory. In an absolute assembler, this will be set by any .ORG directive (0400h in Table 7.1). As each operation mnemonic is encountered, the location counter is incremented by the appropriate number; thus AND.L #LONG_MASK,D0 causes the location counter to advance by six.

As labels are encountered, their name and the state of the location counter are stored in the **symbol table**, which is built up during the first pass. Labels which are explicitly defined, such as LONG_MASK, are of course added to the symbol table without a translation being necessary.

It is necessary to build up a symbol table in the first pass to cope with forward references; thus an instruction BRA NEXT, where NEXT is further on down the source file, cannot be fully translated until NEXT has been encountered and given a value.

Some assemblers may save any translated machine code to speed up the second pass.

During the second pass, the translation is repeated, but this time any references to symbolic names are replaced by the values extracted from the symbol table. With the translation complete, listing, symbol and object files are created in the appropriate format.

In general, assemblers bear a one-to-one relationship to their translated machine code. **Macroassemblers** represent a useful upward extension, by allowing the programmer to define a group of assembly-language instructions as a named macro [6]. This macro can be used repetitively anywhere in the program by simply naming it, followed by a list of operands. The assembler expands this source line to its fundamental components whenever that name is encountered. The programmer can thus emulate more powerful instructions that are not in the MPU's repertoire. As an example, consider the operation where a long-word is to be converted to its modulus (positive equivalent). This can be done by testing for negative and if true negating the target. This sequence is defined in the body of a macro in Table 7.6 between the directives .MACRO and .ENDM. The macro name is LABSOLUTE (long absolute value) and takes one operand, a Data register or address mode. This is indicated in the body of the macro by the dummy ?1 (first operand; this assembler can take up to nine). The numeric label 1$ used in line 8 has the property that its

Table 7.6: A simple macro creating the modulus of the target operand.

```
3    ; Define macro
4                        .macro     LABSOLUTE
5                        tst.l      ?1          ; Is number in ?1 positive
6                        bpl        1$          ; IF so then no action to be taken
7                        neg.l      ?1          ; ELSE negate it
8                1$:     .endm                  ; Continue
9    ;
10   ; Now this macro can be evoked at any time by using its name
11   ; followed by an operand
12   ;
13   ; This fragment converts [D0.L] to an absolute value
14                                              ;~~~~~~~~~~~~~~~~~
15 000400 4A80           LABSOLUTE  d0          ;~ tst.l    d0    ~
       6A02                                     ;~ bpl      1$    ~
       4480                                     ;~ neg.l    d0    ~
16                  1$:                         ;~~~~~~~~~~~~~~~~~
17   ; This fragment converts 20 long words from E100h up to absolute form
18   ;
19 000406 303C0013       move.w     #19,d0
20 00040A 307CE100       move.w     #0E100h,a0  ;~~~~~~~~~~~~~~~~~
21 00040E 4A98    LOOP:  LABSOLUTE  (a0)+       ;~ tst.l    (a0)+ ~
       6A02                                     ;~ bpl      1$    ~
       4498                                     ;~ neg.l    (a0)+ ~
22 000414 51C8FFF8 1$:   dbf        d0,LOOP     ;~~~~~~~~~~~~~~~~~
23   ;
24   ;
```

lifetime only extends to the end of the macro. This is necessary, as macro labels will appear in each expansion; and will thus be defined several times, see Table 7.6 lines 15/16 and 21/22.

The macro is invoked by using its name LABSOLUTE followed by the operand. In Table 7.6 this is done twice, the first specifying a Data register (LABSOLUTE D0) and the second the Address Register Post-increment address mode based on **A0** (LABSOLUTE (A0)+).

A logical progression of this ability to create a new and more powerful instruction set is the evolution of a high-level assembler [7], or even a high-level language.

The 2-pass principle (and the use of macros) apply equally to relocatable assemblers. This time the symbol table cannot be fully resolved, as some symbols appear in other modules. This resolution is the job of a linker program, which is the subject of the next section.

7.2 Linking and Loading

A long program is best implemented by breaking it up into a number of functionally distinct modules, which can be developed separately. Each module is likely to have to cross-reference (XREF) variables from other modules and possibly with a library of standard functions. Full details of these will not be known at the time these modules are designed. Thus there will be a need for a **task builder** to bring all these bits together, filling in these **external symbols** to give a single composite executable program. This task builder is called a linkage-editor or simply **linker** [8, 9]. Assemblers which work in tandem with a linker are known as **relocatable**.

We have already used a relocatable assembler-linker to produce the listings of Tables 6.1 and 6.2. These programs comprised three modules: the display module which did the background task of outputting data from an array to an oscilloscope; the foreground interrupt service routine which updates the array, and the vector table entries. The three modules, Display, Update and Vector, were assembled separately and then linked together to give the composite program.

The most important difference between an absolute and relocatable assembler is in the treatment of symbols. **Symbols** explicitly allocated constant values by the programmer, such as COUNTER in line 11 of Table 6.2(b), are **absolute** and require no further attention by the linker.

Symbols defined implicitly by attaching a label to an instruction are **relocatable**, since their value is only known relative to the *start* of the module, the location of which will be determined by the linker. The label DLOOP in line 23 of Table 6.2(a) is relocatable two bytes after the start of the Display module. Even more vague are symbols referred to but not defined in a module. Such symbols are assumed to be defined in some other module and should be declared so by an **external** declaration, for example the label ARRAY of line 20 in Table 6.2(b).

Besides being tagged Absolute, Relocatable or External, symbols have the attributes of being **Global** (Public) or **Local**. By default local symbols cannot be referenced from an outside module, for example DLOOP in Table 6.2(a). If a symbol is to be globally known then it must be declared as such. In line 16 of Table 6.2(a), ARRAY is declared public by using the .PUBLIC directive. Other assemblers use the GLOBL or XDEF directives. The assembler must pass on the symbol names, tags and attributes to the linker together with the machine code in its output relocatable object file.

Machine code is passed to the linker in streams. The RTS assemblers fundamentally identify two streams, one for program code and the other for data. Programs in Tables 6.1 and 6.2 used the directives .PSECT _TEXT for the former and .PSECT _DATA for the latter, where .PSECT stands for Program SECTion. Most embedded microprocessor systems will require text (which includes tables of constants) in ROM and use RAM for variable data. In certain circumstances the RTS assembler linker can handle two additional data sections, _ZPAGE for data which will lie in the direct/absolute-short memory areas (zero page) of MPUs such as the 6800/9 and 68000 devices and _BSS (Block Symbol Start) frequently used for variables which have no initial value (see Section 10.3).

The Microtec Research Paragon 68K products[1], used later in this section, can handle up to 16 program sections. This is useful where several non-contiguous memory chips are being targeted. For example, initialized variables could be put in a specific segment and placed in ROM. Later, at run time, they can be copied into RAM, where they can be treated as variables; that is changed at will (see Section 10.3).

Some relocatable products do not permit absolute placement of code using the .ORG directive, and in any case this is considered bad practice. The RTS products do permit relocatable ORGs, thus the fragment:

```
START:              -----------
                    -----------
        .ORG        START + 0FFEh
        .WORD       ADDR1
```

will place the data word ADDR1 0FFE:Fh bytes on from START. If you know that the linker will locate START at 0E000h, then this will actually be at 0EFFE:Fh.

That part of the machine code referring to labels, e.g. MOVEA.L #ARRAY,A0 in line 30 of Table 6.2(b), which are relocatable or external is not resolvable at this time. Thus the assembler must parallel the code streams with information relating these bytes to their label. **Object code** also contains headers giving processor information, such as the order of address bytes (most or least significant first), size of processor words, length of symbols, number of machine-code bytes etc. With all this in mind it will be appreciated that relocatable object file formats are much more complex than their absolute counterparts of Table 7.5. As a consequence of this, their structure is very much specific to each product.

As our example for this section, we will follow through the program defined in Table 6.2 but this time using the more sophisticated Microtec Research Paragon

---

[1] Microtec Research Inc., 2350 Mission College Blvd., Santa Clara, CA 95054, USA and Ringway House, Bell Road, Daneshill, Basingstoke, Hampshire, RG24 0FB, UK.

182 *Source to Executable Code*

68K assembler/linker. The instruction mnemonics and address mode representations follow the standard Motorola conventions, but the directives differ considerably from the RTS mnemonics used up to now. Some key directives are:

ident <name>	Gives the module a name (identity).
opt <flags>	Options, such as CASE for case sensitivity.
<name1> equ <value>	Equates a symbol with an absolute value (similar to .define).
sect <n>	Section number; equivalent to .psect.
ds.b/.w/.l <n>	Define Storage; reserve n bytes, words or long words (equivalent to .byte/.word/.double[n]).
dc.b/.w/.l <n,--,m>	Define Constant; puts list of specified bytes, words or long words into section stream (equivalent to .byte n,---,m / .word n,---,m / .double n,---,m).
xdef <name>	Publishes symbol as global, i.e. Cross DEFine (equivalent to .public).
xref <name>	Identifies symbol as external, i.e. Cross REFerence (equivalent to .external).

The source code using this assembler for the Display module is given in Table 7.7(a). I have placed data (ARRAY and X_COORD) in Section 14 and the program text in Section 9. These are the sections chosen for data and text by the Microtec research C compiler, which we will use later; for example see Table 10.9).

The listing file produced after assembly is shown in Table 7.7(b). Both Data and Text sections are shown starting from 00000000h; they will be subsequently located by the linker. This uncertainty also affects machine code relating to labels. Thus in line 29 the value of X_COORD is replaced by its offset from the data segment's zero start value, that is 0200h. Notice that all lines with machine code which contains values to be relocated later are tagged with R. A symbol table is also produced by the Lister utility (as shown at the bottom of the listing), and this shows the section number followed by an offset for each relocatable symbol. Absolute symbols, such as DAC_X, have an absolute value attached.

The output from the Update module source is shown in Table 7.8. Here we have an external symbol which is tagged with E in line 31 and identified in line 21. No value is given for ARRAY in the Symbol table, just **External**.

The Vector code, shown in Table 7.9, similarly has external symbols which are tagged with E. This has been placed in Section 0, so that it can be linked in as the start of the 68000's vector table.

The linker program, depicted in Fig. 7.2(b), has several tasks to perform:

1. To concatenate code from the various input modules in the specified order, to give one contiguous object module.

2. To resolve any intersegment and library symbolic references.

3. To extract code from libraries into the output object module.

4. To generate the object file together with any symbol, listing and link-time error files.

In our example, incoming object modules contain code located in three sectors; 0, 9 & 14 (two for Tables 6.1 and 6.2; _text and _data). The new composite sections are built up by concatenating like streams from the input object files as they come in. Unless otherwise directed, code from Section n simply begins where the last Section n input left off. Thus looking back at Table 6.2, text in the Display module goes from $0400h - 0425h$ and in the Update module from $0426h - 0463h$. However, the programmer can sometimes override this progression by specifying a module's start address independently. This is how the Vector module's text was forced to run from $0000h - 0068h$, as directed in Table 7.10(b).

Table 7.10 shows the invocation of two linker programs. The top one, LOD68K by Microtec Research, is used in our example, whilst the bottom one, LINKX by

Table 7.7: Assembling the Display module with the Microtec Research Relocatable assembler (*continued next page*).

```
        opt         E,CASE
        DISPLAY     idnt
; **********************************************************************
; * Background program which scans array of word data (ECG points)     *
; * Sends out to oscilloscope Y plates in sequence                     *
; * At same time incrementing X plates                                 *
; * so that ARRAY[0] is seen at the left of screen                     *
; * and ARRAY[255] at the right of screen                              *
; * ENTRY : None                                                       *
; * EXIT  : Endless loop                                               *
; **********************************************************************
;
DAC_X:   equ    6000h          ; 8-bit X-axis D/A converter
DAC_Y:   equ    6001h          ; 12-bit Y-axis D/A converter
;
         sect   14             ; Section 14 is Data space
         xdef   ARRAY          ; Make the array global
ARRAY:   ds.w   256            ; Reserve 256 words for the array
X_COORD: ds.b   1              ; and a byte for the X co-ordinate
;
         sect   9              ; Program space
         xdef   DISPLAY        ; Make this program known to the linker
DISPLAY: clr.w  d0             ; Get X co-ordinate byte
DLOOP:   move.b X_COORD,d0     ; expanded to word
         lsl.w  #1,d0          ; Multiply by two to give array index in D0.W
         movea.l #ARRAY,a0     ; Point A0 to ARRAY[0]
         move.w 0(a0,d0.w),DAC_Y ; Get ARRAY[x] to oscilloscope Y plates
         move.w X_COORD,DAC_X  ; Send X co-ordinate to X plates
         addq.b #1,X_COORD     ; Go one on in X direction
         bra    DLOOP          ; and show next sample
         end    DISPLAY
```

(a) Source code for the Display module.

Table 7.7: (*continued*). Assembling the Display module with the Microtec Research Relocatable assembler.

```
                   Microtec Research ASM68008 V6.2a   Page 1 Wed Jan 04 15:59:41 1989

Line Address
  1                             opt     E,CASE
  2                             DISPLAY idnt
  3         ; ***********************************************************************
  4         ; * Background program which scans array of word data (ECG points)     *
  5         ; * Sends out to oscilloscope Y plates in sequence                     *
  6         ; * At same time incrementing X plates                                 *
  7         ; * so that ARRAY[0] is seen at the left of screen                     *
  8         ; * and ARRAY[255] at the right of screen                              *
  9         ; * ENTRY : None                                                       *
 10         ; * EXIT  : Endless loop                                               *
 11         ; ***********************************************************************
 12         ;
 13 00006000          DAC_X:   equ    6000h      ; 8-bit X-axis D/A converter
 14 00006001          DAC_Y:   equ    6001h      ; 12-bit Y-axis D/A converter
 15         ;
 16                            sect   14         ; Section 14 is Data space
 17                            xdef   ARRAY      ; Make the array global
 18 00000000          ARRAY:   ds.w   256        ; Reserve 256 words for the array
 19 00000200          X_COORD: ds.b   1          ; and a byte for the X co-ordinate
 20         ;
 21                            sect   9          ; Program space
 22                            xdef   DISPLAY    ; This program known to the linker
 23 00000000 4240     DISPLAY: clr.w  d0         ; Get X co-ordinate byte
 24 00000002 1039     DLOOP:   move.b X_COORD,d0; expanded to word
          0000 0200 R
 25 00000008 E348              lsl.w  #1,d0      ; x2 to give array index in D0.W
 26 0000000A 207C              movea.l #ARRAY,a0 ; Point A0 to ARRAY[0]
          0000 0000 R
 27 00000010 31F0              move.w 0(a0,d0.w),DAC_Y ; Get ARRAY[x] to Y plates
          0000 6001
 28 00000016 31F9              move.w X_COORD,DAC_X ; Send X co-ord to X plates
          0000 0200 R 6000
 29 0000001E 5239              addq.b #1,X_COORD; Go one on in X direction
          0000 0200 R
 30 00000024 60DC              bra    DLOOP      ; and show next sample
 31                            end    DISPLAY
                         Symbol Table

    Label       Value

    ARRAY      14:00000000
    DAC_X         00006000
    DAC_Y         00006001
    DISPLAY     9 :00000000
    DLOOP       9 :00000002
    X_COORD    14:00000200
```

(b) Resulting listing file before linking.

Table 7.8: Module 2 after assembly.

```
Microtec Research ASM68008 V6.2a  Page 1 Wed Jan 04 16:00:56 1989

Line Address
1                           opt    E,CASE
2                           UPDATE idnt
3     ; **********************************************************************
4     ; * Interrupt service routine to update one array element              *
5     ; * with the latest ECG period, as signalled by the peak detector      *
6     ; * ENTRY : Via a Level1 interrupt                                     *
7     ; * ENTRY : Location of ARRAY[0] is globally known through the linker  *
8     ; * EXIT  : ARRAY[i] updated, where i is a local index                 *
9     ; * EXIT  : MPU state unchanged                                        *
10    ; **********************************************************************
11    ;
12 00009000        COUNTER:  equ    9000h           ; The 16-bit period Counter
13 00009800        INT_FLAG: equ    9800h           ; The external Interrupt flag
14    ;
15                           sect   14              ; Data space
16 00000000        UPDATE_I: ds.b   1               ; for the array update index
17 00000002        LAST_TIME:ds.w   1               ; and for the last reading
18    ;
19                           sect   9               ; Program space
20                           xdef   UPDATE          ; This routine known to the linker
21                           xref   14:ARRAY        ; Get ARRAY from another module
22 00000000 48E7  UPDATE:    movem.l d0/d1/a0,-(sp) ; Save used registers
            C080
23 00000004 4279             clr    INT_FLAG        ; Reset external Interrupt flag
            0000 9800
24 0000000A 3039             move.w COUNTER,d0      ; and get count from the counter
            0000 9000
25 00000010 3200             move.w d0,d1           ; Put in D0.W for safekeeping
26 00000012 9279             sub.w  LAST_TIME,d1    ; Sub frm last cnt for new period
            0000 0002   R
27 00000018 33C0             move.w d0,LAST_TIME    ; & update last counter reading
            0000 0002   R
28 0000001E 4240             clr.w  d0              ; Prepare to get update array indx
29 00000020 3039             move.w UPDATE_I,d0     ; expanded to word size
            0000 0000   R
30 00000026 E348             lsl.w  #1,d0           ; x2 to cope with word ARRAY
31 00000028 207C             movea.l #ARRAY,a0     ; Point A0.L to ARRAY[0]
            0000 0000   E
32 0000002E 3181 0000        move.w d1,0(a0,d0.w)   ; Put new value (D1.W) in ARRAY[I]
33 00000032 5279             addq.w #1,UPDATE_I     ; Move update marker on one
            0000 0000   R
34 00000038 4CDF 0103        movem.l (sp)+,d0/d1/a0 ; Return machine state
35 0000003C 4E73             rte
36                           end    UPDATE

              Symbol Table

Label          Value

ARRAY       14:External
COUNTER        00009000
INT_FLAG       00009800
LAST_TIME   14:00000002
UPDATE       9 :00000000
UPDATE_I    14:00000000
```

186 *Source to Executable Code*

Table 7.9: Module 3 after assembly.

```
Microtec Research ASM68008 V6.2a  Page 1 Wed Jan 04 16:32:28 1989

Line Address
1                        opt    E,CASE
2             VECTOR     idnt
3    ;***********************************************************************
4    ; * Sets up Interrupt and Reset vectors at bottom of ROM                *
5    ; * using globally known labels through the linker                      *
6    ;***********************************************************************
7.   ;
8                        sect   0       ; Use Section 0 for vector table
9                        xdef   VECTOR  ; Make this routine known globally
10                       xref   UPDATE,DISPLAY ; These will be got through linker
11            SSP:
12 00000000   VECTOR: dc.l  0F000h   ; Init value of the System Stack pointer
              0000 F000
13 00000004   PCR:    dc.l  DISPLAY  ; Go to DISPLAY routine on Reset
              0000 0000  E
14 00000008           ds.l  23       ; Other vectors not used here
15 00000064   LEVEL1: dc.l  UPDATE   ; Addr of Level1 IRQ service routine
              0000 0000  E
16                    end   VECTOR

                   Symbol Table

Label      Value

DISPLAY    External
LEVEL1     0:00000064
PCR        0:00000004
SSP        0:00000000
UPDATE     External
VECTOR     0:00000000
```

RTS, was used to generate the code in Table 6.2.

Taking the latter first, LINKX is followed by a Command line comprising a series of flags and file names, by which the programmer directs the action of the linker. The action commanded in Table 7.10(b) is, reading from left to right:

-tb 0000   Start text bias at zero (default)
vector.o   Scan object program vector.o (Table 6.2(c))
-tb 0x400   Next text code starts from 400*h*
-db 0xE000   and data code from E000*h*
display.o   Scan object program display.o

Table 7.10: Linking the three source modules.

```
**************************************************************************
* This is the Command file used for the Microtec Research linker         *
* to combine the three modules previously assembled together             *
* It is called DISPLAY.CMD                                               *
**************************************************************************
* Section 0 is for the Vector table                                      *
* Section 9 is for text                                                  *
* Section 14 is for unitialized variables in RAM                         *
  sect 0 = 0000          * Vector table starts at 0000 for the 68000     *
  sect 9 = 0400h         * Program starts at 0400h (ROM)                 *
  sect 14= 0E000h        * Data starts at E000h (RAM)                    *
  absolute 0,9           * Only put Sections 0 and 9 in the hex file     *
  list d,s,t,x,c         * Options for the Listing file                  *
  load vector.obj        * Load and scan the Vector file first           *
  load display.obj       * Then the background file                      *
  load update.obj        * and finally the Interrupt service file        *
  end
**************************************************************************
```

i: The Command file

LOD68K @display.cmd,display.map,display.abs

ii: Evoking the Microtec linker LOD68K

(a) Linking using the Microtec products.

LINKX -tb 0000 vector.o -tb 0x0400 -db 0xE000 display.o update.o -odisplay.xeq

(b) The equivalent linking process using the RTS products, see Table 6.2.

update.o   and then object program update.o
-o output.XEQ   Create a composite object file named output.XEQ

Note the use of the C language prefix 0x to indicate hexadecimal. The action of these commands can clearly be seen by looking at the addresses of the resulting code of Table 6.2.

In realistic cases the linker command sequence is complex and it is better to use a Command file, which is automatically read at link time. The Command file of Table 7.10(a)(i) is that read by Microtec Research's LOD68K linker to combine our three object modules. Here Program section 0 (used by the Vector module) is commanded to start at 0000h (sect 0 = 0000) whilst Section 9 builds from 0400h up (program) and Section 14 from 0E000h up (data). Modules are scanned in the order given by the Load commands. The Absolute command selects which code sections go into the final absolute hex file (see Table 7.11(b)); I have omitted RAM data (Section 14) from this request.

The Command line of Table 7.10(a)(ii) gives the name of the Command file (prefixed by @), the name of a special Map listing file and the name of the absolute Machine-Code file; the latter two are shown in Table 7.11.

While code is being entered from the various input object files, a composite symbol table is being built up by the linker. For our example this combined symbol table is shown in the Map file produced by LOD68K to give the final location (i.e. map) of all code sections and symbols. There are three types of symbols entered into the linker. Absolute symbols have been given a fixed value by the programmer. These are usually known addresses of external hardware, such as the X and Y digital to analog converters of our Update module. These are marked as ABSCONST under SECTION in the map.

Defined symbols are assigned relative to the beginning of the module they are created. Thus DLOOP is indicated as Section 9 Offset 00000402 in the Display module.

Symbols referred to but not actually defined in a module are usually assigned a value when all the code is in. They must be declared Public where they are defined. When known, the value of a ref (referred to) symbol is substituted in the code where they are referred to. Public symbols are listed separately in the Map file.

The LOD68K linker does not give an Absolute listing file output, unlike the LINKX product (via a utility program ABSX). However, Table 6.2 is indicative of how it would look.

If any ref symbols remain unresolved, the linker will scan such library files as are indicated in the Command line or file. A library file typically comprises a series of object-code programs, each headed up with a name and a code length. Should an unresolved symbol match such a name, the succeeding code is extracted and added to the appropriate Program sections already formed by the linker. Thus, unlike a normal object-code file, only relevant portions of a library file are extracted and used. The linker recognizes a library file from its unique header. A typical evocation of a linker using a floating point mathematics library might be:

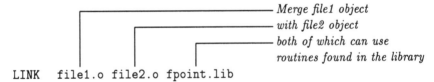

```
LINK    file1.o file2.o fpoint.lib
```

Libraries are typically provided by compiler manufacturers, covering mathematical, string and input/output functions. Compilers and assemblers usually come with a utility program known as a librarian. The programmer uses the librarian to build up his/her own personal libraries or to modify existing ones (see also Section 9.4).

Symbols which remain unresolved after the linking process are indicated as errors. Whether they are depends on the format of the output object code. If this is in the same relocatable mode as the input, then the resulting file can be subsequently linked again with other object files. An absolute format, such as shown

Table 7.11: Output from the Microtec linker (*continued next page*).

```
Microtec Research Lod68K V6.2a        Thu Jan 05 10:17:34 1989

OUTPUT MODULE NAME:    display
OUTPUT MODULE FORMAT:  MOTOROLA S2

SECTION SUMMARY
---------------

SECTION   ATTRIBUTE            START      END        LENGTH     ALIGN

0         NORMAL DATA          00000000   00000067   00000068   2 (WORD)
9         NORMAL CODE          00000400   00000463   00000064   2 (WORD)
14        NORMAL DATA          0000E000   0000E205   00000206   2 (WORD)

MODULE SUMMARY
--------------

MODULE         SECTION:START        SECTION:END      FILE

VECTOR         0:00000000           0:00000067       vector.obj
DISPLAY        14:0000E000          14:0000E200      display.obj
               9:00000400           9:00000425
UPDATE         14:0000E202          14:0000E205      update.obj
               9:00000426           9:00000463

LOCAL SYMBOL TABLE
------------------

SYMBOL                       ATTRIB    SECTION    OFFSET      MODULE:FUNCTION

ARRAY                        ASMVAR    14         0000E000    DISPLAY:
COUNTER                      ASMVAR    ABSCONST   00009000    UPDATE:
DAC_X                        ASMVAR    ABSCONST   00006000    DISPLAY:
DAC_Y                        ASMVAR    ABSCONST   00006001    DISPLAY:
DISPLAY                      ASMVAR    9          00000400    DISPLAY:
DLOOP                        ASMVAR    9          00000402    DISPLAY:
INT_FLAG                     ASMVAR    ABSCONST   00009800    UPDATE:
LAST_TIME                    ASMVAR    14         0000E204    UPDATE:
LEVEL1                       ASMVAR    0          00000064    VECTOR:
PCR                          ASMVAR    0          00000004    VECTOR:
SSP                          ASMVAR    0          00000000    VECTOR:
UPDATE                       ASMVAR    9          00000426    UPDATE:
UPDATE_I                     ASMVAR    14         0000E202    UPDATE:
VECTOR                       ASMVAR    0          00000000    VECTOR:
X_COORD                      ASMVAR    14         0000E200    DISPLAY:

PUBLIC SYMBOL TABLE
-------------------

SYMBOL                       SECTION        ADDRESS        MODULE

ARRAY                        14             0000E000       DISPLAY
DISPLAY                      9              00000400       DISPLAY
UPDATE                       9              00000426       UPDATE
VECTOR                       0              00000000       VECTOR
```

(a) Map file.

Table 7.11: (*continued*) Output from the Microtec linker.

S00600004844521B
S20C0000000000F00000000400FF
S2080000640000042669
S21400040042401039000E200E348207C0000E00093
S21400041031F00000600131F90000E200600052395E
S20A0004200000E20060DCB3
S21400042648E7C0804279000098003039000900006
S21400043632009279000E20433C00000E204424033
S21400044630390000E202E348207C0000E0003181FB
S212000456000052790000E2024CDF01034E73F4
S5030009F3
S804000426D1

(b) Hex file.

in Table 7.5, precludes any further processing of this nature. Some Linkers, such as the RTS product, naturally produce the former and require a utility program, often called a Hexer, to extract an absolute file.

From Fig. 7.2 we see that the end of the translation chain is the **Loader** program. The purpose of a Loader is to take the object code and place it in memory, from where it can be run. The operation of the Loader depends somewhat on the relationship between the computer doing the translation (i.e. assembly and linkage) and the target system that will actually run the generated code.

In the situation where they are the same, as depicted in Fig. 7.3(a), the Loader will frequently be part of the computer's operating system. Such a Loader can usually deal with both relocatable and absolute object files produced by a Linker. In the former case the operating system decides on the location of the various code streams; in the latter the programmer can influence this decision through the Linker. In such a resident system, the user program in its object form normally resides on disk. When the operator decides to run the program, the operating system first loads and locates the code, then proceeds directly to execution; that is load and go. Some configurations, mainly mainframe, combine the linkage and loading operations in one Linker-Loader program.

Although it is possible to interface devices to a computer and use a resident configuration, in engineering applications the **cross-target** arrangement depicted in Fig. 7.3(b) is the more usual. Here the microprocessor-oriented hardware is distinct from the computing apparatus doing the code conversion. Indeed it is unlikely that they even use the same processor. In this situation the assembler is known as a **cross-assembler** as opposed to a **resident assembler**. Where the user hardware is a dedicated controller with its software in ROM, the Loader must be in the target system. This may well be part of the operating software of an intelligent EPROM programmer, into which absolute object code is downloaded into a RAM buffer for later programming. The blown EPROM is then moved by hand to the target. Alternatively during development the Loader may be in an

Linking and Loading 191

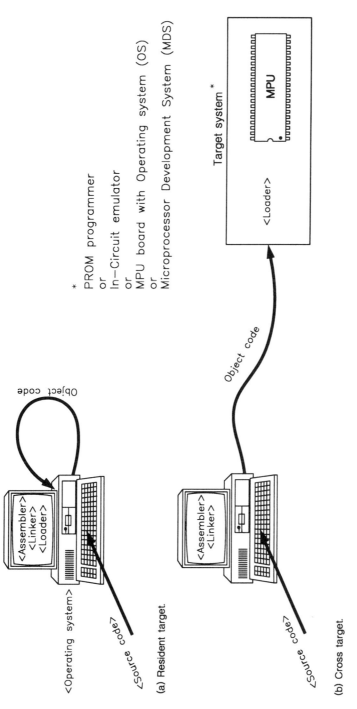

Figure 7.3: Assembly environment.

in-circuit emulator interface package or the operating system of a microprocessor development system (MDS). In all of these cases it is likely that the Loader will act on absolute object code, such as depicted in Table 7.5. Absolute Loaders are somewhat less complex than their relocatable counterparts, which are used in resident configurations.

The cross environment is necessary because dedicated microcontrollers rarely have the facilities necessary to develop their own software. The additional software and hardware resources necessary for this purpose cannot be integral to the system as they must be easily jettisoned when their use is over. Targets of this form, without their own general-purpose operating system, are often referred to as **naked systems**. The use of a general purpose computer with an in-circuit emulator can be thought of as supplying these resources to a naked system in a form that can be readily disengaged when no longer needed.

7.3 The High-Level Process

We have defined a **high-level language** as a code which is modelled more on the algorithm of the problem rather than on the underlying machine which will actually solve it. The level of a language can be quantified as a function of the 'distance' it is removed from its ultimate machine code. The **compiler** is then the system program which translates from one language to another [10]. Strictly this includes programs which convert between high-level languages such as Pascal to C (PTC) or BASIC to C (BASTOC). In this book, we use the term compiler in its narrower sense, to denote translation to the target processor machine-level language.

In principle an assembler and compiler carry out a similar task, but clearly the latter has a much more onerous burden to discharge. There are two parts to a compilation process: analysis and synthesis. The analysis part separates the source text into the constituent parts of which the language is composed. This is akin to the verbs, nouns, adjectives etc. of human language. The structural relationship of these elements, called leximes, must then be ascertained. The synthesis part generates the desired code from the intermediate representation created by the analysis phases.

All this is easily stated, but the details are of necessity rather complex and of little relevance to other than compiler designers; interested readers are directed to references [11, 12]. However, it is instructive to expand a little on the process of compilation.

Lexical analysis, or parsing, subdivides the source language into its fundamental chunks or tokens, and identifies each token, whether operator, constant, variable etc. Thus the two characters >= may be recognized as a relational operator of type Greater or Equal, and this could be coded as REL_OP GE. Taking a slightly more meaningful example, consider the expression:

```
sum = (n + 1) * n/2;
```

which evaluates the sum of all integers up to n. A Lexical analysis would produce something like that shown in Table 7.12, where each chunk is parsed into a token and an attribute. For instance, the variable sum is an identity (the name of a variable is commonly known as its identity) and its attribute is an address or pointer into the Symbol table.

Table 7.12: A possible Lexical analysis of sum = (n+1)*n/2;

Source Expression	Token	Attribute	Comment
sum	id	Pointer to Symbol table entry for sum	Identity sum
=	assign-op	—	Assignment operator
(	par-op	L	Parenthesis, Left
n	id	Pointer to Symbol table entry for n	Identity n
+	add-op	—	Addition operator
1	const	1	Constant value 1
)	par-op	R	Parenthesis, Right
*	mul-op	—	Multiply operator
n	id	Pointer to Symbol table entry for n	Identity n
/	div-op	—	Divide operator
2	const	2	Constant value 2
;	end-op	—	End-of-statement op

This example is fairly simple to decompose into its constituent elements. As a more difficult situation, consider the fragment of C source code:

n+++m

where the ++ operator following a variable means Auto-Increment after use, and prior to a variable means Auto-Increment before use. A single + in the normal way means Add. Are the tokens n++ +m or n+ ++m?

Actually, the former is the correct interpretation, as C compilers analyze using the 'maximal munch' strategy [13]. Here the parser moves from left to right, biting off the longest possible token; hence ++ first followed by +. Thus the expression means add to m the post-incremented value of n.

A Lexical analysis says nothing about the relationships between the various leximes. For this, a subsequent Syntactic analysis must be performed. The interrelationship and order of operators, constants and variables for our example is shown in the Syntax tree of Fig. 7.4. The expression to the right of the assignment operator is evaluated from the most distant parts up: that is, first add n + 1 then multiply by n and then divide by 2. The variable sum is finally overwritten by this value. This process is governed by the precedence order defined by the language (e.g. multiplication has a higher precedence than addition), direction of evaluation (e.g. right to left) — see Table 8.4 for an example of both of these —, parenthesis, brackets, loop constructs etc. More elaborate Syntax trees are often called Parse trees.

Parse trees are in turn subjected to Semantic analysis. This gathers type information relevant to the coming code-generation phase. In particular the type and

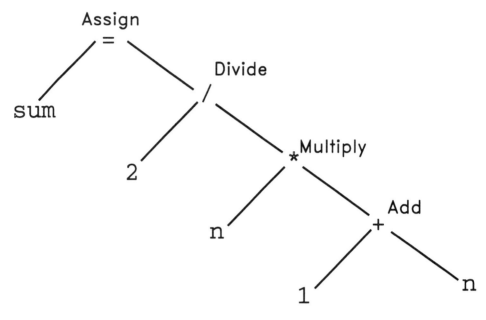

Figure 7.4: Syntax tree for sum = (n+1) * n/2;

size of variables and constants need to be checked and altered according to the rules of the language. For example in C if we have an expression of the form:

Z = X + Y;

where X has been declared an integer (say 32 bits) and Y a short integer (say 16 bits), then Y must be expanded to 32 bits before the addition is performed. Other type conversions include signed and unsigned combinations, floating and fixed-point mixes etc. Errors may be reported during this phase as well as all previous phases.

The output of the Semantic analysis is a type of Intermediate code. Although Intermediate code is independent of a real machine, it nevertheless reflects the type of operation available in the target. The synthesis of real machine code involves the determination of storage requirements and addressing algorithms for the variables and the expansion of the Intermediate code statements to sequences of machine-specific instructions. Intermediate code for our example may be something like:

1. Move integer n in from memory.

2. Put in multiplier location.

3. Add one.

4. Move to the multiplicand location.

5. Multiply them.

6. Divide the returned value by two.

7. Move out to where sum is.

The actual machine code produced by the Cosmic 6809 C cross-compiler V3.1 for this example is shown in Table 7.13. Notice how closely it mirrors this pseudo code.

The front end of the compiler covering the analysis phases through to the production of Intermediate code is mainly a function of the source language and largely independent of the target machine. The back end of the compiler includes those portions of the compiler that generate the specific target language. In theory the target may be changed by replacing only the back end components.

The code produced by the compiler illustrated in Table 7.13 is in normal assembly-level format. To complete the production of machine code, it can be passed on through the chain of Fig. 7.2, that is the assembly process. Other modules from high-level language programs, assembly-level programs and libraries can be linked to generate the final executable code.

Table 7.13: 6809 target code for sum = (n+1) * n/2;

```
; Compilateur C pour MC6809 (COSMIC-France)
            .processor m6809
            .psect    _text
L3_n:       .byte     0
L31_sum:    .byte     0,0
;   unsigned int sum_of_n()
;   {
            .psect    _text
_sum_of_n:
;   static unsigned char n;
;   static unsigned int sum;
;   sum = (n+1)*n/2;
            ldb       L3_n      (Move integer n in from memory        )
            clra                (Expand to integer in (D)             )
            tfr       d,x       (Move to multiplier location (X)      )
            addd      #1        (Add one.  In multiplicand location (D) )
            jsr       c_imulx   (Multiply them                        )
            ldx       #2        (Prepare to divide returned value by 2 )
            jsr       c_idivx
            std       L31_sum   (Move out to where sum is stored      )
;   return(sum);
            rts                 (Returned in Accumulator D            )
;   }
            .public   _sum_of_n
            .external c_idivx
            .external c_imulx
            .end
```

The compilation process used to produce the code in Table 7.13 is shown in Fig. 7.5. The Whitesmiths Group series of compilers use separate programs to implement the various processes discussed above [14]. These are:

pp
   The preprocessor implements the Lexical analysis. Also expands out #include file and #define substitutions and macros.

p1
   Performs Syntax and Semantic analysis to produce intermediate code.

p2nn
   Generates source code for machine nn's assembler. For example, p209 synthesizes source code for a 6809 MPU, p280 for 8080/8085 processors.

optnn
   This is an optional peephole optimizer which eliminates redundant instructions generated by p2nn.

Not shown in the diagram is the Listing utility which produces interleaved listings of assembly-level statements as comments. Also the optional front-end Pascal to C (PTC) translator used when Pascal source is desired.

Splitting up the compiler into separate programs has the advantages of flexibility and requires less in the way of memory capacity of the computer. A single composite compiler program, as used in most commercial products, is much faster but does demand more of the translating engine. In both situations, various options are selected by following the program(s) with flags in the form of command lines or files, much as depicted in Table 7.10.

The top right process box in Fig. 7.5 is labelled Peephole optimizer. Depending on the sophistication of the compiler translation, a variable percentage of the machine code produced is either inefficient or redundant. For example, as each high-level code statement is processed separately, a subsequent translation may not be aware that a variable that it requires is already down in a Data register. Accessing an array element in a loop may require the use of a complex address mode to calculate its location, (see line 27 of Table 7.7 as an example). However, if the array elements are going to be accessed sequentially, it will be faster to load an Address register prior to entering the loop with the address of the array element to be accessed on the first loop iteration. Thereafter, indirect addressing with automatic increment/decrement can be used. This is known as strength reduction. There are of course obvious faux pas such as using multiplication for the function X * 1!

Peephole optimization is a method of improving the quality of the machine code by moving a small window over the target program looking for redundancies. This window is typically 30–100 code lines. In general, the window-sized scan will be repeated until no further improvements can be made.

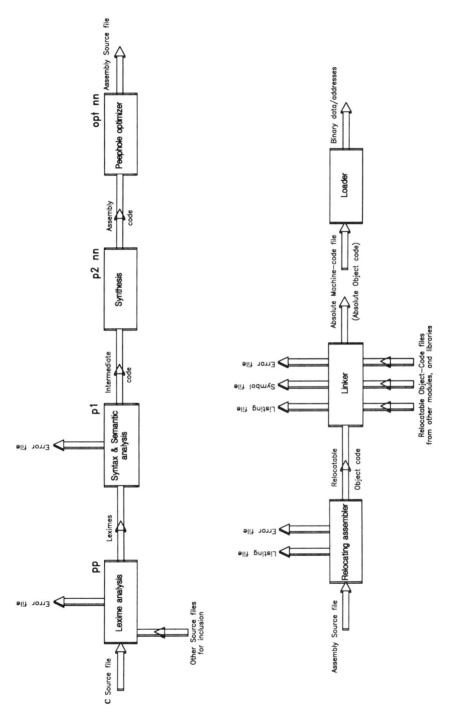

Figure 7.5: The Whitesmiths C compiler process.

198   *Source to Executable Code*

There are many different types of techniques for optimization transformations. Reference [15] gives an overview of this area. For example, the Microtec Research Paragon MCC68K Version 3 C cross-compiler has five optional optimizations.

Some optimizations can be dangerous in certain situations. The classical case involves a program testing the data in an external peripheral's Control register, perhaps the Transmit_Data_Register_Empty_flag in a UART. The compiler will translate this into a loop which repetitively brings in the Transmit Control register (TCR) state, checks the appropriate bit and repeats unless True. However, the optimizer may decide that once the variable is down in the MPU's Data register, then it is a shame to keep bringing it down each loop pass; why not bring it down just once before the loop is entered. Of course the optimizer does not realize that the variable TCR can be altered, seemingly spontaneously, by some agency outside the sphere of influence of the software. ANSII C has a type of variable known as volatile, which warns optimizers to leave well alone.

Table 7.14: Passing a simple program through the compiler of Fig. 7.5 (*continued next page*).

```
unsigned int sum_of_n()
    {
    static unsigned int n;
    static unsigned int sum;
    sum=0;
    while(n>0)
        {
        sum=sum+n;
        n=n-1;
        }
    return(sum);
    }
```

(a) C source.

```
; Compilateur C pour MC6809 (COSMIC-France)
            .processor m6809
            .psect    _data
L3_n:       .byte     0,0
L31_sum:    .byte     0,0
            .psect    _text
_sum_of_n:  clra                          (sum=0000      )
            clrb
            std       L31_sum
L1:         ldx       L3_n
            cmpx      #0                  (n>0?          )
            jbeq      L11                 (Exit IF true  )
            ldd       L31_sum             (Get sum       )
            addd      L3_n                (Add n to it   )
            std       L31_sum             (= new sum     )
            ldd       L3_n                (Get n         )
            addd      #-1                 (Subtract 1    )
            std       L3_n                (n=n-1         )
            jbr       L1                  (Repeat while  )
L11:        ldd       L31_sum             (Return sum    )
            rts
            .public   _sum_of_n
            .end
```

(b) Resulting assembly source code.

Table 7.14: Passing a simple program through the compiler of Fig. 7.5 (*continued next page*).

```
; Compilateur C pour MC6809 (COSMIC-France)
            .processor m6809
            .psect    _data
L3_n:       .byte     0,0
L31_sum:    .byte     0,0
; unsigned int sum_of_n()   ; {
            .psect    _text
; static unsigned int n;    ; static unsigned int sum;    ; sum=0;
_sum_of_n:  clra
            clrb
            std       L31_sum
; line 6    ; while(n>0)
L1:         ldx       L3_n
;***** cmpx #0   (Removed by optimizer)
            jbeq      L11
;           {                ;             sum=sum+n;
            ldd       L31_sum
            addd      L3_n
            std       L31_sum
;           n=n-1;
            ldd       L3_n
            addd      #-1
            std       L3_n
;           }
            jbr       L1
; line 10   ; return(sum);
L11:        ldd       L31_sum
            rts
; }
            .public   _sum_of_n
            .end
```

(c) Optimized, with C source interspersed.

```
1  ; Compilateur C pour MC6809 (COSMIC-France)
2                       .processor m6809
3                       .psect    _data
4   0000  0000 L3_n:    .byte     0,0
5   0002  0000 L31_sum: .byte     0,0
6  ;    unsigned int sum_of_n()
7  ;    {
8                       .psect    _text
9  ;    static unsigned int n;
10 ;    static unsigned int sum;
11 ;    sum=0;
12  E000   4F   _sum_of_n:  clra
13  E001   5F               clrb
14  E002   FD0002           std       L31_sum
15 ; line 6; while(n>0)
16  E005   BE0000 L1:       ldx       L3_n
17 ;***** cmpx #0
18  E008   2714             jbeq      L11
19 ;       {
20 ;       sum=sum+n;
21  E00A   FC0002           ldd       L31_sum
22  E00D   F30000           addd      L3_n
23  E010   FD0002           std       L31_sum
24 ;       n=n-1;
25  E013   FC0000           ldd       L3_n
26  E016   C3FFFF           addd      #-1
27  E019   FD0000           std       L3_n
28 ;       }
29  E01C   20E7             jbr       L1
30 ; line 10; return(sum);
31  E01E   FC0002 L11:      ldd       L31_sum
32  E021   39               rts
33 ; }
34                          .public   _sum_of_n
35                          .end
```

(d) Object listing.

Table 7.14: (*continued*) Passing a simple program through the compiler of Fig. 7.5.

```
:20E000004F5FFD0002BE00002714FC0002F30000FD0002FC0000C3FFFFFD000020E7FC00AD
:02E020000239C3
:0400000000000000FC
:00E000011F
```

(e) Machine code.

As an example, Table 7.14(a) shows a simple C program passed through the chain of Fig. 7.5. We will look at this program in more detail in the next chapter, but essentially it comprises a loop adding a decrementing 8-bit integer n to the 16-bit integer variable sum (see Table 2.1). Assembly source code produced by the compiler's second pass in Table 7.14(b) is sent through the optimizer, which removes or changes relevant instructions. These are normally shown as comments in the output listing. In Table 7.14(c), line 17 has been commented out from the original source code. The optimizer has recognized that the previous line, which loads the variable n into the X Index register, also sets the Z flag if n is zero. Thus the subsequent comparison with zero (to test the condition n > 0?) is redundant. If the while operand had been anything other than zero, for example (n > 1), then the generated instruction CMPX #1 would be valid.

Not all compilers produce assembly-level code for assembly and linkage, although most Fortran and C translators do. A load-and-run compiler may directly generate machine code, load it and execute.

The most common alternative to compilation is **interpretation**. In this situation the source code is not translated but run as it is. An interpreter program must be resident with this source at run time, as it translates each line 'on the fly' and then executes it. This process is of course very slow compared to running a machine-code program directly. However, developing such programs is faster, as a recompilation is not necessary after each change in the source. For small programs, an interpreter represents a large overhead, as it must be accommodated in target memory in tandem with the source. Nevertheless, high-level source code is more compact than its equivalent machine code and very large interpreted programs may actually require less storage, even with the resident interpreter. The BASIC language is usually run under an interpreter, although compilers are available for this language and may be used once the interpreter-based development has been completed. C interpreters are also available as a development aid, but are rarely used.

A compromise is sometimes effected, where a compiler produces an intermediate code and at run time a much simplified interpreter 'executes' this code as its source. Pascal is traditionally used in this manner (the compiler producing p-code).

References

[1] Barron, D.W.; *Assemblers and Loaders*, MacDonald and Jane's, (UK), 3rd ed., 1978, Chapters 1-4.

[2] Calingaert, P.; *Assemblers, Compilers and Program Translation*, Computer Science Press, Springer-Verlag, 1979, Chapter 2.

[3] Fischer, W.P.; Microprocessor Assembly Language Draft Standard, *Computer*, **12**, no. 12, Dec. 1979, pp. 96-109.

[4] *Standard for Microprocessor Assembly Language*, ANSI/IEEE Standard 694-1985, IEEE Service Center, Publications Sales Dept., 445 Hoes Lane, POB 1331, Piscataway, NJ 08855-1331, USA.

[5] Wakerly, J.F.; *Microcomputer Architecture and Programming: The 68000 Family*, Wiley, 1989, Section 6.3.

[6] Barron, D.W.; *Assemblers and Loaders*, MacDonald and Jane's, (UK), 3rd ed., 1978, Chapter 6.

[7] Walker, G.; Towards a Structured 6809 Assembler Language, Parts 1 and 2, *BYTE*, **6**, nos. 11 and 12, Nov. and Dec., 1981, pp. 370-382 and 198-228.

[8] Barron, D.W.; *Assemblers and Loaders*, MacDonald and Jane's, (UK), 3rd ed., 1978, Chapter 5.

[9] Barron, D.W.; *Assemblers and Loaders*, MacDonald and Jane's, (UK), 3rd ed., 1978, Chapter 8.

[10] Aho, A.V.; *Compilers*, Addison-Wesley, 1986, Chapter 1.

[11] Aho, A.V.; *Compilers*, Addison-Wesley, 1986, Chapters 3-9.

[12] Calingaert, P.; *Assemblers, Compilers and Program Translation*, Computer Science Press, Springer-Verlag, Chapters 6 and 7.

[13] Koenig, A.; *C Traps and Pitfalls*, Addison-Wesley, 1989, Section 1.3.

[14] Reid, L. and McKinlay, A.P.; Whitesmiths C Compiler, *BYTE*, **8**, no. 1, Jan. 1983, pp. 330-343.

[15] Aho, A.V.; *Compilers*, Addison-Wesley, 1986, Chapter 10.

CHAPTER 8

Naked C

In the beginning there was CPL (Combined Programming Language), a language developed jointly by Cambridge and London universities in the mid 1960s. BCPL (Basic CPL) was a somewhat less complex but more efficient variant designed as a compiler-writing tool in the late 1960s [1]. At around that time, Bell System Laboratories were working on the UNIX operating system for their DEC PDP series of minicomputers. Early versions of UNIX were written in assembly language [2]. In an attempt to promote the spread of this operating system to different hardware environments, some work was done with the aim of rewriting UNIX in a portable language. The language B [3], which was essentially BCPL with a different syntax (and was named after the first letter of that language), was developed for that purpose in 1970 [4], initially targeted to the PDP-11 minicomputer.

Both BCPL and B used only one type of object, the integer machine word (16 bits for the PDP-11). This typeless structure led to difficulties in dealing with individual bytes and floating-point computation. C (the second letter of BCPL) was developed in 1972 to address this problem, by creating a range of objects of both integer and floating-point types. This enhanced its portability and flexibility. UNIX was reworked in C during the summer of 1973, comprising around 10,000 lines of high-level code and 1000 lines at assembly level [5]. It occupied some 30% more storage than the original version.

Although C has been closely associated with UNIX, over the intervening years it has escaped to appear in compilers running under virtually every known operating system, and targeted to mainframe CPUs down to single-chip microcontrollers. Furthermore, although originally a systems programming language, it is now used to write applications programs ranging from CAD down to the intelligence behind microwave ovens and smart egg-timers!

For over ten years, the official definition of C was the first edition of *The C Programming Language*, written by the language's originators Brian W. Kernighan and Dennis M. Ritchie [6]. It is a tribute to the power and simplicity of the language, that over the years it has survived virtually intact, resisting the tendency to split into dialects and new versions. In 1983 the American National Standards Institute (ANSI) established the X3J11 committee to provide a modern and comprehensive definition of C to reflect the enhanced role of this language. It took nearly a decade to finally approve the resulting definition, known as Standard or

ANSII C.

In the event, the philosophy of the standard was to alter Old C as little as possible, and that such changes should allow existing programs to compile with, at most, minor changes. The two major changes were the tightening up of the syntax for declaring and defining functions, so that the compiler can report errors due to mismatched arguments. The original specification did not define the libraries accompanying C; although many such functions became de facto standards, there were many portability problems. ANSI C has a standard library, which is a specified part of the language running in a hosted environment; that is with an operating system in situ.

Within the scope of this book, it is impossible to do more than survey the elements of programming in C. There are many excellent texts devoted entirely to this end, some of which are listed at the end of this chapter [7, 8, 9, 10, 11]. To reduce the size of this summary, aspects of C which are unlikely to be of interest to non-hosted environments, that is naked MPU-based systems, have been omitted, for example, file and terminal I/O functions. In addition I have concentrated on the newer ANSI C language, which I have given the generic term C. Where the original specification is alluded to, the term old C has been used. At the time of writing, virtually all compilers are implementing ANSI C.

8.1 A Tutorial Introduction

Let us begin by taking a simple but functional C program and dissect it line by line.

Table 8.1: Definition of function sum_of_n().

```
1   /******************************************************************
2    * Function sums all integers up to n (maximum 65,535)            *
3    * ENTRY : n passed as unsigned short                             *
4    * EXIT  : sum returned as unsigned int                           *
5    ******************************************************************/
6   unsigned int sum_of_n(unsigned short int n)
7   {
8   unsigned int sum;
9   sum=0;
10  while(n>0)         /* For as long as n is greater than zero */
11      {
12      sum=sum+n;     /* add n to the sum                      */
13      n=n-1;         /* and decrement n                       */
14      }
15  return(sum);
16  }
```

The program itself, shown in Table 8.1, is a slightly modified version of Table 7.14(a). It is written in the form of a subroutine (known in C as a function) with the variable n being passed to it from the calling program, and **sum** being returned to the caller. The algorithm continually adds n to the initially cleared sum, as n is decremented to zero.

1–5: These five lines are comments. Any characters between delimiters /* and */ are regarded as a single space by the compiler. Comments can be anywhere where whitespace (the collective term for blank, tab or newline) can appear. Thus lines 10, 12 and 13 have comments after the executable part. Generally, whitespace is used as a matter of style to make the code easier to read. The language itself is entirely freeform, provided that the various statements etc. can be distinguished by the preprocessor.

6: This line names the function **sum_of_n** and declares that it returns an unsigned integer value and acts on an unsigned short integer variable n passed to it by the calling program. Setting out the function parameters like this is known as prototyping. Objects of type int mean that such variables have fixed-point (as opposed to real floating-point) values. A short integer size is typically 16 bits whilst a plain integer is typically 32 bits (see Fig. 8.3). Here they are to be treated as unsigned numbers.

7: A left brace thus { is equivalent to **begin** in Pascal. All **begins** must be matched with an **end**, or in **C** a right brace }. It is good programming style to indent each **begin** from the column of the immediately preceding line(s) and to ensure that **begin** and **end** braces line up. In this case line 16 is the corresponding **end** brace. Between lines 7 and 16 is the body of the function sum_of_n().

8: There is only one variable which is local to our function. Its name and type are **defined** here. In C all variables (unless external) must be defined before they are used. Conventionally, all variables are defined at the beginning of the function. A definition tells the compiler what properties the named variable has, for example size, so that it can allocate suitable storage. Several variables of the same type may be defined in the one statement, for example:

```
int   var1, var2, var3;
```

The line is terminated by a semicolon ; as are all statements in **C**.

9: Here we assign (=) the value 0 to the variable **sum**, that is clear it. A definition and an initializing assignment can frequently be combined; thus:

```
unsigned int sum = 0;
```

is a legitimate statement combining lines 8 and 9.

10: In evaluating sum we need to repeat the same process for as long as n is greater than zero. This is the purpose of the while loop introduced in this line. The body of this loop, that is the statements which appear between the following left and right braces of lines 11 and 14, is continually executed for as long as the expression in the parentheses evaluates to non-zero (**True** in C). This test is done *before* each pass through the body. Thus in our example, on entry the expression n > 0 is evaluated. If True, then n is added to sum, n is decremented, and the loop test repeated. In this case, eventually n reaches zero. Then the expression n > 0 evaluates to **False** (zero), and the statement *following* the closing brace is entered (line 15).

An alternative is while(n), which will also terminate when n reaches zero (False). This is similar to the difference at assembly level between the Test and Compare operations.

11: The begin brace defining the while body. Notice that for style it is indented.

12: The right expression to the assignment is evaluated, sum + n, and the resulting value (r_value) given to the left variable (l_value), sum. The expression sum += n; in C is equivalent and means increment sum by n.

13: Here one is subtracted from n and the result becomes the new n. The decrement operator -- can also be used giving the expression n--;.

14: The end brace. Notice the style with the opening brace (in line 7) and closing brace vertically aligned. This reduces the chance of an error in complex expressions. Braces are used to surround **compound statements**; that is sequences of single statements. Such blocks can be treated in exactly the way a single statement is dealt with. Except where they surround the body of a function, braces may be omitted when the block has only one statement (a **simple statement**). In our example, lines 9–14 could be replaced by:

```
while(n) sum+=n--;
```

which reads: while n is non-zero, add n to sum and decrement n. C can be written in a terse style like this, but the result can be difficult to read. The style used in this book would be:

```
while(n)
    {sum += n--;}
```

where I have used (the optional) braces for clarity.

15: Only one value can be returned directly from a function and this is specified by the return operator. The type of this parameter was given in the prefix of the function declared in the prototype of line 6. The value of the function is the value of this variable. Thus if we had a function that returned the square root of a constant passed to it, then the expression in the calling function:

```
x = sqr_root(y);
```

would assign the value of `sqr_root(y)`, that is its returned value, to **x**.

16: The closing brace for function `sum_of_n()`.

At a simple level, our dissection has given us a feeling for the basic architecture of a C function. C programs are normally structured in a modular fashion with a central function, conventionally named `main()`, calling up a series of functions, some of which may be from a library. Functions can of course be nested. This structure is shown in Fig. 8.1.

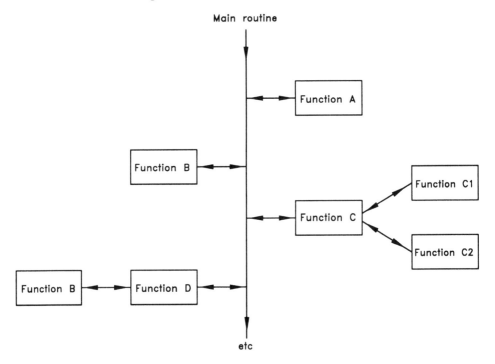

Figure 8.1: Structure of C programs.

We will spend the remainder of this and the next chapter exploring the basic concepts informally introduced here, and enlarging our repertoire of C operations and constructions.

8.2 Variables and Constants

C has a rich variety of elementary objects, upon which are based the more complex groupings of strings, arrays and structures. Objects have properties such as size, structure, range and scope, some of which are summarized in Fig. 8.2. We will discuss these properties at some length in this section, except for scope which is deferred until Section 9.1.

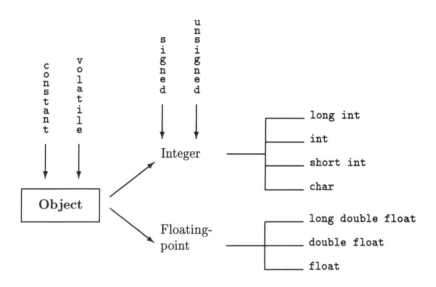

Figure 8.2: Properties of simple object types.

Simple objects are based on a fixed set of basic types, which are illustrated in Fig. 8.3. The fundamental division is between **real** and **integer** forms. The former are valued in terms of floating-point numbers, with sign, magnitude and exponent parts. Three real types are specified, namely **float**, **double float** and **long double float**. C does not guarantee that the three types will in any given implementation differ in precision, only that a `double float` object will never be of lower precision than a plain `float` equivalent, and similarly that a `long double float` will never be of lower precision than a `double float` equivalent. The actual format is implementation dependent, but typically conforms to the ANSI Standard 754-1985 [12] shown in Fig. 8.3.

Most microprocessor-target implementations treat `long double` objects as the default `double`. Some also permit the optional situation where all real types are treated as single-precision `float` objects. This gives faster processing at the expense of precision, especially when real operations are not implemented in a mathematics co-processor. Even when only one or two precisions are actually implemented, it is not considered an error to declare an object of an unimplemented size. For example, where an implementation only supports a single- and double-precision format, an object declared `long double float` is represented in the double precision format, the best possible. `double float` can be abbreviated to just **double**, likewise `long double float` to `long double`.

208 *Naked C*

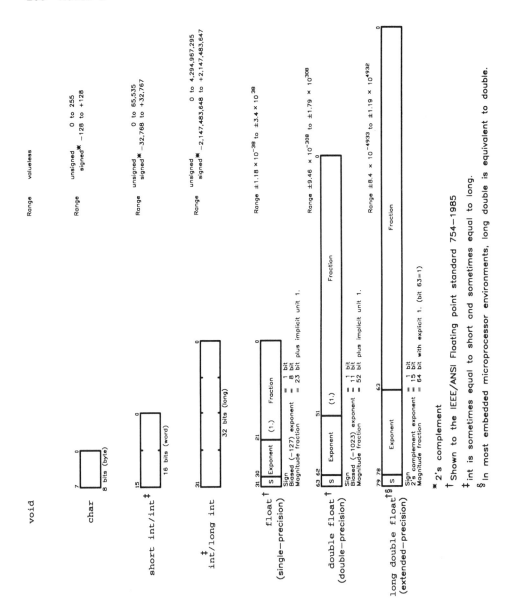

Figure 8.3: Basic set of C data types.

Four integer types representing objects in a fixed-point format are also specified. They are in rising order of range: char, short int, int and long int. Their range is implementation dependent, and typically only three actual sizes are supported. The **char** type is nearly always an 8-bit byte, and is named for an object just big enough to hold a single character (typically, but not always, ASCII-coded).

However, the char type object is a true integer and can be used to define a basic (byte) unit of memory or 8-bit I/O port. A plain int object is supposed to be the size most comfortably handled by the target processor, but is guaranteed to never be less than 16 bits wide. Typically it is 16 bits for an 8-bit MPU and 32 bits for 16/32-bit devices. Some compilers permit the option of either size. int objects are **signed**, that is both positive and negative magnitudes are represented. A 2's complement representation is usual for MPU targets, but others are in use.

A **short int** object is warranted never to be of a greater size than a plain **int** and conversely a **long int** is never smaller than int. Normally with MPU targets there are only two distinct sizes, with **short** being 16 bits and **long** being 32 bits. The int size may be either. short and long int are also signed representations, covering the range $-2^{(n-1)}$ to $+2^{(n-1)} - 1$ in a 2's complement implementation, where n is the number of bits. The qualifier **short int** can be shortened to just **short**; similarly **long int** shortens to **long**.

Although all int types default to a signed representation, they may be prefixed by the (redundant) qualifier **signed** for clarity. The **unsigned** qualifier can be used to give an object which is positive only, and covering the range 0 to $2^n - 1$. Thus the declaration:

unsigned long int sum;

defines an object called sum which can range from 0 to 4,294,967,295 (assuming 4-byte size).

char types are not guaranteed either way. The qualifier **signed** or **unsigned** must be used if such objects are reliably to partake in mathematical operations with other integer types, see Section 8.4.

A **void** object does not exist and does not (naturally) take up any space. It is normally used to declare that a function does not return a variable back to the caller or that no variable is passed by the caller to the function. This is illustrated in Section 9.1. void could properly be said to be a pseudo type.

One of the major properties of an object is where it will be stored: in a register, in an absolute memory location or in a relative memory location in a stack-based frame. The programmer can use the qualifiers **register**, **static** or **auto** to declare which storage class the named variable is to be assigned.

In order to illustrate how these high-level attributes map down to assembly level, I have compiled three versions of a slightly modified version of the sum-of-integers C program of Table. 8.1. The output of this compiler is shown in Table 8.2. This shows the original C source code statements as comments (the syntax of the assembler uses a * to denote a comment) interspersed with their resulting assembly-level instructions. I have manually added comments in parentheses to clarify what is happening at this assembly level; the compiler cannot do this.

By default, all variables are automatically assigned locations in a frame when they are defined on entry to the function in which they operate. They are then accessed relative to the Frame Pointer, as illustrated in Figs 5.6 and 5.7. This is the situation shown in Table 8.2(a) where the variables are declared type **auto** (line C3). The Frame is made eight deep (LINK A6,#-8) as described in Section 5.2. The 4-byte variable n is located at A6-4:3:2:1, and can be fetched using MOVE.L -4(A6),D7, where **A6** is the Frame Pointer. Similarly the resulting 4-byte sum is located at A6-8:7:6:5, hence the operation MOVE.L D7,-8(A6). Although the qualifier auto is shown in this example, it is usually omitted as the default.

Once a function or compound statement has been completed, the frame for any internally defined variables (that is variables defined inside the braces) is closed and its contents lost. Thus if that code is re-entered at some time in the future, no sensible use can be made of an auto variable's previous incarnation. Thus an auto variable's lifetime is simply from where it is defined to its corresponding closing brace. It is unknown outside this region, that is its **scope** is **local** to that of the braces within which it was defined.

A **static** variable is permanently allocated storage, rather than residing inside a transient frame. Thus in Table 8.2(b), both variables n and sum are given room in the data program section (.data is the directive used for the Whitesmith's assembler as equivalent to .psect _data, and similarly .text for the text section), and the compiler names them L3_n and L31_sum respectively. Now to fetch n we have the instruction MOVE.L L3_n,D7. Similarly, to update sum we have MOVE.L D7,L31_sum. Both L3_n and L31_sum of course translate to absolute addresses after linking. Absolutely located variables usually take longer to fetch and return to memory as opposed to stack-based (i.e. auto) storage.

Internally defined static variables have the same scope as auto variables, that is they are local to the function or compound statement in which they are defined. Their lifetime is however that of the program run. Thus if the code is re-entered, the last value of that static variable will still be known. static variables can be declared outside a function, in which case they are globally known from their definition point onwards. This will be discussed in Section 9.1.

Variables have to be brought down to a register to be processed, and then returned to their abode in memory (either to a fixed or relative address) afterwards. All these toings and froings are time consuming and take up program space. In processors with a copious supply of registers, some can be reserved to keep variables in situ for longer periods. This is especially valuable in a loop situation where, otherwise, variables would have to be continually swapped in and out of memory.

The programmer can designate any number of auto variables as candidates for register storage, by using the keyword **register**. The compiler does not have to take any notice of this, and if ignored, such variables are treated as auto types. The Whitesmith 68020 C cross-compiler V3.2, used to generate the code shown in Table 8.2(c), reserves three Data and three Address registers for this purpose. Such register variables are widened to 32 bits (int) when fitted into the designated register. Floating-point variables cannot be designated register types, but pointers (addresses) of such objects can. The scope and lifetime of register variables is identical to that of auto types; indeed they behave in an equivalent way to these, except that their address cannot be taken (see Section 9.2).

Variables can be given a value at any time by simple assignment, for example sum = 0, in line C4 of Table 8.2. It is possible to **initialize a variable** at the time

Table 8.2: Variable storage class (*continued next page*).

```
*    1   sum_of_n()
*    2   {
             .text
             .even
_sum_of_n: link     a6,#-8         (Open frame for n and  sum  )
*    3   auto unsigned int n,sum;
*    4   sum=0;
             clr.l    -8(a6)        (sum in -8(a6) = 0           )
*    5   while(n>0)
L1:          tst.l    -4(a6)        (n in -4(a6) checked for =0? )
             beq.s    L11           (Exit if true                )
*    6       {sum=sum+n--;}
             move.l   -4(a6),a1     (Get n                       )
             subq.l   #1,-4(a6)     (Decrement it in its lair    )
             move.l   a1,d7         (Move old n to d7            )
             add.l    d7,-8(a6)     (Add old n to sum            )
             bra.s    L1            (and repeat                  )
*    7   return(sum);
L11:         move.l   -8(a6),d7     (Sum returned in d7          )
             unlk     a6
             rts
             .globl   _sum_of_n
*    8   }
```

(a) Variables stored in relative memory (38 bytes).

```
             .data
             .even
L3_n:        .byte    0,0,0,0       (4 bytes for n at L3_n       )
             .even
L31_sum:     .byte    0,0,0,0       (and 4 for sum at L31_sum    )
*    1   sum_of_n()
*    2   {
             .text
             .even
*    3   static unsigned int n,sum;
*    4   sum=0;
_sum_of_n:   clr.l    L31_sum       (sum = 0                     )
*    5   while(n>0)
L1:          tst.l    L3_n          (Check for n=0?              )
             beq.s    L11           (Exit if true                )
*    6       {sum=sum+n--;}
             move.l   L3_n,a1       (Get n                       )
             subq.l   #1,L3_n       (Decrement it in its lair    )
             move.l   a1,d7         (Move old n to d7            )
             add.l    d7,L31_sum    (Add old n to sum            )
             bra.s    L1            (and repeat                  )
*    7   return(sum);
L11:         move.l   L31_sum,d7    (sum returned in d7          )
             rts
             .globl   _sum_of_n
*    8   }
```

(b) Variables stored in absolute memory (44 bytes).

Table 8.2: (*continued*) Variable storage class.

```
*    1   sum_of_n()
*    2   {
             .text
             .even
_sum_of_n:   movem.l   d5/d4,-(sp)    (Save used registers        )
*    3   register unsigned int n,sum;
*    4   sum=0;
             moveq.l   #0,d4          (Sum in d4.l                )
*    5   while(n>0)
L1:          tst.l     d5             (Check for n=0?             )
             beq.s     L11            (Exit if true               )
*    6       {sum=sum+n--;}
             move.l    d5,d7          (move n to d7               )
             subq.l    #1,d5          (Decrement it in its lair   )
             add.l     d7,d4          (Add old n to sum           )
             bra.s     L1             (and repeat                 )
*    7   return(sum);
L11:         move.l    d4,d7          (Sum returned in d7         )
             movem.l   (sp)+,d5/d4    (Restore all used regs      )
             rts
             .globl    _sum_of_n
*    8   }
```

(c) Variables stored in registers (26 bytes)

of its definition; thus we may have:

int x=5, y=10, z=-3;

defining x as a (signed) integer with an initial value of +5, y likewise at +10 and z starting off life as −3. How this is done at machine code level, and the resulting effects at high level, depends on whether the variable has a permanent storage location (i.e. is static) or temporary (i.e. auto or register).

Variables that are static as viewed from the high-level perspective are given their initial value *before* the program begins execution. This is obvious when the assembly code of a static initializing definition is examined, as shown in Table 8.3(a). Each static variable has its location reserved for it in data space with the constant in situ, by using a .BYTE (or DC) directive. Thus when the program is put into memory prior to execution by using a loader, these constants will be placed at the appropriate addresses. Loading is a one-off procedure, and no matter how often the definition code is executed, the contents of these locations will not be re-initialized.

Notice from the listing that c has been given an initial value of zero. The language specification guarantees that all uninitialized static variables will be zero (see also lines 4 and 5 of Table 7.14(d)).

Relying on initial static variable states is dangerous when ROMable **C** code (that is code destined to be located in ROM) is executed. This is because there is

Variables and Constants

Table 8.3: Initializing variables.

```
            .processor  m6809
            .psect    _data
L3_a:       .word     5       ; a is predefined as 5
L31_b:      .word     23      ; b is predefined as 23
L32_c:      .byte     0,0     ; No explicit initialization gives an initial zero value
;      1  main()
;      2  {
            .psect    _text
_main:
;      3  static int a=5, b=23, c;
;      4  c=a+b;
            ldd       L3_a    ; Get a
            addd      L31_b   ; Add b
            std       L32_c   ; = c
;      5  }
            rts
            .public   _main
            .end
```

(a) Compile-time initialization.

```
            .processor m6809
            .psect    _text
;      1  main()
;      2  {
_main:      pshs      u       ; Open frame
            leau      ,s
            leas      -6,s    ; Six deep
;      3  auto int a=5, b=23, c;
            ldd       #5      ; Make a 5
            std       -2,u
            ldd       #23     ; Make b 23
            std       -4,u
;      4  c=a+b;
            ldd       -2,u    ; Get a
            addd      -4,u    ; Add b
            std       -6,u    ; = c
;      5  }
            leas      ,u      ; Close frame
            puls      u,pc
            .public   _main
            .end
```

(b) Run-time initialization.

no loading action before execution, the program being permanently stored in ROM. Data in RAM will be garbage, as the power-up state for such memory is unspecified. This is discussed further in Section 10.3.

Variables that are auto or **register** can be initialized in their definition using the same syntax, but the effects are very different from the previous situation. As can be seen from Table 8.3(b) such a definition leads to executable code identical to that produced by the sequence:

```
auto int a, b, c;
a = 5;
b = 23;
```

Such code is executed at each pass through the function; that is the constants a and b are re-initialized each time. auto and register variables are said to be **initialized at run time**, as opposed to static variables which get their primary values at **load time**. Uninitialized auto and register variables have no predictable value, as their locations will either hold the random power-up state of volatile memory or a value generated by some other code which used the same locations previously.

The const and volatile type modifiers are new to ANSII C [13]. An object declared const must not be changed by the compiler subsequent to any optional pre-initialization. Code such as:

```
int a, b;
const int c;
c = a + b;
```

will be flagged by the compiler as erroneous.

The const qualifier is particularly useful in generating fixed look-up tables and arrays. Objects declared both static and const are normally placed by the compiler in the same program section as text. In a dedicated system, this is usually in ROM.

The const modifier can also be used together with the **volatile** modifier to declare a peripheral register as read-only. Specifically the volatile qualifier warns the compiler that the specified variable may be altered by some outside agency not known by the program; that is its value is subject to spontaneous and random change. Thus for example, an input port will reflect an external event not under the program's control. Also the compiler should never try to modify an input port's contents; it is read-only.

The classical example is monitoring a bit in a Status register, waiting for an event to happen, for example:

```
unsigned char i;                            /* i is an ordinary variable           */
volatile unsigned char status;              /* status is the Status register       */
const volatile unsigned char in_port;       /* in_port is the read-only input      */
while (!(status & 0x80))                    /* As long as bit 7 is False (0)       */
    {;}                                     /* Do this (a null statement)          */
i = port;                                   /* When bit 7 is True (1), read in_port */
```

Here the Status register is continually ANDed (**&**, the bitwise AND operator) with the mask $10000000b$ ($80h$ = 0x80). If bit 7 (the flag which says an event has happened out there) is 0 or False, then the expression !(status & 0x80) returns

!(False). As ! is the logic NOT operator, this yields NOT False or True, and the body of the while construction is executed. The single ; statement terminator is used to give a null body. When bit 7 is high !(status & 0x80) returns !(True) = False and the polling terminates.

If the volatile qualifier is not used, then the compiler may well **optimize** the situation by reading in status to a register. The compiler will then continually test this *copy*, to save regularly bringing it down into the MPU. This would only make sense if status were an ordinary variable whose value could *only* be changed by the compiler. Note that the port peripheral register has been declared both const and volatile. This means it is a read-only object (the compiler should not try and alter it) and its value can only be modified by an outside agency. The object descriptor pair of modifiers unsigned char are the equivalent to saying that the qualified object is byte sized, as illustrated below. Another example of volatile is given in the listing of Table 9.6. Normally objects of this kind are pointers to (i.e. addresses of) hardware ports or fixed memory locations, rather than the objects themselves. We will discuss pointers in Section 9.2.

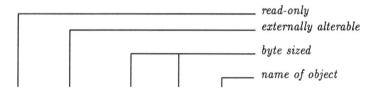

```
const volatile unsigned char in_port
```

Object **identifiers** (that is their names) can be any combination of case sensitive alphanumerics and the underscore character _. The first character must not be a number. With one exception, the initial 31 characters are guaranteed to be significant. Longer identifiers can be used, but the additional characters may be ignored; for example the two variables:

```
base_emitter_bias_resistor_R1150
base_emitter_bias_resistor_R1159
```

may be treated as the same.

The exception is variables which are declared **extern**. These are not defined anywhere in the program but are assumed to be added later through the linker, for example calls to library functions. As the C specification does not include the linker, only the first six characters can be relied on to be significant, and case distinctions may be ignored for such variables. External variables are discussed in Section 9.1.

Integer **constants** can be written in decimal, hexadecimal or octal. Thus the statements:

```
i = 255;          /* Decimal     */
i = 0xFF;         /* Hexadecimal */
i = 0377;         /* Octal       */
```

are identical, with the 0x prefix indicating hexadecimal and 0 for octal. Be careful, 377 (decimal) and 0377 (octal) are very different in C. Character constants are indicated by single quotes; thus:

```
i = 'a'           /* Same as i = 0x61; if ASCII */
```

See lines 4 and 5 of Table 8.5 for another example.

Constants are normally regarded by the compiler as plain integer, or if floating-point, then double precision. The default type can be overridden by using the suffix L for a long integer or long double floating-point, U for an unsigned integer, UL for an unsigned long integer and (F) for a single-precision floating-point constant.

If an integer constant is too large to fit into a plain integer size, then the compiler will **promote** it according to the rules:

```
int → long → unsigned long                        (if plain decimal)
int → unsigned int → long → unsigned long         (if hex or octal)
```

Similarly real constants may be promoted to long double precision.

All integer constants are regarded as positive, with the assignment i = -255; being treated as a positive constant of 255, modified by the unary operator - (minus).

Some typical floating-point assignments are:

```
i = 255.    ;   /* Note use of decimal point */
i = 255.0   ;
i = 2.55E2  ;
i = .0255E+4;
i = 2550.E-1;
```

which are all identical.

8.3 Operators, Expressions and Statements

We have already casually introduced several C operators without much discussion. These were + for Addition, - for Subtraction, * for Multiplication, / for Division, ++ and -- for Increment and Decrement, & for bitwise AND, ! for logic NOT and () to pass parameters to a function. In all, there are 45 defined operators in C as listed in Table 8.4.

Operators are used to combine operands into expressions, for example:

```
x + y * z
```

Here we have a problem, will z be multiplied by the sum of x and y, or will x be added to the product of y and z? The outcome will differ considerably between the two cases. Parentheses can be used to force the way an expression is put together; thus we can have:

Table 8.4: C operators, their precedence and associativity (*continued next page*).

Operator	Operation	Example
Top priority		
Direction (associativity) $\Rightarrow$		
()	Function call	sqr()
[]	Array element	x[6]
.	Structure element	PIA1.CRA
->	Structure element using a pointer	
Unary operators		
Direction (associativity) $\Leftarrow$		
!	Logical NOT	!x
~	Inversion (1's complement)	~x
-	Negative	y=-x
+	Unary plus	y=x- +(y+z)
++	Increment	x++ or ++x
--	Decrement	x-- or --x
&	Address of	&x
*	Contents of address	*address
(type)	Cast	(long)x
sizeof	Size of object in bytes	sizeof x
Arithmetic		
Direction (associativity) $\Rightarrow$		
*	Multiplication	z=x*y
/	Division	z=x/y
%	Remainder	z=x%y (Integer types only)
+	Addition	z=x+y
-	Subtraction	z=x-y
Shift		Integer types only
Direction (associativity) $\Rightarrow$		
>>	Shift left	z=x>>3
<<	Shift right	z=x<<3
Relational operators		Boolean objects
Direction (associativity) $\Rightarrow$		
<	Less than	while (x<3)
<=	Less than or equal	while (x<=3)
>	Greater than	while (x>3)
>=	Greater than or equal	while (x>=3)
==	Equivalent	while (x==y)
!=	Not equivalent	while (x!=0)

Table 8.4: (*continued*) C operators, their precedence and associativity.

Operator	Operation	Example
Bitwise logic		Integer types only
Direction (associativity) ⇒		
&	AND	x&0xFE (Clear bit 0)
^	Exclusive-OR	x^0x01 (Toggle bit 0)
|	OR	x|0x01 (Set bit 0)
Objectwise logic		Boolean objects
Direction (associativity) ⇒		
&&	Logical AND	x&&y is True if both x and y are True
||	Logical OR	x||y is True if both or either x and y are True
?:	Conditional	x=(y>z)?5:10 x=5 if y>z True else x=10
Assignment		
Direction (associativity) ⇐		
=	Simple	x=3
+=	Compound plus	x+=3 e.g. (x=x+3)
-=	Compound minus	x-=3 e.g. (x=x-3)
*=	Compound multiply	x*=3 e.g. (x=x*3)
/=	Compound divide	x/=3 e.g. (x=x/3)
%=	Compound remainder	x%=3 e.g. (x=x%3)
&=	Compound bit AND	x&=3 e.g. (x=x&3)
^=	Compound bit EX-OR	x^=3 e.g. (x=x^3)
|=	Compound bit OR	x|=3 e.g. (x=x|3)
<<=	Compound shift left	x<<=3 e.g. (x=x<<3)
>>=	Compound shift right	x>>=3 e.g. (x=x>>3)
Direction (associativity) ⇒		
,	Concatenate	if(x=0,y=3;x<10,x++)
Lowest priority		

```
x + (y * z)
(x + y) * z
```

as C follows the usual rules of computing the contents of parentheses first (innermost outwards for nested parentheses).

The way in which an expression is combined is obviously of critical importance. In C, operators are graded in order of their **precedence**. Table 8.4, which lists operators in descending order of precedence, shows that multiplication is of a higher precedence than addition, and so will be implemented first. Thus the first form of the parenthesized expression above is equivalent to our original statement.

This still leaves us with the problem of mixing operators at the same level of precedence. For example:

x/y/z

Is this (x/y)/z or x/(y/z)? The outcomes are very different. Most operators associate from left to right, thus the equivalent here is:

(x/y)/z

An example of right to left association is the statement:

f = x = y = z = 0;

What value will f have? The answer here is 0, as assignment operators associate from right to left. Firstly z will be assigned to 0, then y to z (i.e. 0), then x to y (i.e. 0) then f to x (i.e. 0); so all variables will be set to 0, that is:

f = (x = (y = (z = 0)));

We have illustrated this situation using a statement as opposed to an expression. C programs are made up of a series of statements or actions, each terminated by a semicolon ;. The majority of these are **expression statements**, where combinations of variables and constants are linked by one or more operators. **Compound statements** may be formed by enclosing a series of simple statements in braces, as shown in Table 8.1, lines 11–14. Anywhere it is legitimate for a simple statement to appear, can be filled by a complex statement.

Expressions have values, for example in the peculiar looking statement:

x = 12 * (y = z + 5) + 2;

the expression y = z + 5 assigns the value z + 5 to y and takes this outcome for itself. Thus it is equivalent to the compound statement:

```
{
y = z + 5;
x = 12 * y + 2;
}
```

Most of the operators listed in Table 8.4 are intuitive and will not be covered in any detail here. Apart from functions, arrays and structures, discussions of which are deferred to Chapter 9, the **unary operators** have the highest priority. Unary operators attach to a single object, for example ~x inverts all bits in x (1's complement). Most operators are **binary** in that they connect two objects, for example x + y. Unaries bind very tightly to their object due to their high priority; thus:

a = b + ~x;

and

a = b + (~x);

are the same, as ~ has a higher priority than the binary addition operator.

Care must be taken when inverting C objects, as all zeros *implicit* in the variable become ones. Consider:

```
int i = 0xA9, j;
j = ~i;                 /* j = 0xFF56 or 0xFFFF56 */
j = ~i & 0xFF;          /* j = 0x0056 or 0x000056 */
```

Although i is assigned constant 0xA9, its bit pattern will be 0000 0000 1010 1001b or 0000 0000 0000 0000 0000 0000 1010 1001b, depending on whether int is 16 or 32 bits. On inversion, all the implicit zero bits will become one as shown. Bit ANDing (the && operator) by 1111 1111b will clear these, as this int constant has implicit leading digits of zero. & has a lower priority than ~, so no parentheses are required. In a 2's complement machine the unary - operator acts in a similar way to ~, that is -a is the same as ~a + 1 (2's complement is invert plus 1). As you would expect, unary - simply changes the sign of the object.

Consider the statement:

```
f = a + (b-c);
```

You might think that the expression (b-c) would be evaluated first and then a added to it. In fact C will ignore the parentheses, deeming them unnecessary, as the binary addition (+) and subtraction (-) operators have the same level of priority. Then, according to the table, evaluation occurs from left to right; that is a + b and then -c. If it is important to you to add (b-c) to a and not b alone (perhaps because you are afraid of overflow) then the unary + operator will ensure this happens; that is:

```
f = a + +(b-c);
```

Unary + forces evaluation of its operand, as this has a higher priority than the binary + Addition operator (see Table 8.4).

Although C guarantees the way an expression is put together according to the rules of precedence and associativity, it says nothing about the sequence in which component sub-expressions are produced. Consider the following (convoluted) statement:

```
f = a + (z = z+4) + 3*z;
```

where the writer hoped that the parentheses would force the variable z to be incremented by four first, then multiplied by 3 and finally, left to right, a added to the new value of z and then added to three times the new value of z. But what if the compiler took it into its head firstly to multiply z by three (i.e. old z) and store the answer away somewhere, then evaluate (z+4) and store it away, and then add a to the new value of z plus three times the old value! This type of occurrence is known as a **side effect**, as it is usually caused by using an assignment, increment, decrement or function that changes the value of an object that appears elsewhere in the expression. C makes no promises that side effects will occur in a predictable order within a single statement [14]. A safer sequence would be:

```
z = z + 4;
f = a + z + 3*z;
```

or

```
f = a + 4*(z + 4);
```

Unary operators normally tag their object to the left, the possible exception being the Increment ++ and Decrement -- unaries. These can be before or after the identifier; their effect being subtly different. A left Increment/Decrement unary operator means *first* change the object and *then* use it. A right unary means *first* use the (old) value in the calculations and *then* change the object. For example:

```
sum = sum + n--;     /* Add n to sum, then decrement n      */
sum = sum + --n;     /* Decrement n first, then add n to sum */
```

The former is clearly shown in line C6 of Table 8.2(b). First n is fetched from memory into internal storage (MOVE.L L3_n,A1), then the original object out there in memory is decremented (SUBQ.L #1,L3_n). Finally the original value is used for the addition (MOVE.L A1,D7; ADD.L d7,L31_sum).

Because of side effects, care must be taken that Incremented/Decremented objects do not appear elsewhere in the same statement; for example:

```
z = 6*n-- + a/n;
```

Will the n used in the denominator have the old or new value? You will be at the mercy of the vagaries of your compiler in writing such code.

Rather confusingly, some of the unary operators have the same symbols as binary operators, with very different meanings; particularly **address of** (&) and **contents of** (*). The compiler normally has no difficulty distinguishing between the unary and binary from their context (see Section 9.2 for these unary operators).

We have already given an example of & as a bitwise binary operator. Also available are OR (|), Exclusive-OR (^) and the unary bitwise NOT (~). In a binary bitwise operation, each bit of the integer object (not floating-point types) is affected by the corresponding bit of the integer operand. This latter right-hand operator can be a constant or variable.

Bitwise logic operations are identical in action to their assembly-code cousins (e.g. see Table 4.4). Other 'bit-banging' operations are Shift Right (>>) and Shift Left (<<). As before only integer type objects are permitted. The following code:

```
BCD_LOW  = (packed_BCD & 0x0F) + '0';
BCD_HIGH = (packed_BCD >> 4) + '0';
```

separates the 8-bit object `packed_BCD` into its two 4-bit constituent BCD digits in their ASCII form. The low digit is obtained by clearing the upper four bits with a bitwise AND, whilst the high digit is separated out by shifting right four times. Adding ASCII 0 (i.e. 0x30) converts to the appropriate ASCII code. Parentheses are used, as & and >> are of lower priority than +. Table 8.5 shows how these operations translate to 68000 code.

222 *Naked C*

Table 8.5: Bitwise AND and Shift operations.

```
        .text
        .even
        link    a6,#-4      ; Make frame
*   3   unsigned char packed_BCD, BCD_LOW, BCD_HIGH;
*   4   BCD_LOW=(packed_BCD & 0x0f)+'0';
        moveq.l #15,d7      ; [D7] = 000000FFh
        moveq.l #0,d6
        move.b  -1(a6),d6   ; [D6] = 000000[packed_BCD]
        and.l   d6,d7       ; [D7] = 000000[packed_BCD&000000FFh]
        moveq.l #48,d6      ; [D6] = 30h; that is  '0'
        add.l   d7,d6       ; [D7] = 000000[packed_BCD&000000FFh + '0']
        move.b  d6,-2(a6)   ; Assigned to BCD_LOW
*   5   BCD_HIGH=(packed_BCD >> 4)+'0';
        moveq.l #0,d7       ; [D7] = 00000000h
        move.b  -1(a6),d7   ; [D7] = 000000[packed_BCD]
        asr.l   #4,d7       ; [D7] = 000000[packed_BCD] >> 4
        moveq.l #48,d6      ; Once again add '0'
        add.l   d7,d6
        move.b  d6,-3(a6)   ; Assigned to BCD_HIGH
        unlk    a6          ; Close frame
        rts
```

The Shift Left operation always feeds in zeros. Shifting right is more problematical. If the object is unsigned, then a Logic Shift Right is generated, with zeros moving in. The situation is confused when a signed object is being acted upon. Most compilers will emit an Arithmetic Shift Right, where the sign bit is propagated along. However, this is not guaranteed. If a Logic Shift Right is desired, then the object can temporarily be treated as unsigned; for example:

Temporary cast

z = (unsigned int)a >> 6;

where I have assumed a is a **signed int** type. The unary operator **(type)** used to force the variable a is known as a **cast**.

C has a range of relational and logic operations which treat objects as **Booleans**, that is having only two values, True (non-zero) and False (zero). We have already used the Greater Than (>) operator in line 5 of Table 8.2. Here the value of n is compared to 0. If Greater Than, then the outcome of the expression n > 0 is 1 (i.e. True); otherwise the outcome is 0 (False). Actually in this case the construction while (n) would do the same thing. Unary logic NOT (!) simply changes the truth value of the object; for example:

```
while ((!n && m) || (n && !m) )         /* while this is true */
    {do this, that and the other}       /* loop body          */
```

executes the loop body if n is False (!n True) AND m is True OR ELSE n is True AND m is False. In other words, only if one of m or n is False (i.e. 0) will the loop body be executed. Notice the use of && and || for logic AND and OR, as opposed to the bitwise & and | operator symbols.

All logic (Boolean) expressions are guaranteed to be evaluated left to right, and this evaluation ceases as soon as an overall result can be ascertained. Thus in the example above, if n were False and m were True, the sub-expression (n && !m) would not be executed. Thus fancy programming such as:

(!n && m)||(n && !m++)

would be dangerous as the m++ increment would only happen if n was True and/or m was False. In this case, the first expression would be False and the compiler would move onto the second expression.

Mixing up the logic equivalent operator == and assignment operator = is a major source of error [15] (not helped by most texts calling == equal). Compare the following two statements:

```
if (a == b)  {do this;}       /* Correct   */
if (a = b)   {do this;}       /* Dangerous */
```

In the former case the value of a is compared to that of b. If they are the same, (True) the value of the expression is 1, and {this;} is executed. If they differ, the result is 0 (False) and {this;} is skipped (see page 227). Neither a nor b are changed by this process. In the latter case a is assigned the value of b, and the value of the expression is b. If b is non-zero then {this;} is done, and if zero, skipped. It is unlikely the programmer meant to do this, and if he/she did, then it should be done in a less obscure fashion.

As a final example, consider the problem of determining the state of the most significant bit of an unsigned int object x. This simply requires ANDing by 2^{n-1}, where n is the number of bits in the object. Unfortunately an int object can have 16 bits in some implementations and 32 bits in others (other values are also possible but rare). If the software is to be written in a portable form, then one of the two masks 2^{15} (1000000$0b$) and 2^{31} (10000000000000000b) has to be chosen.

C has a unary operator called sizeof, which operates on a type designator or object, and which returns its size in bytes. This also applies to composite objects such as arrays and structures. Using this, a possible sequence might be:

```
if (sizeof(x) == 4)           /* Has x got 4 bytes? */
    {mask = 0x80000000;}      /* If True            */
else
    {mask = 0x8000;}          /* If not True        */
msb = mask & x;
```

Notice the use of == to compare the size of x with 4.

A rather more ingenious coding is given by:

```
msb = ((sizeof(x)==2) * 0x8000 + ((sizeof(x)==4) * 0x80000000)) & x;
```

where we rely on a Boolean expression returning 0 if False and 1 if True.

Where a variable is to be assigned to one of two values depending on the truth of an expression, C provides a compact ternary operation using the ?: pair. Repeating the above now gives us:

```
msb = (sizeof(x)==2 ? 0x8000 : 0x80000000) & x;
```

where the expression in parentheses evaluates to 0x8000 if sizeof(x)==2 evaluates to True, else 0x80000000 if False. Try rewriting the statement using the NOT Equivalent (!=) operator.

Besides Addition and Subtraction, the basic arithmetic operations of Multiplication, Division and Modulus (%) are provided. Division of two integral objects yields a truncated integral quotient, thus 6/4 = 1. The Modulus operation of two integral objects gives the remainder, thus 6%4 = 2. Truncation direction and modulus sign are implementation dependent with negative objects. Modulus only operates on integral objects.

As an example, consider the following code which converts an 8-bit binary variable to a hundreds, tens and units BCD digit:

```
unsigned char  binary, hunds, tens, units;
units  = binary%10;     /* e.g. 253%10 = 3 (units)    */
binary = binary/10;     /* Residue = 25 after this    */
tens   = binary%10;     /* %10 gives 5 (tens)         */
hunds  = binary/10;     /* /10 gives 2 (hundreds)     */
```

In Section 9.2, we repeat this example for larger binary numbers, using an array data structure.

Consider the statement above:

```
binary = binary/10;
```

In C this can be written in a compressed manner as:

```
binary /= 10;
```

using the /= compound assignment function. This could be read as divide **binary** by 10.

Apart from compound assignments' concise notation, there can be advantages in the size of machine code emitted where complex objects are involved. As an example, consider a 2-dimensional byte array (see Section 9.2) of 100 rows and 12 columns. If, say, we wish to multiply an element 5 rows down and 3 columns across, we could write:

```
x[5][2] = x[5][2] * n;
```

using simple assignment. The compiler knows where the start address of the array is, so to get x[5][2] it must multiply the number of rows (5) by the *maximum* number of columns (i.e. 12). Finally add the *actual* number of columns (2 across). This is the number of bytes on from the start (62), see Fig. 9.3(b), and would then be used as part of some Indexed address mode to give the effective address (ea). Once x[5][2] was down, it would be multiplied by n. The compiler would then move to the left side of the assignment, and if not very bright would again calculate the ea (probably previously thrown away) to determine the target address for the Store/Move. This takes lots of wasted time and code.

The alternative compound assignment is written:

```
x[5][2] *= 2;
```

The compiler now knows that the ea has only to be calculated once, which consequently produces a superior coding.

Using this notation, line 6 of Table 8.2 could be replaced by:

```
sum += n;
```

or even, using the comma operator (,), lines 5 and 6 could be combined as:

```
while (sum += n--, n > 0)     {;}
```

The **comma operator**, shown at the bottom of Table 8.4, allows expressions to be concatenated. Each such expression is guaranteed to be evaluated from left sub-expression to right, with the value being that of the rightmost sub-expression. Thus, in the example above, sum += n-- will be executed and *then* the test n > 0. The value (True or False) of this latter is the one acted upon by the while instruction. Notice the use of {;} to indicate a null statement (i.e. do nothing). The braces are optional. It is normally recommended that the comma operator be used with caution.

A close scrutiny of the code produced in Table 8.5 shows that the three objects packed_BCD, BCD_LOW and BCD_HIGH are stored in memory as bytes (at [A6]-1, [A6]-2 and [A6]-3 respectively), as expected by their declaration as char. However, when brought down into a MPU register, they are converted into 32-bit ints. For example:

```
MOVEQ.L  #0,D6         ; Clears all 32 bits
MOVE.B   -1(A6),D6     ; packed_BCD occupies lower 8 bits
```

shows the promotion of the unsigned char packed_BCD to 32-bit status by making the upper 24 bits zero. If packed_BCD had been signed, then a Sign Extension would have been used (e.g. EXT for the 68000 MPU). This promotion to int is the reason why an Arithmetic Shift Right (ASR) was used to implement >> in line C5, as opposed to the expected LSR, as int is signed and the compiler sensibly uses Arithmetic Shift operations for signed numbers.

In general, C prefers to do all its fixed point arithmetic in int form. Thus, as shown by the thick arrow in Fig. 8.4, all objects declared **signed** or **unsigned char**, **signed** or **unsigned short** are automatically made int for the duration of their stay in the processor. Some compilers give the option of disabling this widening, which can be useful for 8-bit MPUs which have difficulty in this area. However, this extension facility is non-standard. In a similar manner, C prefers to

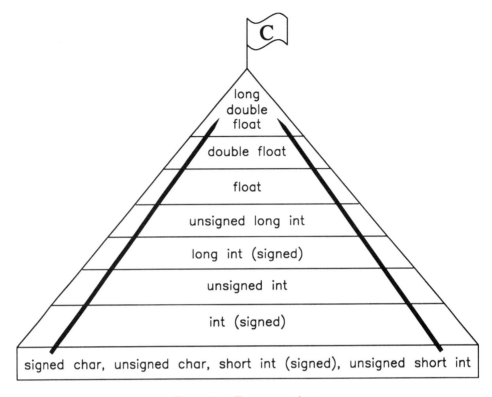

Figure 8.4: Type promotions.

do its floating-point operations in double float form. This too may sometimes be changed to the non-standard single-precision float size, to save time and storage.

C permits arithmetic with mixed types. Consider the following example:

```
short z;
int   x;
unsigned long y;
float a;
a = x + y/z;
```

What type will the right-hand side end up with, and how will that equate with the left-hand type?

Well, firstly object z will be promoted to unsigned long to match the numerator, and the result will be unsigned long. Then x will be promoted to unsigned long to match, and added to give an unsigned long right-hand value. Finally this is converted to float, which is the value assigned to the left-hand variable.

In general, in a mixed type operation, the objects involved migrate upwards to the highest commonalty, as defined in the hierarchy of Fig. 8.4, with int being the base integral type.

One point that needs watching is the notion that an unsigned integral type is of a higher order than its signed counterpart. This is because an unsigned quantity can hold a larger magnitude for the same size, see Fig. 8.3. This can cause strange outcomes when mixing unsigned and signed types together. For example, in the statement above, if x was −1 on a 2's complement machine, it would be stored as 0xFFFFFFFF (for 32 bits). Now because of y, it must be converted to **unsigned long**, and in this case it will be treated as a positive number (4,294,967,295). In some situations, this can lead to spectacular results, although it will work out correctly in this case. In general, if possible do not mix signed and unsigned numbers.

In an assignment, the right-hand value (r_value) is converted to the l_value type, in this example, the **float** equivalent to the **unsigned long** r_value. Where the l_value type is further down the hierarchy, then truncation or other unspecified shortening will occur, and unless the actual value can be fitted into the lower type, an erroneous result will be recorded.

As a final example of what can go wrong consider the code fragment:

```
long int sum;      /* Reserve 32 bits for sum   */
unsigned int n;    /* and a 16-bit n            */
sum = (n+1)*n/2;   /* Sum of all integers up to n */
```

compiled with a 16-bit **int** and 32-bit **long** compiler model. All arithmetic is done at **unsigned int** level (i.e. 16-bit precision). However, if n is large enough, overflow will occur; for example if n is 256, then (n+1)*n will give 256 and not 65,792 (256 is $65,792 - 65,536$)! The fact that sum is defined as **long** will not save the situation, as this means only that the final (erroneous) r_value will be promoted to 32 bits. If values of sum greater than 65,535 are expected, then the variable n may be treated as a 32-bit object by using the cast operator (i.e. (long)n), which will force 32-bit arithmetic thus:

```
sum = ((long)n+1) * n/2;
```

Why didn't I bother to cast the second n? Why is the code of Table 7.13 safe?

8.4 Program Flow Control

The flow control instructions specify the structure of the computation process. Primarily they provide the means whereby the MPU can bypass, alternate or repeat a specified block of statements based on the outcome of an expression. In C this outcome is defined as False if the value returned is 0, otherwise it is True.

The most fundamental decision structure is shown in Fig. 8.5, where the truth outcome of an expression used as the argument of the **if instruction** is used to decide between a 2-way branch. In (a), an expression returning True forces the execution of the do this; statement, otherwise nothing. The **else instruction** can be used in conjunction with if, that is **if-else**, to force one statement on True and another on False.

228   *Naked C*

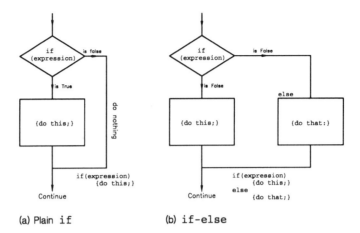

Figure 8.5: Simple 2-way decisions.

As an example of a straight if decision, the following statement returns the positive equivalent (the modulus) of the variable x:

```
if (x<0)   {x= -x;}
```

The statement following the if (*expression*) is executed when x<0 is True (i.e. x is negative), otherwise it is skipped. The braces surrounding the if body are optional when it comprises a single statement. Compound statements must be braced as usual. As a matter of style, in this text braces are normally used irrespectively.

An if-else construction is used in the following code snippet, which converts an ASCII-coded digit in the range '0' to '9' and 'A' to 'F' into its equivalent decimal value 0 to 15.

```
if (ascii <= '9')
    {decimal = ascii - '0';}
else
    {decimal = ascii - '0' - 7;}
```

Here the ASCII code for '0' (i.e. $30h$) is subtracted if the digit lies between '0' and '9' ($30h-39h$) and $37h$ is subtracted if it does not (which assumes that it must be between 'A' and 'F', $41h-46h$).

if instructions may be nested, although care must be taken in using braces to force the proper association. As an example, consider a Real-Time Clock function entered via an interrupt once a second. We will discuss how this might be accomplished in a C program in Section 10.2. Once in the function, we have three variables: Seconds, Minutes and Hours. The logic for the update is:

1. Add one to the Seconds count.

2. If this gives 60 then zero Seconds and increment Minutes.

3. If this gives 60 then zero **Minutes** and increment **Hours**.

4. If this gives 24 then zero **Hours**.

As shown in Table 8.6, the **Seconds** variable is first incremented and then compared for greater than 59 in line 4 (note the ++ operator *before* the variable **Seconds**). If this is not True then the following complex statement, delineated by the braces of lines 5 and 15 is skipped and the function exits. Otherwise, this complex statement is entered, **Seconds** are zeroed in line 6 and the next **if** instruction executed. This does the same thing with **Minutes**, and if the result is not greater than 59 its body, delineated by braces in lines 8 and 14, is skipped and the function terminated. Finally the third-level nested **if** increments and checks **Hours**. If the result is not greater than 23, then its body is skipped to the brace in line 13 and thence the exit point at line 16.

Table 8.6: A nested **if** Real-Time Clock interrupt service routine.

```
1:   unsigned char Seconds,Minutes,Hours;
2:   void clock(void)
3:   {
4:   if(++Seconds>59)
5:       {
6:       Seconds=0;
7:       if(++Minutes>59)
8:          {
9:          Minutes=0;
10:         if(++Hours>23)
11:            {
12:            Hours=0;
13:            }
14:         }
15:      }
16:   return;
17:   }
```

Notice how the **if** instructions are indented, and how the different nesting levels' braces line up. It is essential to take care with constructions like this to avoid error.

Nesting **if**s with **else**s can cause errors, as any **else** will associate itself to the nearest unattached **if**, thus:

```
1: if (n > 0)                 /* IF n is above zero THEN                    */
2:     if (n > max) {n = max;} /* restrict to no more than max              */
3: else n = 0;                /* Otherwise ensure it never goes negative */
```

The writer of this code fragment meant to restrict the variable n to the range 0−**max**, limiting it to these boundary values if beyond. Thus the logic was:

1. Check n above zero, IF False then make n = 0.

2. Check n above **max**, IF True make n = **max**, ELSE do nothing.

230   *Naked C*

What actually happens is the `else` of line 3 will attach itself to the `if` of line 2, not that of line 1; thus if n is lower than zero, all of lines 2 and 3 are bypassed. Furthermore, if n is not above `max`, then n will be made zero! The situation is solved by proper use of braces:

```
if (n > 0)
   {
   if (n > max) {n = max;}
   }
else n = 0;
```

or better still use the **else-if** instruction:

```
if (n > 0)            {n = 0;}
else if (n > max)     {n = max;}
```

Although nested ifs can be utilized to make multiple decisions, their use is not very elegant, and, as we have seen, error prone. The **else-if** construction illustrated in Fig. 8.6 is the more structured approach. The several expressions are evaluated in order until the first True result. The statement associated with this expression is executed, and the rest of the chain is by-passed. An optional final `else` can be used at the end to give a default action.

As an example, let us redo the Real-Time Clock function, this time using an `else-if` construction. In Table 8.7, line 4 is a plain `if`, which checks the state of `Seconds` after incrementing. Should `Seconds` be less than 60 the dummy null statement `{;}` is executed and the rest of the structure bypassed. If not, then the `Minutes` variable must in turn be incremented and checked. However, first `Seconds` must be zeroed. I have used the comma concatenate operator to implement a compound expression doing both in line 5. As far as the `else-if` operator is concerned, the rightmost expression value is utilized in determining its action. Similarly the `Hours` variable is checked in line 6. A plain `else` gives the final fall-through option which happens only once a day, when going from 23:59:59 to 00:00:00 hours.

Table 8.7: An `else-if` Real-Time Clock interrupt service routine.

```
1:   unsigned char Seconds,Minutes,Hours;
2:   void clock(void)
3:   {
4:   if(++Seconds<60)                    {;}
5:   else if (Seconds=0,++Minutes<60)    {;}
6:   else if (Minutes=0,++Hours  <60)    {;}
7:   else                                {Hours=0;}
8:   return;
9:   }
```

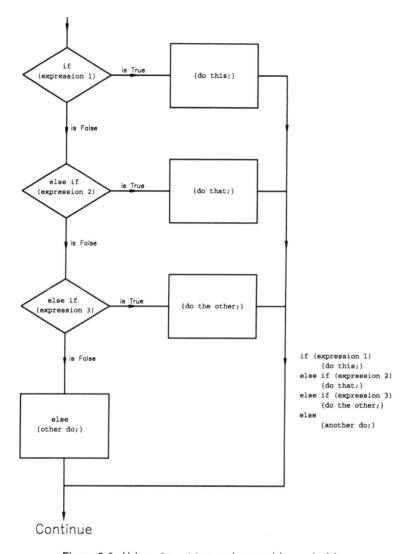

Figure 8.6: Using `else-if` to make a multi-way decision.

One final `else-if` example evaluates the factorial of an unsigned object valued between 0 and 12. If outside this range a zero is returned to indicate an error situation. Table 8.8 is self-explanatory. Remember that as soon as a True outcome is met, the associated expression is evaluated and the whole of the rest of the structure bypassed. As we shall see, there are more efficient ways to implement this function.

The **switch-case** instruction gives an alternative multi-way decision structure, as shown in Fig. 8.7. This time a single expression, which must return an integral type result, is evaluated at the head of the structure. A series of case expres-

Table 8.8: Generating factorials using the `else-if` construct.

```
 1:    unsigned long factor(int n)
 2:    {
 3:    unsigned long factorial;
 4:    if((n==0)||(n==1))      {factorial=1;}
 5:    else if (n==2)          {factorial=2;}
 6:    else if (n==3)          {factorial=6;}
 7:    else if (n==4)          {factorial=24;}
 8:    else if (n==5)          {factorial=120;}
 9:    else if (n==6)          {factorial=720;}
10:    else if (n==7)          {factorial=5040;}
11:    else if (n==8)          {factorial=40320;}
12:    else if (n==9)          {factorial=362880;}
13:    else if (n==10)         {factorial=3628800;}
14:    else if (n==11)         {factorial=39916800;}
15:    else if (n==12)         {factorial=479001600;}
16:    else                    {factorial=0;}          /* Error condition */
17:    return(factorial);
18:    }
```

sions compare this result with a constant. On finding equality, the accompanying statement is executed. If no equality is found, an optional `default` statement is allowed.

The `switch-case` structure is much less flexible than its `else-if` multi-way cousin. Only one expression is evaluated, and actions are not mutually exclusive. We see from Fig. 8.7 that if, say, the expression produced a value equal to constant B, then not only is the `do that;` statement executed, but also `do the other;` and the default `other do;` as well. Normally the `switch-case` decision tree is implemented at machine level by using the result of the expression as a pointer into a look-up table holding a series of Jump to Subroutines, that is the case routines. As these are stored consecutively in memory, once a statement is entered, it will be executed and on return will immediately jump to the next subroutine. Thus it will fall through all following case routines until the terminating RETURN FROM SUBROUTINE instruction is reached. As compensation for `switch-case`'s lack of flexibility, the resulting machine code is normally more compact, and execution is quicker than an `else-if` equivalent. The code of Table 8.8 yielded 264 bytes on a 68000 MPU compiler, while that of Table 8.9 took 212 bytes. Larger structures produce proportionately greater savings.

Two things are noticeable concerning Table 8.9. Firstly `case` statements can be stacked, as in line 6 where the outcome is the same if n is 0 or 1. Secondly each `case` statement is compound, ending with the `break` instruction. This forces the execution to bypass *all* remaining statements down to the `return` of line 20. Leaving out `break` is not a syntax error, but is rarely what the programmer meant to do [16].

`switch-case` structures are frequently used in conjunction with a keyboard to

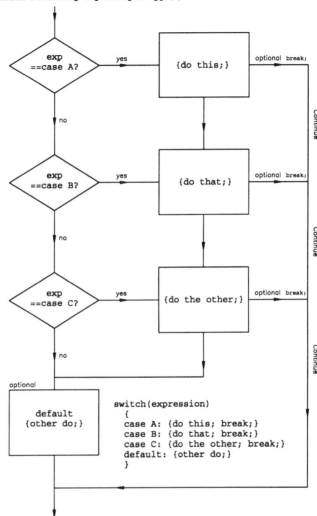

Figure 8.7: switch-case multi-way decision.

select an appropriate response to each keypress, usually by jumping to a subroutine. Thus, if key M is pressed, do a memory examine; if V is pressed, view a block; etc.

The **loop** structure is the standard technique for repeating a process a number of times, either on a single object or on an array or block of related objects. We have already extensively used this approach at assembly level, for example Table 5.7. C has three statements specifically handling loops: while, do-while and for.

Initially, let us see how we could handle a loop without using specific looping instructions. Consider the following code fragment, which evaluates the factorial

Table 8.9: Generating factorials using the `switch-case` construct.

```
 1:   unsigned long factor(int n)
 2:   {
 3:     unsigned long factorial;
 4:     switch(n)
 5:     {
 6:       case 0: case 1:  {factorial=1;           break;}
 7:       case 2:          {factorial=2;           break;}
 8:       case 3:          {factorial=6;           break;}
 9:       case 4:          {factorial=24;          break;}
10:       case 5:          {factorial=120;         break;}
11:       case 6:          {factorial=720;         break;}
12:       case 7:          {factorial=5040;        break;}
13:       case 8:          {factorial=40320;       break;}
14:       case 9:          {factorial=362880;      break;}
15:       case 10:         {factorial=3628800;     break;}
16:       case 11:         {factorial=39916800;    break;}
17:       case 12:         {factorial=479001600;   break;}
18:       default:         {factorial=0;}
19:     }
20:     return(factorial);
21:   }
```

by repetitive multiplication of a decrementing n.

```
         factorial = 1;
LOOP:    if (n>1)            {factorial *= n--; goto LOOP;}
```

This uses the **goto** instruction, together with a label, to repeat the **if** test on each pass of the loop body, in a similar fashion to an assembly language implementation (see Table 4.13).

The **goto** instruction can be used to force an unconditional branch to a label anywhere within a function. However, its use is frowned upon (it has been stated that the quality of programmers is a decreasing function of the density of **goto** instructions in the programs they produce [17]), as used without care it can lead to spaghetti (unstructured) code. Nevertheless, its use is sometimes virtually indispensable, particularly when trying to escape to the outside world from within nested loops. Use with caution.

We have already met the **while** loop back in Table 8.1. Here the body — lines 11 to 14 — was repetitively executed as long as the test n>0 was True. On False the code following the body, that is line 15, is entered.

Three elements present in any loop should be noted. Firstly, variables must be set to their initial state before the loop proper is entered, see line 9 in Table 8.1. Then, in a **while** construct, a test is made, as shown in Fig. 8.8(a). If the outcome is True, the loop body is executed. Finally, some change must be made to the test variables, so that this test will eventually have a False outcome, and execution will

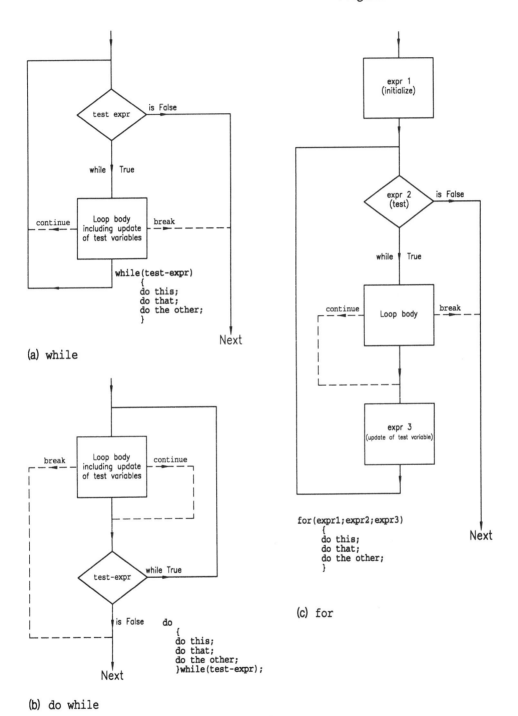

Figure 8.8: Loop constructs.

go on to the next code section. Sometimes this change is explicit, as in line 13 of Table 8.1, and sometimes implicit in the loop body.

Table 8.10: Generating factorials using a while loop.

```
unsigned long factor(int n)
{
unsigned long factorial;
factorial=1;
while(n>1)
    {
    if(n>12)   {factorial=0; break;}
    factorial *= n--;
    }
return(factorial);
}
```

Two keywords are used in conjunction with while. A break forces an immediate exit from within the loop, usually on some exceptional situation. In Table 8.10, this occurs if n>12, which is the error condition demanding a return of zero. This is done by testing for greater than 12 and breaking if True in line 7. In the case of a nested loop, breaking will move the execution only to the next outer level.

The continue keyword forces an early repeat of the test by jumping over the rest of the loop body. As an example, consider an array of signed elements (see Section 9.2). The following code totalizes only array members that are positive:

```
     sum = 0, x = 0;
     while (x < MAX)
         }
         if (array[x] < 0)   {continue;}
         sum += array[x++];
         }
```

Sometimes it is necessary to go through the loop body *first* before testing for exit. This ensures that at least one pass will be performed irrespective of the outcome of the test. The structure of this **do-while** (repeat-until) loop is shown in Fig. 8.8(b). A break can be used in a similar way as in while, and continue causes a drop down to the test (rather than upwards). The do-while loop is the least used of the three kinds of C loop constructions.

The most versatile of the three is the **for loop**. This is similar to while but combines the initialization, test and loop variable update as its arguments; thus:

```
for(expr1; expr2; expr3)
    {loop body}
```

causes expression 1 to be evaluated *once* at the beginning. Expression 2 is tested next, and while True enters the loop body. The third expression is evaluated *after* each loop iteration. Thus normally, but not exclusively, expression 1 is used to initialize variables, expression 2 for the while test and expression 3 to change the tested expression. The while equivalent is:

```
expr1;
while(expr2)
    {
    loop body
    expr3;
    }
```

As an example, lines 9 – 14 of Table 8.1 can be replaced by:

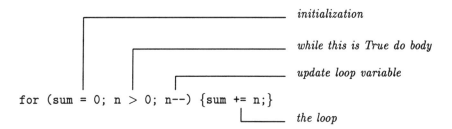

See also Table 10.14.

Expressions can be compounded using the concatenate , operator; so several variables can be initialized together, and fairly complex processing can be done directly in expression 3. The body of the loop must be enclosed by braces, unless it is a single statement, and there *must* be a body; for example:

```
for(sum = 0; n>0; sum += n--)   {;}
```

has a null statement as its body (braces are optional).

Any of the three expressions may be omitted, but the semicolons must stay. An omitted expr2 always returns a True result giving an endless loop. This may be deliberate, for instance traffic lights must be controlled continuously irrespective. In this case we could have the structure:

```
main ()
    {
    for (;;)                    /* Forever do */
        {Control traffic lights}
    }
```

Similarly while(1) can be used to delineate a continuous loop, as the expression is always True. See line 7 of Table 9.5(b).

break as usual causes an immediate exit from the loop, as in line C6 of Table 8.11. continue passes control to expr3, which usually updates the loop variable before returning to the test, see Fig. 8.8(c).

Table 8.11: Generating factorials using a for loop.

```
unsigned long factor(int n)
{
unsigned long factorial;
for(factorial=1; n>1; n--)
   {
   if(n>12)   {factorial=0; break;}
   factorial*=n;
   }
return(factorial);
}
```

(a) Source code.

```
*    1  unsigned long factor(int n)
*    2  {
         .text
         .even
_factor:
         link    a6,#-4          * Open frame
*    3  unsigned long factorial;
*    4  for(factorial=1; n>1; n--)
         moveq.l  #1,d0
         move.l   d0,(-4,a6)     * factorial = -1 (at A6-4:-3:-2:-1)
L1:      cmpi.l   #1,(8,a6)      * n (at A6+8:9:A:B) > 1?
         ble.s    L11            * IF not THEN terminate
*    5     {
*    6     if(n>12)       {factorial=0; break;}
         cmpi.l   #12,(8,a6)     * n > 12?
         ble.s    L14            * IF yes THEN go on
         clr.l    (-4,a6)        * ELSE return 0 as an error marker
         bra.s    L11            * and break
*    7     factorial*=n;
L14:     move.l   (-4,a6),d7     * Get factorial
         mulu.l   (8,a6),d7      * Long (32x32) multiply by n
         move.l   d7,(-4,a6)     * and return it
*    8     }
         subq.l   #1,(8,a6)      * n--
         bra.s    L1             * Repeat
*    9  return(factorial);
L11:     move.l   (-4,a6),d7     * Return factorial in D7.L
         unlk     a6             * Close frame
         rts                     * and terminate function
         .globl   _factor
*   10 }
```

(b) Resulting 68020 MPU assembler code (64 bytes) with annotated comments.

References

[1] Richards, M.; BCPL : A Tool for Compiler Writing and Systems Programming, *Proc. AFIPS SJCC*, **34**, 1969, pp. 557–566.

[2] Richards, M.; The Typeless Survivor, *.EXE* (UK), **6**, no. 6, Nov. 1991, pp. 74–81.

[3] Ritchie, D.M. and Thompson, K.; The UNIX Time-Sharing Systems, *Bell Systems Technical Journal*, **57**, no. 6, part 2, pp. 1905–1929.

[4] Johnson, S.C. and Kernighan, B.W.; The Programming Language B, *Comp. Sci. Tech. Ref.*, no. 8, Bell Laboratories, Jan. 1973.

[5] Ritchie, D.M. et al.; The C Programming Language, *Bell System Technical Journal*, **57**, no. 6, part 2, July/Aug. 1978, pp. 1991–2019.

[6] Collinson, P.; What Dennis Ritchie Says; Part 1, *.EXE* (UK), **5**, no. 8, Feb. 1991, pp. 14–18.

[7] Kernighan, B.W. and Ritchie, D.M.; *The C Programming Language*, Prentice-Hall, 1978.

[8] Kernighan, B.W. and Ritchie, D.M.; *The C Programming Language*, 2nd. ed., Prentice-Hall, 1988.

[9] Banahan, M.; *The C Book*, Addision-Wesley, 1988.

[10] Kelly, A. and Pohl, I.; *A Book on C*, Benjamin Cummings Publishing Co., 2nd. ed., 1989.

[11] Gardner, J.; *From C to C*, Harcourt Brace Jovanovich/Academic Press, 1989.

[12] *Standard for Binary Floating-Point Arithmetic*, ANSI/IEEE Standard 754-1985; IEEE Service Center, Publications Sales Dept., 445 Hoes Lane, POB 1331, Piscataway, NJ 08855-1331, USA.

[13] Jaeschke, R.; The Proposed ANSI C Language Standard, *Programmer's Journal*, **5**, part 4, pp. 38–40, 1987.

[14] Kernighan, B.W. and Ritchie, D.M.; *The C Programming Language*, 2nd. ed., Prentice-Hall, 1988, Section 8.3.

[15] Koenig, A.; *C Traps and Pitfalls*, Addison-Wesley, 1988, Section 1.1

[16] Koenig, A.; *C Traps and Pitfalls*, Addison-Wesley, 1988, Section 2.4

[17] Dijkstra, E.W.; Goto Statement Considered Harmful, Letters to the Editor, *Communications of the ACM*, March 1968, pp. 147–148.

CHAPTER 9

More Naked C

Here we discuss functions as the building block of C programs, data structures of various kinds, libraries and headers. In keeping with our definition of naked C in Chapter 8, we continue to studiously ignore hosted input/output and file handing operations.

9.1 Functions

The **function** in C is the direct equivalent to the subroutine at assembly level, and is directly translated as such. A function encapsulates an idea or algorithm into a named structure. It can be used in an expression as a normal variable, by naming it together with any parameters that are being passed. All functions, except **void**, return a value as defined by the **return** instruction. This is the value that is substituted for the function in the calling expression. For example, if we have the function defined in Table 8.11, then the code fragment:

```
x = 4;
y = 1/factor(x);
```

will make y the reciprocal of 4!, that is $\frac{1}{24}$. The value of factor(4) is of course 24, as returned in line 9 of that table.

In this section, we specifically need to look at how functions are declared and defined, how parameters are passed back and forth, and the scope of objects declared inside and outside functions.

C programs are structured as a collection of external objects. These objects are mainly global variables and functions. This is graphically shown in a much simplified form in Fig. 8.1. The main function, conventionally called main(), acts as a central spine calling up the various ancillary functions in the appropriate order, usually with a minimum of processing itself. In a hosted environment, main() interacts with the operating system, from which it can obtain and sometimes return information. In a naked environment it is normally entered via an assembly-level startup routine, and frequently runs forever in an endless loop. More details are given in Section 10.1.

Although main() is regarded as a little special in C, in reality the compiler treats it in the same fashion as any other function. The layout of Fig. 9.1 shows this, with

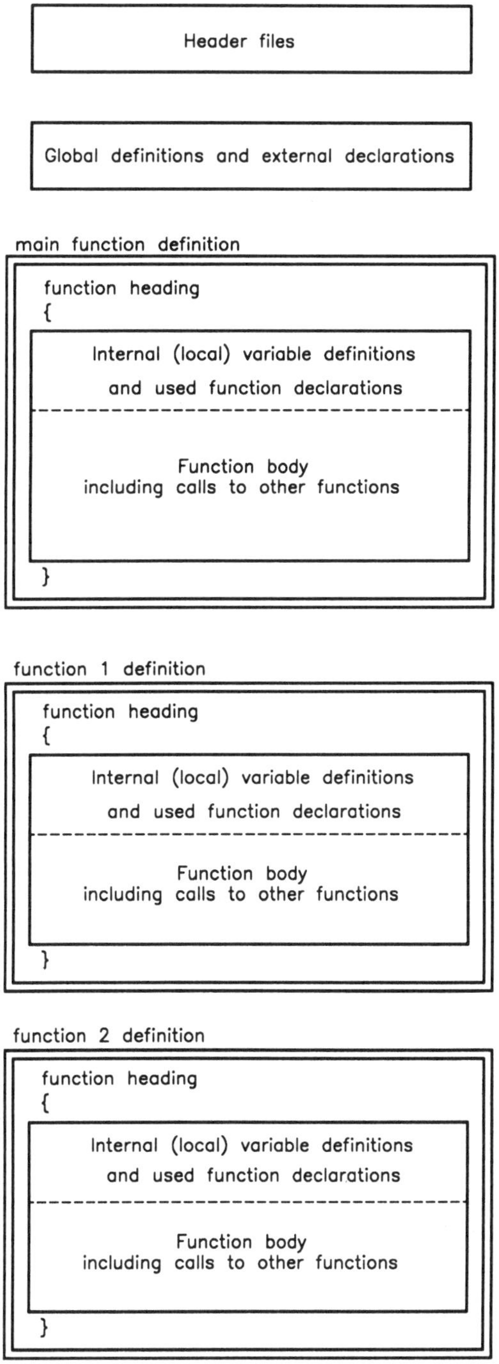

Functions

```
┌─────────────────────┐
│    Header files     │
└─────────────────────┘

┌──────────────────────────────────────────────┐
│ Global definitions and external declarations │
└──────────────────────────────────────────────┘
```

main function definition
```
┌────────────────────────────────────────────────┐
│ function heading                               │
│ {                                              │
│   ┌──────────────────────────────────────────┐ │
│   │  Internal (local) variable definitions   │ │
│   │     and used function declarations       │ │
│   │ - - - - - - - - - - - - - - - - - - - -  │ │
│   │              Function body               │ │
│   │     including calls to other functions   │ │
│   │                                          │ │
│   └──────────────────────────────────────────┘ │
│ }                                              │
└────────────────────────────────────────────────┘
```

function 1 definition
```
┌────────────────────────────────────────────────┐
│ function heading                               │
│ {                                              │
│   ┌──────────────────────────────────────────┐ │
│   │  Internal (local) variable definitions   │ │
│   │     and used function declarations       │ │
│   │ - - - - - - - - - - - - - - - - - - - -  │ │
│   │              Function body               │ │
│   │     including calls to other functions   │ │
│   └──────────────────────────────────────────┘ │
│ }                                              │
└────────────────────────────────────────────────┘
```

function 2 definition
```
┌────────────────────────────────────────────────┐
│ function heading                               │
│ {                                              │
│   ┌──────────────────────────────────────────┐ │
│   │  Internal (local) variable definitions   │ │
│   │     and used function declarations       │ │
│   │ - - - - - - - - - - - - - - - - - - - -  │ │
│   │              Function body               │ │
│   │     including calls to other functions   │ │
│   └──────────────────────────────────────────┘ │
│ }                                              │
└────────────────────────────────────────────────┘
```

Figure 9.1: Layout of C programs.

main() being one of three functions in the figure. Each of these functions must be defined. A **function definition**, typical examples of which are shown in Tables 8.1 and 8.6 – 8.11, consists of a prototype heading followed by local variable definitions and any called function declarations, and then by the body of the function. This **body** is a series of any legal C statements enclosed in braces. Unusually, these braces must be present even if the body comprises a single statement; for example:

```
{
return (x*x);
}
```

is a legitimate function body (squaring x, which has been passed to the function). The `return` instruction is the mechanism whereby the function is assigned a value, as seen by the caller. If `return` is omitted, the function will still exit back to the caller (i.e. there will be an RTS or equivalent at the end of the subroutine), but the value seen by the caller will be undefined. Functions which do not return a value, that is are `void` (e.g. Table 8.6), can either omit this statement or as a matter of style include a null `return;`. Parentheses are optional around `return`'s expression, but are frequently used for clarity.

The **function prototype** at the head of the body simply names and indicates the type of the function (i.e. its return value) and the types of any parameters passed to the function. Doing this, we have for our squaring function definition:

```
int square(signed char x)      /* Prototype head */
    { return (x*x); }          /* Body           */
```

The prototype declares the name of the function as `square()`, returning an `int` value and accepting a `signed char` variable, here named x. In the body of the function, the formal parameter x behaves as an `auto signed char` variable. It is local to this function, that is it is unknown outside. Note that in this case we can return an `int`, as we know that x will be promoted to `int` type for the calculation, see Fig. 8.4. If the type of the expression returned is not that indicated in the prototype, it will be converted using the normal C rules.

At the assembly level, `return` is normally implemented by evaluating the expression and putting it in a register prior to RTS. Thus in the line labelled L11: of Table 8.11, `factorial` is placed in **Data register_D7**. In line L11: of Table 7.14, sum is returned in **Accumulator_D**. The **AX** register is normally used for 80x86 family returns (see Table 10.14(b)) Registers may be concatenated for return types larger than a single register capacity, for example **X:D** in 6809 implementations for `long` or `float` returns.

Unlike some languages, such as Pascal, function definitions are not allowed inside other functions; that is, each definition must be self-standing, as shown in Fig. 9.1 and Table 9.1. Any function can of course be called up from any other function (or even from itself for recursive operations). When a function is going to be used, it needs to be declared in a similar manner to any other object.

As our example for this section, consider a function that will return the integral power of a signed integral variable, that is y^{exp}. This is shorthand for a repetitive multiplication of 1 by **y**, **exp** times, which covers the case where **exp** = 0. Thus, for example, $2^3 = 1 \times 2 \times 2 \times 2$.

The software implementation of Table 9.1(a) uses this algorithm, but recognizes that overflow will occur for certain combinations of **y** and **exp**, and returns 0 for

Table 9.1: The C program as a collection of functions (*continued next page*).

```
main()
    {
    signed char n;                      /* Define variable n          */
    unsigned char x;                    /* Define variable x          */
    register int p;                     /* Define variable p          */
    int power(signed char y, unsigned char exp); /* declare power()   */
    n=25;   x=3;
    p=power(n,x);                       /* p = 25^{3}                 */
    }
/* Here follows the definition of power()                             */
int power(signed char y, unsigned char exp) /* Generates y^{exp}      */
    {
    int result, old_result, abs(int);
    for(result=1; exp>0; exp--)
        {
        old_result = result;
        result*=y;                      /* Repetitive multiplication by y */
        if(abs(result)<=abs(old_result)) {return 0;} /* Overflow error */
        }
    return result;
    }
/* Here follows the definition of abs()                               */
int abs(int z)
    {return (z>=0 ? z:-z);}
```

(a) C source code.

```
*       1   main()
*       2       {
1           .text
2           .even
3  _main:   link    a6,#-2
4           movem.l d5/d0,-(sp)
*       3       signed char n;              /* Define variable n       */
*       4       unsigned char x;            /* Define variable x       */
*       5       register int p;             /* Define variable p       */
*       6       int power(signed char y, unsigned char exp); /*declare power() */
*       7       n=25;   x=3;
5           move.b  #25,(-1,a6)
6           move.b  #3,(-2,a6)
*       8       p=power(n,x);               /* p = 25^{3}              */
7           moveq.l #0,d5
8           move.b  (-2,a6),d5       ; Copy x from [A6]-2, extended to int
9           move.l  d5,(sp)          ; Put on Stack
10          move.b  (-1,a6),d5       ; Copy n from [A6]-1
11          extb.l  d5               ; Sign extended to int
12          move.l  d5,-(sp)         ; and push out on Stack
13          jsr     _power           ; Go do ftn power(n,x), returning in D7.L
14          addq.l  #4,sp            ; Restore SP from last push
15          move.l  d7,d5            ; p lives in D5.L (register variable)
16          movem.l (sp)+,d5/d0      ; Move it to D7.L
17          unlk    a6               ; and return
18          rts
*       9       }
*      10
```

Table 9.1: (*continued*) The C program as a collection of functions.

```
*      11   /* Here follows the definition of power() */
*      12
*      13   int power(signed char y, unsigned char exp) /* Generates y^{exp} */
*      14   {
19          .even
20  _power: link    a6,#-16
*      15      int result, old_result, abs(int);
*      16      for(result=1; exp>0; exp--)
21          moveq.l  #1,d0
22          move.l   d0,(-4,a6)        ; result (living in [A6]-4) = 1
23  L1:     tst.b    (15,a6)           ; exp (living in [A6]+15) > 0?
24          beq.s    L11               ; Break if True
*      17      {
*      18         old_result = result;
25          move.l   (-4,a6),(-8,a6)   ; old_result at [A6]-8 = result
*      19         result*=y;               /* Repetitive multiplication by y */
26          move.l   (-4,a6),d7        ; Get result
27          move.b   (11,a6),d6        ; and y as passed by caller in [A6]+11
28          extb.l   d6                ; 68020 style sign extension byte to int
29          mulu.l   d6,d7             ; 68020 type 32x32 multiplication
30          move.l   d7,(-4,a6)        ; put away as new result
*      20         if(abs(result)<=abs(old_result)) {return 0;}/* Overflow error */
31          move.l   (-8,a6),(sp)      ; Copy old_result and put into stack
32          jsr      _abs              ; Get absolute value back in D7.L
33          move.l   d7,(-12,a6)       ; Put away for safekeeping in [A6]-12
34          move.l   (-4,a6),(sp)      ; Repeat for result
35          jsr      _abs
36          cmp.l    (-12,a6),d7       ; Compare abs(result) with abs(old_result)
37          bgt.s    L12               ; IF greater THEN continue
38          moveq.l  #0,d7             ; ELSE return a zero to indicate error
39          unlk     a6
40          rts
41  L12:    subq.b   #1,(15,a6)        ; Decrement exp char
42          bra.s    L1                ; and repeat for loop
*      21      }
*      22      return result;
43  L11:    move.l   (-4,a6),d7        ; Exit with power in D7.L
44          unlk     a6
45          rts
*      23   }
*      24
*      25   /* Here follows the definition of abs() */
*      26
*      27   int abs(int z)
*      28      {return (z>=0 ? z:-z);}
46          .even
47  _abs:   link     a6,#0
48          tst.l    (8,a6)            ; Test z passed in [A6]+8
49          blt.s    L01               ; IF less, then prepare to negate
50          move.l   (8,a6),d7         ; ELSE exit with z unchanged
51          bra.s    L21
52  L01:    move.l   (8,a6),d7         ; Get z
53          neg.l    d7                ; negate it
54  L21:    unlk     a6
55          rts                        ; Return with abs(z) in D7.L
56          .globl   _main
57          .globl   _abs
58          .globl   _power
```

(b) Resulting 68020 MPU assembly code (line numbers added for clarity).

this situation. This is determined when the result of the kth multiplication is less than the $(k-1)$th. However, as y can be signed, it is the modulus of these results that must be compared, rather than their actual value.

The program is structured as three functions. The obligatory main() is for demonstration only, and terminates with the value of 25^3. Of interest to us is the declaration in line C6 of the function power(). This declaration is in prototype form, where the function return type (int here) is followed by its name and the type of the two objects to be passed (a signed char and unsigned char). For clarity the formal parameters y and exp are used in the declaration. This is optional, and the declaration:

```
int power(signed char, unsigned char);
```

is acceptable.

This line is a **declaration**, unlike lines C3–5 where variables are **defined**. The difference is that a definition gives the properties of the referenced object as well as reserving storage. On the other hand a declaration only makes known the object's properties; no storage space is assigned.

A declaration function prototype is not mandatory, and the alternative:

```
int power();
```

is accepted by the compiler without complaint. However, without a prototype declaration, the compiler will not be able to check that the writer has sent the correct number and types of variables to the function. This was an endless source of error in old C, where prototyping was not featured. Indeed out of deference to old C, the compiler will accept no function declaration at all, and will assume an int type return as default. Not recommended! I have given main() an old style name-only header rather than the prototype void main(void), as a function not returning any type and not receiving any parameters. Table 8.6 is another example of the use of void.

Function power() is defined in lines C12–23 in the normal way, with a prototype head and body in braces. The formal parameters y and exp are known only down to the closing brace and so their names can be reused by any other function, although for clarity it is better not to reuse parameter names. The actual (as opposed to formal) parameters sent are of course n and x.

Function power() makes use of function abs() twice, so this is declared as an object known to it in line C15, together with other variable definitions. The definition of abs() is in lines C27 and C28 (see the ?: operator on page 224).

Any number of **parameters** may be passed to a function, although only one can be returned (a function can have only one value at a time!). Parameters are passed by value (i.e. copied) in a manner which is implementation dependent, but normally on the System stack, as illustrated in Fig. 5.4 and described in Section 5.2. The mechanism can clearly be seen by inspecting the assembly code of Table 9.1(b). Each function has its own private frame, in which its local (i.e. auto) variables are stored. For main() this frame is two bytes deep (line 3) and holds x at the bottom and n at the top. To send the value of x to power() it is first copied into **D5** and put on the System stack (lines 7–9) and the process repeated for n (lines 10–12). Finally a JSR is made to _power: in the normal way to transfer to the subroutine.

Several implementation-specific properties of this compiler (Whitesmiths V3.2 68020 C cross-compiler) cloud this issue. Firstly, all functions identified as name() commence at _name: at assembly level. This compiler passes chars (and shorts) promoted to ints, although they are treated according to their proper type in the function (lines 27 and 41). This is probably to cope with old-style non-prototype declarations, where parameters are promoted to int or double [1]. Finally, and rather obscurely, the compiler always puts the *first* variable passed (the rightmost) away using the plain Address Register Indirect address mode, whereas further variables (moving leftward) use the Address Register Indirect with Pre-Decrement mode for a proper Push action. Thus we have:

```
9    MOVE.L   D5,(SP)     ; A straight move to [A7]
12   MOVE.L   D5,-(SP)    ; A proper Push to   -[A7]
```

Why is this? Well, the former is quicker, especially as the System stack does not have to be restored to its original value (cleaned up) on return (e.g. line 14) to compensate for its decrementation. This is useful, as the majority of function calls only pass a single parameter, but of course whatever was on the System stack before will be overwritten. This compiler gets around this either by having created a frame which is overly large (see Fig. 9.2) or else when registers were saved on the System stack after the frame was opened (line 4) a sacrificial register was put away. In either case the stack content is irrelevant at the time when the parameters are sent out, and can be overwritten. This compiler specifies that registers **D3, D4, D5, A3, A4, A5, A6** are not to be altered on return from a function. Thus, as main() uses **D5**, it is saved in line 4 together with **D0**, whose value on return is unspecified; that is the sacrificial register.

Compiler-specific details like these are irrelevant at the higher level. However, they are important when mixing C and assembly-level subroutines, as described in Section 10.1, and when debugging, see Chapter 15.

Some compilers pass one or more variables in a register. For example the Cosmic/Intermetrics V3.3 6809 C cross-compiler will normally put the first copied value in Accumulator_D if char, short or int, and pass copies of any subsequent variables through the System stack in the normal way. In either case the objects are always known only to the called function, living either in the local frame or register set. Fig. 9.2 illustrates the frame as seen by power(). Notice how the passed variables are referenced *above* the local Frame Pointer **A6**, that is y at A6'+11 and exp at A6'+15, whereas the internal variables are *below* at A6'-8 for result and A6'-12 for old_result. Notice the sacrificial long word at the bottom of the stack, which is overwritten in lines 31 and 34 by the single parameter passed to abs().

The concept of **scope** as the lifetime of an object was introduced in Section 8.2. Let us look at this in more detail. Objects defined *inside* braces are **local** to that multiple statement only. Thus, in the code fragment:

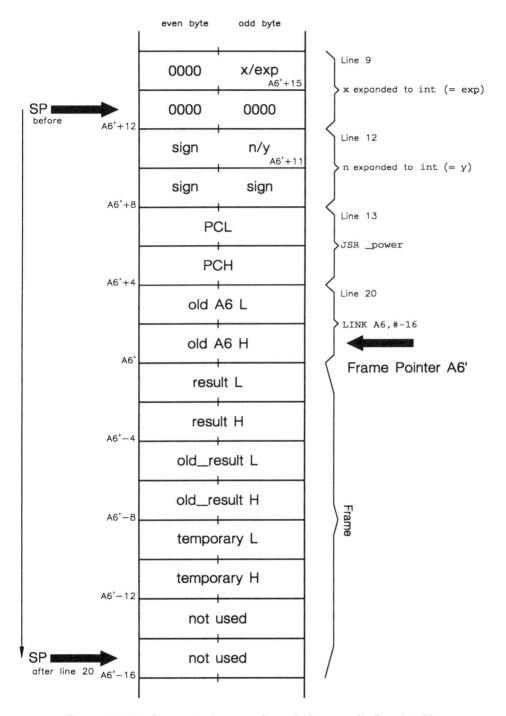

Figure 9.2: The System stack as seen from within power(), lines 21–38.

```
function()
{
int i;
    {
    do lots of things with i;
    }

    {
    int i;
    do more things with a different i;
    }
    etc
}
```

The two is are different, and the lower i is known only down to its local }. Outside this redeclared area the first i is known.

Generally, variables are declared at the opening function brace and disappear from view at the closing brace. When the function is re-entered, static variables (placed in absolute memory) will have kept their last value, whereas auto variables (assigned space in a stack-based frame) have lost theirs.

Unless a function is declared static [2], its identifier is broadcast as an external object; that is it is declared public or global at assembly level, and as such is known to the linker. In lines 56–58 of Table 9.1(b), the three labels _main, _abs and _power are declared .GLOBL (similar to .PUBLIC), as would be expected. This means that any other file which has been separately compiled for future linking can use, say, function power(), as the assembly line JSR _POWER will be recognized by the linker (see Section 7.2). However, the declaration:

`extern int power(signed char y, unsigned char exp);`

must appear in this separate file either before the first function (in which case its scope is the entire file) or else in functions which call it. This tells the compiler that the function power() will be found elsewhere through the linker. The **extern** qualifier in C will generate a .EXTERNAL or XREF directive at assembly level, for example:

`.external   _power`

Variables can also be made publicly known. If an object is defined outside a function, then it is globally recognized. For example, the variables Seconds, Minutes and Hours in Table 8.6 are not only known to clock() but to any function defined afterwards. Usually global variables are defined at the head of the file, and so are known to main() and everything else. Table 9.5(b) shows the resulting assembly code for an external object (Array[] in line C1). By convention, such variables are identified by a leading capital letter. Where they are used by separately compiled files through the linker, they must be announced by using the keyword **extern**, for example:

```
extern unsigned char Seconds, Minutes, Hours;
```

Such a statement is considered a declaration — as no storage is granted — not a definition.

Like `static` variables, global (known as `extern`) variables are allocated absolute memory locations (see Table 9.5(b)). They can be initialized where they are defined and behave in the same way; that is, the initial values are given at load time. If no initialization value is given, C specifies an implicit zero value.

Rather confusingly, a variable declared outside a function can be qualified by the `static` keyword. If this is done, the variable is known from that point on only *within* the file in which it appears. It will not be passed through the symbol table to the linker as a global object. `static` externally defined objects are stored and initialized in the same way as ordinary (i.e. across files) defined objects.

In review, it is important to distinguish between a `static` variable declared inside and one declared outside a function. They are both stored in absolute memory (i.e. not in a frame) and are both initialized in the assembler by using data storage directives (e.g. .BYTE, .DS etc). The former is only locally known within its function. The latter is known throughout its file (if declared at the top) but not beyond it. Leaving out the `static` qualifier on an externally defined variable broadcasts its name to all files through the linker. However, to use such a variable in an outside file its name and properties must be declared in such outside files qualified by the `extern` keyword.

Functions too are globally known from their declaration onwards and to external files. However, qualifying a function definition by the keyword `static` restricts its scope to its local file [2]. Thus replacing line C13 in Table 9.1 by:

```
static int power(signed char y, unsigned char exp)
```

will not generate assembler line 58, that is the identifier _power is not broadcast as public.

9.2 Arrays and Pointers

As far as C is concerned, an **array** is a set of objects of the same type. Although it does provide the specific array operator [], no special array-oriented procedures are supported.

Arrays must be defined in the same way as all other C objects; some examples are:

```
static unsigned long arr[1024];
auto int fred[256];
const static unsigned char table_7[10];
```

The first defines an array named arr[], comprising 1024 consecutive long-words in absolute memory. At assembly level the reservation will be labelled something like `_arr: .double[1024]`. Thus `arr` is actually the address of the first element of the array (e.g. see Table 9.5(b), _Array:).

The second definition reserves 256 units in the frame. Although these locations are in relative memory, the root name `fred` in the C source still refers to the address of element `fred[0]`.

The final statement defines ten consecutive bytes of constants, beginning at `_table_7:`, which are in absolute memory, probably destined for ROM. In practice, this is useless as it stands as an initial value must be given as part of its definition, otherwise it will be filled with zeros by the compiler — in the normal way for `static` objects — which will generate an assembly-level line something like:

```
_table_7:    .byte 0,0,0,0,0,0,0,0,0,0
```

As the array is `const`, no subsequent change can be made to any element.

The size of an array must be determinable by the compiler, which in practice means the use of a constant dimension specifier or its equivalent during its definition. Incidentally the `sizeof` operator works with arrays, and indeed any C object. Thus `sizeof(fred)` will yield 1000 for a 4-byte `int` implementation.

An array can hold any type of object, and can have any of their attributes. However, it is unlikely that the compiler will pay any attention to a `register` qualifier. Each element will have the characteristics expected of it according to its type, including initialization properties. Thus an `auto` array is initialized at run time on each entry to its local sphere of influence, whilst `static` and global arrays are set up once and for all at load time, otherwise are zero.

Some definitions of initialized arrays are:

```
int factor[5] = {1,1,2,6,24};
static unsigned const char square[] = {0,1,4,9,16,25,36,49,64,81,100};
```

In the latter case the dimension of the array was not given, the compiler taking it as the number of initializers (i.e. eleven); and array elements `square[0]` to `square[10]` will have the values shown. The dimension n, specified either explicitly or implicitly in a definition, is the number of elements. However, as element 0 is the first, the final element is $n-1$. Thus the first example really means `factor[0] = 0`, `factor[1] = 1`, `factor[2] = 4`, `factor[3] = 9`, and `factor[4] = 16`. There is *no* `factor[5]`! The compiler will not warn you of errors like this, and using undefined elements is a fruitful source of obscure errors. If the explicit array size parameter is greater than the number of initializers, all unspecified elements are assumed to be zero, unless an `auto` storage-class array. Old C did not permit `auto` array initialization, but this is not true of ANSII C.

At any time an element m of an array can be referred to by following the root name by `[m]`. The resulting object can then be treated like any other C object of the same type. As an example, consider the following code fragment which applies the 3-point low-pass filter transformation, defined on page 112, to an array of 256 elements:

```
for (i = 255, i > 1, i--)
    {array[i] = array[i]/2 + array[i-1]/4 + array[i-2]/2;}
```

Thus $\text{array}[255] = \frac{1}{2}\text{array}[255] + \frac{1}{4}\text{array}[254] + \frac{1}{2}\text{array}[253]$ etc.

A **look-up table** is a synonym for an array, usually, but not always, of constants. Table 9.2 shows this technique used to generate our old friend, the factorial. Here an array of 13 elements hold the equivalents of 0! to 12!. The array, or table, is declared in line C3 as `static` (i.e. stored in absolute memory) and `const`, together with the appropriate values. The `const` qualifier forces the compiler to place these values in the text program section along with the program, both of which will be in ROM in an embedded implementation. This storage is simply thirteen sequential long-words beginning at L5_array:, in the same manner as in Fig. 9.3(b).

The resulting assembly-level program itself uses the array index n multiplied by four (to match the element size) to give the offset from the base address. Putting this in an Address register (**A1**) and adding the base address L5_array to it gives the position of `array[n]`. This is then transferred to **D7.L** for return.

Multi-dimensional arrays can be implemented in C, although their use is rare. Some example definitions are:

```
unsigned char calendar[12][31];    /* 12 months of 31 days each    */
int x[100][12];                    /* 100 rows of 12 columns       */
char x[3][2] = {7,6},{9,13},{0,5}; /* 3 rows of 2 columns
    x[0][0]=7, x[0][1]=6, x[1][0]=9, x[1][1]=13, x[2][0]=0, x[2][1]=5 */
```

Higher-order arrays are treated as arrays of array objects. In this manner `calendar[12][31]` can be considered as 12 objects (months), each of which comprises an array of 31 days holding `unsigned char` objects. Noting from Table 8.4 that [] associates left to right, the 2-dimensional array `x[3][2]` can be written as `(x[3])[2]`, that is an array called `x` of three objects, each of which contain two elements. Fig. 9.3(b) shows the memory organization of a 2-dimensional array. As can be seen, the rightmost subscript varies fastest as we go up in memory, with one complete pass for an increment of the next left index. In accessing any element [m][n], the relationship $(m \times COL \times w) + n$ must be calculated, where m is the row, COL is the number of columns, w is the element size in bytes and n is the column co-ordinate. Adding this offset to the base address of `array[0][0]` gives the element address. The principle applies for any number of dimensions.

We have seen in Section 9.1 that variables are normally passed to functions by copying their value into a stack prior to the call. In general it is feasible to copy an entire array through a stack, but this technique is not used in C because of the overheads in time and space. For example, the array `x[100][12]` would need 1200 Copy and Push actions prior to the call, which is hardly efficient. Instead the *address* of the base element is passed through the stack. With access to this pointer, the function can reference any element as described. Careful consideration of this shows that passing the address permits the function to change the actual array elements and not copies, as is the case where simple objects are involved.

As an example of a function acting on arrays of data, consider a block copy operation where a number of elements from [0] to [length-1] are to be copied from one array to another. The calling program must pass three parameters: the

Table 9.2: Generating factorials using a look-up table.

```
unsigned long factor(int n)
    {
    static unsigned const long array[13] =
    {1,1,2,6,24,120,720,5040,40320,362880,368800,39916800,479001600};
    if(n>12)   {return 0;}
    return(array[n]);
    }
```

(a) C source code.

```
          .text
          .even
L5_array: .long    1               ; array[0]
          .long    1               ; array[1] etc.
          .long    2
          .long    6
          .long    24
          .long    120
          .long    720
          .long    5040
          .long    40320
          .long    362880
          .long    368800
          .long    39916800
          .long    479001600       ; array[12]
*    1 unsigned long factor(int n)
*    2 {
          .even
_factor:  link     a6,#-4
*    3    static unsigned const long array[13] =
          {1,1,2,6,24,120,720,5040,40320,362880,368800,39916800,479001600};
*    4    if(n>12)      {return 0;}
          cmpi.l   #12,8(a6)       ; Compare n living in [A6]+8 with 12
          ble.s    L1              ; IF lower or equal THEN continue
          moveq.l  #0,d7           ; ELSE exit with 0 in D7.L
          unlk     a6
          rts
*    5    return(array[n]);
L1:       move.l   8(a6),d7        ; Get n
          asl.l    #2,d7           ; Multiply by 4 to match array element size-long
          move.l   d7,a1           ; and put in A1.l
          adda.l   #L5_array,a1    ; Add it to the address of array[0]
          move.l   (a1),d7         ; which then points to array[n]. Get it
          unlk     a6              ; and return
          rts
          .globl   _factor
*    6 }
```

(b) Resulting 68000 assembly code.

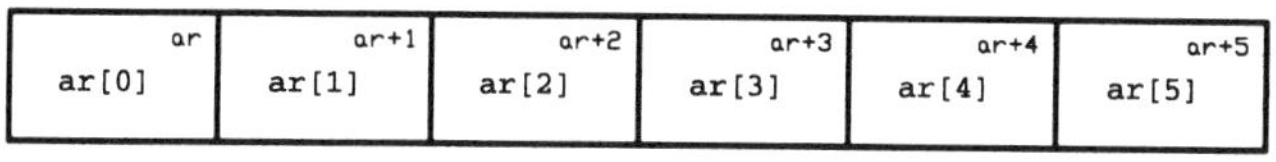

ar ar[0]	ar+1 ar[1]	ar+2 ar[2]	ar+3 ar[3]	ar+4 ar[4]	ar+5 ar[5]

(a) A 1-dimensional array ar[n] of six elements.

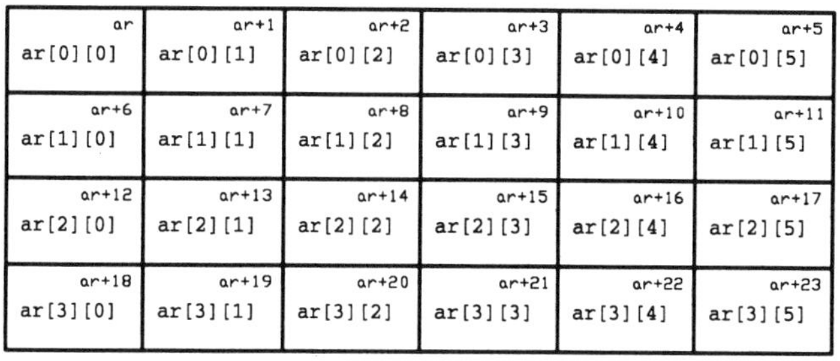

ar ar[0][0]	ar+1 ar[0][1]	ar+2 ar[0][2]	ar+3 ar[0][3]	ar+4 ar[0][4]	ar+5 ar[0][5]
ar+6 ar[1][0]	ar+7 ar[1][1]	ar+8 ar[1][2]	ar+9 ar[1][3]	ar+10 ar[1][4]	ar+11 ar[1][5]
ar+12 ar[2][0]	ar+13 ar[2][1]	ar+14 ar[2][2]	ar+15 ar[2][3]	ar+16 ar[2][4]	ar+17 ar[2][5]
ar+18 ar[3][0]	ar+19 ar[3][1]	ar+20 ar[3][2]	ar+21 ar[3][3]	ar+22 ar[3][4]	ar+23 ar[3][5]

Address of ar[p][q]
= ar+(p*6+q)*w
where w = element size in bytes.

(b) A 2-dimensional array ar[m][n] with 4 rows of 6 columns, 24 elements.

Figure 9.3: Array storage in memory.

base addresses of array1 and array2, and length. As the base address of an array is just its root name, the calling routine will include lines similar to these:

caller()

```
long n;                          /* Length parameter                */
char block_x[256], block_y[256]; /* Two arrays of 256 bytes         */
```

```
void block_copy(char array1[], char array2[], long length);
/* Declaration prototype of function block_copy()                   */
```


```
block_copy(block_x, block_y, n);  /* Call up function block_copy()  */
```

This would result in the first n elements of array block_y[] being physically replaced by the first n elements of array block_x[].

The code itself, reproduced in Table 9.3(b), shows that the address of array2[]'s base was passed in a long-word 12–15 bytes above the Frame Pointer (**A6**) and that of array1[] in locations 8–11 bytes above. Parameter length is passed by copy in [A6]+16/19. Array element i's address is calculated as i (stored in **D5.L**) plus the relevant base address. This is a pity, as the sequential nature of the process would

Table 9.3: Altering an array with a function.

```
void block_copy(char array1[],char array2[],long length)
   {
   register long i;
   for(i=length-1;i>=0;i--)
       {array2[i] = array1[i];}
   return;
   }
```

(a) C source code.

```
*.   1   void block_copy(char array1[],char array2[],long length)
*    2   {
             .text
             .even
     _block_copy:
             link    a6,#0
             move.l  d5,-(sp)
*    3       register long i;
*    4       for(i=length-1;i>=0;i--)
             move.l  16(a6),d5  ; i (in D5.L) equated to length passed in [A6]+16
             subq.l  #1,d5      ; minus one
     L1:     tst.l   d5         ; i>=0?
             blt.s   L11        ; IF not THEN pass on
*    5          {array2[i] = array1[i];}
             movea.l d5,a1      ; i to A1.L
             adda.l  12(a6),a1  ; added to array2's base address passed in [A6]+12
             movea.l d5,a2      ; i to A2.L
             adda.l  8(a6),a2   ; added to array1's base address passed in [A6]+8
             move.b  (a2),(a1)  ; array1[i] moved to array2[i]
             subq.l  #1,d5      ; i--
             bra.s   L1         ; and repeat
     L11:    move.l  (sp)+,d5
             unlk    a6
             rts
             .globl  _block_copy
*    6       return;
*    7   }
```

(b) Resulting 68000 code.

suit the use of the Address Register Indirect with Post-Increment mode to creep up (walk) through the block, such as in lines 23 and 25 of Table 5.7. However, the code size of 40 bytes compares well with that of 39 in the hand-assembled version.

Notice how an array is denoted in the function prototype by the root name and empty square brackets, for instance array1[]. A size is not necessary, although it can be added for clarity if desired. Multi-dimensional arrays must give the size of each dimension, except the leftmost. As we have seen, this is in order to calculate the address of any element. Care must be taken, as the compiler does not check for overrun; thus reference to `array[9]` in an array defined to have four elements is accepted, and the contents of memory location `array + 9 × w` actually fetched or, worse still, changed!

What if we wanted to copy the contents of a ROM into a RAM, as was the case in Table 5.7? The function will be exactly the same, but this time we must pass the start address of the two chips. We have seen that we can determine the address of an array by just referring to its root name; can we extend the principle? The affirmative answer to this leads us to one of C's strengths, the use of **pointers**.

A pointer is a constant or variable object holding information relating to *where* another object is stored. In MPUs with linear addressing techniques, such as most 8-bit devices and the 68000 family, this is just the absolute address. In segmented architectures, as exhibited by the 8086 family, typically near and far pointers exist; the former holding the address within the current segment (usually two bytes) and the latter the segment:address (usually four bytes).

Pointers may be taken of any C object, except a `register` type, by using the **address-of** unary operator `&` (see Table 8.4). For example, if a variable `x` exists, then its address can be assigned to the pointer variable `ptr` thus:

```
ptr = &x;
```

where `ptr` has previously been suitably defined.

Conversely, if we have a pointer, then we can get the object it points to by using the indirection unary operator **contents-of** (`*`). Thus if we have a pointer `ptr` then:

```
y = *ptr;
```

which reads from left to right, `y` is assigned the contents of address `ptr`.

We now have the problem of what value will a pointer have if the pointed-to object is bigger than one byte, say a 4-byte long-word variable or 100-word array? And how would the construction `ptr+1` be interpreted? In assembly/machine code, the address is normally the lowest byte address of the object, for example:

```
MOVE.B   D0,0C000h    ; [C000]                   -> D0(7:0)
MOVE.L   D0,0C000h    ; [C000:C001:C002:C003] -> D0(31:0)
```

and C uses the same convention. Thus, the value of a pointer to the 100 short-element array (`&ar[0]`) stored in memory at 0xC000 — 0xC063 will simply be 0xC000. In the case of an array, we can use the root name as the base address; hence:

```
ptr = &ar[0];
ptr = ar;
```

are the same, and `ptr` will be a pointer to whatever kind of object the array comprises, provided of course that it has been previously defined as such.

Although the storage size of the base address is fixed, and is independent of the object, pointers do take on the type of their referred-to object. Thus, for example, we can have pointer-to-int and pointer-to-float entities. This, and other properties, are bequeathed to the pointer variable during its definition. In common with any other object, all pointer variables must be defined before use. Some examples are:

```
char * port;            /* port is a pointer variable to a char object */
float * pvar1, * pvar2; /* pvar1 & pvar2 are pointers to float objects */
void * point;           /* point is a generic pointer                  */
```

In the first instance, the pointer variable port is brought into being and declared to be addressing a char object. This can be read as 'the contents of port is a char'. Alternatively, the * indirection operator can be transcribed as 'pointer to', if read right to left; that is 'port is pointer-to a char object'. The second example creates two pointer-to float objects, namely pvar1 and pvar2. This reads (from right to left) pvar1 is a pointer-to a float, pvar2 is a pointer-to a float.

The final definition is rather enigmatic, as it appears to be saying that point is a pointer-to nothing! A **pointer to void** type is treated as a generic pointer (pure address) and can be cast to any real type and back without any loss of integrity.

The concept of **different types of pointers** is important when dealing with **pointer arithmetic**. Pointers can be incremented/decremented, added/subtracted with constants and pointers of the *same kind* and compared with pointers of the *same kind*. Consider a pointer pvar having a value at some instant of 0xC000, then:

```
pvar += 4;
```

will have what value? If pvar is a pointer-to char (or void) then 0xC004 will be the answer, but if pvar is a pointer-to long (or float) then 0xC010 is the answer. Thus in pointer arithmetic, constants indicate objects, and are multiplied by their sizes for the purposes of any calculation.

A more sophisticated example of pointer arithmetic is given by the example:

```
for (i=0; i<100; i++)    {*(ar+i) = 0;}
/* Contents of object pointed to by ar+i is cleared; i = 0 to 99, step 1*/
```

which is similar to:

```
for (i=0; i<100; i++)    {ar[i] = 0;}
```

both of which clear an array of 100 elements.

If the compiler is to make sense of the pointer operation ar + i, then it must know what type of object ar points to. In this case it will know from the prior definition of the array ar[]. Thus, if this was long ar[100], then ar + i would actually be calculated as ar + 4 × i.

The root array name can be used as the parameter of the sizeof operator. Thus sizeof(ar) in the example above will return 400 as the storage used by the array. If the sizeof operand were just an ordinary pointer (i.e. not an array root), then the size of the pointer itself would be generated, typically 2 or 4 bytes.

Pointer types can be to any object of any complexity. As an example consider our 2-dimensional array of chars, calendar[12][31]. This has 12 arrays each comprising 31 char elements. We know that calendar is the address of the element calendar[0][0] and that it is a pointer-to char type. Following this argument, what then is calendar[10]? Actually it is the address of November, that is where the 11th array of 31 days begins. What will the compiler make of the statement calendar[10] + 1;? The result of the addition will be the address of December, that is 31 chars on, or calendar[11]. Thus, calendar[10] is a pointer-to array-of-31-chars type! If a pointer variable is to be assigned to such a type, then it must be defined accordingly, for example:

```
char (* month)[31];    /* Declare month as pointer-to-array-of 31 chars */
month = calendar[0] + i;
```

The complex definition of the pointer month reads from inside the parentheses going left and then right: month is a pointer-to / an array of 31 / chars. Parentheses must be used, as [] is of higher precedence than *. Pointer variable month can then participate in pointer arithmetic where the other pointer variables are of the *same* type.

The pointer variable itself may be given properties, such as being const or static (i.e. stored in absolute memory locations, see Table 14.8), for example:

```
static unsigned char * const port;
```

which says that port is a const pointer-to an unsigned char object and is stored statically. Besides giving port its arithmetic properties, the compiler has been told to store it in absolute memory and flag any attempt to change it (typically placed in ROM in an embedded system).

In dealing with the interface to hardware, the software designer must be able to direct data to and from ports at known addresses. Thus in C, we must be able to assign specific addresses to pointers. In ANSII C a pointer can only be assigned a value if their types match; thus the statement:

```
char * port = 0x9000;          /* !!!!!!!!!!!! Incorrect */
```

in an attempt to make port point to 9000h will fail, as 0x9000 is considered an int type constant (old C, however, permitted this cross-type assignment). The way around this problem is to use a cast to convert the int constant to a pointer-to type; thus:

Casts 0x9000 *to a pointer-to-char type*

```
char * port = (char*)0x9000; /* Correct */
```

which now states that port is a pointer-to a char type variable, and its value is 0x9000. The cast reads (right to left) pointer-to-char type.

On this basis, if we wished to call up the function of Table 9.3 to copy a ROM starting from E000h to RAM starting at 2000h of length 1000h bytes, then the caller would include the following:

```
call()                                      /* The caller function    */
----------------------
----------------------
char * const ROM_start = (char*)0xE000;     /* ROM_start is a pointer */
char * const RAM_start = (char*)0x2000;     /* as is RAM_start        */
----------------------
----------------------
block_copy(ROM_start, RAM_start, 0x1000);   /* Invoke the copy function */
----------------------
```

or even

```
call()
----------------------
----------------------
block_copy((char*)0xE000, (char*)0x2000, 0x1000);
----------------------
```

but the former is more readable.

We have already noted that * can be translated as contents-of (read from left to right), and can be used as such to access the contents of any object of which we have an address. Thus the statement:

```
z = *port;    /* Same as z = *(char*)0x9000; - i.e. contents of 9000h */
```

says that z is assigned the contents of the variable pointed to by port. As we have previously (page 257) defined port as a pointer-to a char at 0x9000, then effectively we are assigning z to the byte stored in 9000.

Consider the situation depicted in Fig. 9.4, where one of several byte (char) ports drives 7-segment displays. We need to write a function to interface this port, which will accept a number from 0 to 9, and the address of the destination port. Functions interfacing to hardware input/output are often known as drivers. Calling this driver function outch() (for output character), we would have as its declaration:

```
int outch(unsigned char number, unsigned char * const port);
```

This prototype identifies number as an unsigned char (i.e. byte) and the address variable port as a const (fixed) pointer to an unsigned char object. The function is declared as returning int, as it is proposed to return −1 to indicate an error (defined as n > 9), otherwise 0. Based on this declaration, to display 7 at the digit located at 0x9000, we would use the call:

Casts 0x9000 *to pointer-to-*unsigned char *type*

```
outch(7, (unsigned char*)0x9000);
```

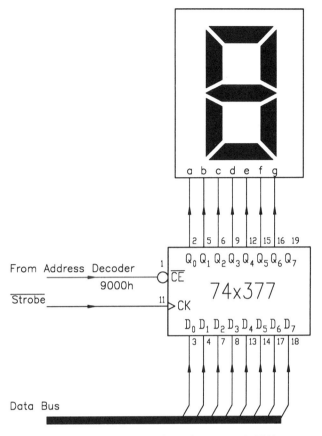

Figure 9.4: A simple write-only port at 0x9000.

Notice the cast (`unsigned char*`), which is necessary to match the constant address to the same pointer-to type as `port`.

The program given in Table 9.4 is straightforward. The code conversion is done by means of a look-up table, as described in Table 9.2. The extracted value is moved to `port` in line C5, by the simple expedient of saying that the contents of `port` are the nth entry of the table. Notice that in line C3 the table is an array of statically stored constant characters (i.e. bytes), and has thus been assigned to the _text program section. In an embedded system this will be in ROM. Objects which are statically stored are considered to have been given their initial values at compile time (i.e. during loading), not run time. At assembly level this appears as .BYTE (or equivalent) directives. The `const` qualifier ensures that an attempt to modify the table values will be flagged as erroneous, which is sensible if the table is in ROM! Thus these initial values are the only values that the table will ever have. The compiler evaluates the size of the array from the number of initializers.

The example of Table 9.4 shows that it is as easy to pass a pointer through to a function as a copy of an object. This can be exploited to change an object itself

Table 9.4: Sending out a digit to a 7-segment port.

```
int outch(unsigned char number, unsigned char * const port)
    {
    static const unsigned char table_7[] =
            {0x20,0x79,0x24,0x30,0x19,0x12,2,0x78,0,0x10};
    if(number>9)    {return -1;}      /* Return error                          */
    *port = table_7[number];          /* or *port = *(table_7+number);         */
    return 0;                         /* Return success                        */
    }
```

(a) C source listing.

```
; Compilateur C pour MC6809 (COSMIC-France)
            .processor m6809
            .psect    _text
L5_table_7: .byte     32,121,36,48,25,18,2,120,0,16 ; 7-segment values
;    1 int outch(unsigned char number, unsigned char * const port)
;    2 {
_outch:     pshs      u               ; Open frame (not needed!)
            leau      ,s
;    3      static const unsigned char table_7[]
;             = {0x20,0x79,0x24,0x30,0x19,0x12,2,0x78,0,0x10};
;    4      if(number>9)    {return -1;}    /* Return error            */
            ldx       4,u             ; Get number passed in [U]+4
            cmpx      #9              ; >9?
            jble      L1              ; IF not THEN continue from L1:
            ldd       #-1             ; ELSE return with -1 in D
            jbr       L4              ; L4 is the exit point
;    5      *port = table_7[number];  /* or *port = *(table_7+number); */
L1:         ldx       #L5_table_7     ; Point X to bottom of table
            ldd       4,u             ; Get number again
            ldb       d,x             ; Get word at number+table_7 bottom ([D]+[X])
            stb       [6,u]           ; Store at address passed in [U]+6; i.e. port
;    6      return 0;                 /* Return success                */
            clra
            clrb
L4:         leas      ,u              ; Close frame
            puls      u,pc
;    7 }
            .public _outch
            .end
```

(b) Resulting assembly code.

in a function, rather than the copy. Just send a reference to the object and not the copy (e.g. &x, not x). The *contents* of that object can then be changed at will; thus (rather trivially) to add ten to an int object x we have:

```
void add_ten(int * pvar)        /* pvar is a pointer to an int object */
    {*pvar += 10;}              /* Get inside it and increment by ten */
```

which is called up as:

`add_ten(&x);`

The variable *pvar (i.e. x) can be manipulated in exactly the same way as any 'ordinary' variable. The Contents-Of operator *, in common with all other unary operators, has a high precedence and thus parentheses need rarely be used; for example:

```
z = *(pvar1)/5 + *(pvar2)*7;
z = *pvar1/5   + *pvar2*7;
```

are equivalent. However, care must be taken when the ++ and -- unary Increment and Decrement operators are used, as these have the same precedence. As unaries read from right to left, we have as an example:

```
x = *pvar++;    /* Increment pointer and take contents of */
x = (*pvar)++;  /* Increment contents of pvar             */
x = ++*pvar     /* Same as above                          */
```

As we have already observed, incrementing a pointer is not the same as incrementing a normal variable. Instead of one being added, a constant equal to the size of the object being pointed to is summed. Thus, incrementing a pointer to a **long** variable will usually add four. In a similar manner, decrementation, addition and subtraction are scaled. Only pointers of the *same* kind can be compared, added or subtracted. As we have seen, this same-type rule also applies to assignments, including constants. Assignments and comparisons with **void** pointers are also permitted.

There is some exception to the assignment rule, as a pointer can always be assigned to or compared with 0. ANSII C guarantees that no object ever lives at 0, thus a function returning a pointer can use this to indicate an error situation. Care needs to be taken in processors that can physically use address $0000h$, such as the 6809, to avoid storing any variable there.

To declare a function returning a pointer, we use the declaration syntax:

`int * fred(parameter list);   /* fred() returns a pointer to int    */`

Pointers to a function (i.e. where the function begins) are also possible (see also Section 10.1), thus:

`int (*fred)(parameter list); /*fred is a pointer to the function fred()*/`

Hence, it is possible to store a table of pointers to functions (see page 286), frequently seen in assembly-level programs as jump tables [3]. Pointers to pointers can be defined ad infinitum, if rarely used.

We introduced pointers by noting that arrays were handled using addresses, especially when being passed to functions. Thus, by inference it is possible to use pointer rather than array notation in such functions. We did this as one way of clearing a 100-element array. Another example is given in lines C3 and C5 of Table 9.4, which can be replaced by:

```
unsigned const char * table_7 = {0x20, ......, 0x10}; /* 7-segment table  */
*port = *(table_7 + n);                               /* Send out nth entry */
```

Pointer notation is of course more comprehensive than array notation, and as such is more flexible. Most texts state that the use of the former will often lead to better code production by the compiler. However, it is the author's opinion that with modern compilers this is rarely so; thus whichever notation is clearest should be used. For example, in the case of a memory copy, the prototype of Table 9.3 (line C1) would be more obviously presented as:

```
void block_copy(char * ROM_start, char * RAM_start, long length)
```

whereas array notation is more relevant to Table 9.4.

As our final example, we will repeat the array scan and update procedure of Table 6.2, this time coded in C. Algorithm and hardware details are given on page 154. Object identifiers have been kept the same to facilitate a comparison between the hand-assembled and C versions.

The array of 256 `short ints` is defined outside a function to give it global status, as can be seen by the `.GLOBL _Array` directive at the bottom of Table 9.5(b). Conventionally, such objects are identified by an initial capital letter.

In the `display()` function, the objects `dac_x` and `dac_y` are declared as constant pointers to a `char` (byte) and `short` (word) respectively, and given the value 0x6000 and 0x6002 respectively. An endless loop then sends out the values `Array[x_coord]` and `x_coord` to the Y and X digital to analog converters. We rely on `x_coord` being a byte-sized (`unsigned char`) object and folding over after 255 (FFh).

The `update()` function is entered via an interrupt (see Table 10.2) and resets the external interrupt flag (see Fig. 6.6) before reading the counter. Both `counter` and `int_flag` are declared constant pointers to their respective addresses/object types in lines C20 and C21. The variables `last_time` and `update_i`, respectively holding the counter reading from the last interrupt and the array index to put the new difference in, are defined as `static`, so that they can remember their previous value. Notice how they are assigned the fixed locations L3_last_time and L31_update_i in the .data program section. Although they will reside in RAM just before the 256 array words, they have not been declared global.

Updating the array element simply means taking the contents of `counter` (see Fig. 6.1), that is `*counter`, and subtracting the last reading. This difference (beat-to-beat variation) is assigned to `Array[update_i++]` in line C23, which also increments the array element. As in `x_coord`, use is made of the byte nature of `update_i` to provide wraparound at 256.

Finally, a comparison of the listings of Tables 6.2 and 9.5 gives 99 bytes for the former and 152 for the latter. This excludes the data, vector table and the latter's strategy for dealing with interrupts.

Table 9.5: Displaying and updating heart rate (*continued next page*).

```
short Array[256];                        /* Global array of 256 words         */
/* The background routine is defined here                                     */
void display(void)
    {
    register unsigned char x_coord;      /* The x co-ordinate                 */
    char *const dac_x = (char*)0x6000;   /* 8-bit X-axis D/A converter        */
    short *const dac_y = (short*)0x6002; /* 12-bit Y-axis D/A converter       */
    while(1)                             /* Do forever                        */
        {
        *dac_y = Array[x_coord];         /* Get array[x] to Y plates          */
        *dac_x = x_coord++;              /* Send X co-ordinate to X plates    */
        }
    }

/* The foreground interrupt service routine is defined here                   */

void update(void)
    {
    static unsigned short last_time;     /* The last counter reading          */
    static unsigned char update_i;       /* The array update index            */
    short * const counter = (short*)0x9000; /* The counter is at 9000/1h      */
    char * const int_flag = (char*)0x9080;  /* The external interrupt flag    */
    *int_flag = 0;                       /* Reset external interrupt flag     */
    Array[update_i++] = *counter-last_time; /* Difference is new array value  */
    last_time = *counter;                /* Last reading is updated           */
    }
```

(a) C source code.

```
*    1   short Array[256];                       /* Global array of 256 words     */
*    2   void display(void)
*    3       {
             .text
             .even
_display:    link      a6,#-10      ; Open frame
*    4       register unsigned char x_coord;         /* The x co-ordinate        */
*    5       char *const dac_x = (char*)0x6000;   /* 8-bit X-axis D/A converter */
             move.l    #6000h,-6(a6) ; Store address 6000h in [A6]-6
*    6       short *const dac_y = (short*)0x6002;/* 12-bit Y-axis D/A converter*/
             move.l    #6002h,-10(a6) ; and address 6002h in [a6]+10
*    7       while(1)                             /* Do forever                  */
*    8           {
*    9           *dac_y = Array[x_coord];         /* Get array[x] to Y plates    */
L1:          movea.l   -10(a6),a1   ; Point A1 to 6002h
             moveq.l   #0,d7
             move.b    -1(a6),d7    ; Get x_coord in [A6]-1
             movea.l   d7,a2        ; Put in A2
             adda.l    a2,a2        ; Crafty way of multiplying by 2 for word array
             adda.l    #_Array,a2   ; Add Array base address
             move.w    (a2),(a1)    ; and move Array[x] to dac_y (6002h)
*   10           *dac_x = x_coord++; /* Send X co-ordinate to X plates */
             movea.l   -6(a6),a1    ; Point A1 to dac_x (6000h)
             move.b    -1(a6),d7    ; Get x_coord
             addq.b    #1,-1(a6)    ; Increment it
             and.l     #255,d7
             move.b    d7,(a1)      ; Send it out
*   11           }
             bra.s     L1
```

Table 9.5: (*continued*) Displaying and updating heartbeat.

```
*   13
*   14  /* The foreground interrupt service routine                         */
*   15
*   16  void update(void)
*   17  {
*   18        static unsigned short last_time;  /* The last counter reading */
*   19        static unsigned char update_i;    /* The array update index   */
            .data
            .even
L3_last_time: .=.+2              ; Reserve 2 bytes for last_time
L31_update_i: .=.+1              ; Reserve one byte for update_i
*   20     short * const counter = (short*)0x9000; /* The counter is at 9000:1 */
            .text
            .even
_update:    link     a6,#-8       ; Make frame
            move.l   #9000h,-4(a6) ; Use A6-4 to store address 9000h (counter)
*   21     char * const int_flag = (char*)0x9080; /* The external interrupt flag*/
            move.l   #9080h,-8(a6) ; and A6-8 for 9080h (int_flag)
*   22     *int_flag = 0;                   /* Reset external interrupt flag  */
            movea.l  -8(a6),a1    ; Point A1 to int_flag
            clr.b    (a1)         ; and reset it
*   23     Array[update_i++] = *counter-last_time;/* Diff is new array value  */
            move.b   L31_update_i,d7 ; Get update_i
            addq.b   #1,L31_update_i ; Meanwhile inc it in memory for next time
            and.l    #255,d7      ; Extend to {\tt int} (32 bits)
            movea.l  d7,a1        ; Original value in byte form to A1
            adda.l   a1,a1        ; Multiplied by two for word array
            adda.l   #_Array,a1   ; Add the array base; points to Array[update_i]
            movea.l  -4(a6),a2    ; A2 points to counter
            move.w   (a2),d7      ; Get it
            ext.l    d7           ; in long form
            moveq.l  #0,d6
            move.w   L3_last_time,d6 ; Get last_time in long form
            sub.l    d6,d7        ; Subtract them
            move.w   d7,(a1)      ; Put away in Array[update_i]
*   24     last_time = *counter;            /* Last reading is updated        */
            movea.l  -4(a6),a1    ; Point A6 to counter
            move.w   (a1),L3_last_time ; Get and put it away as new last_time
            unlk     a6           ; Close frame
            rts
            .globl   _update
            .globl   _display
            .data
            .even
_Array:     .=.+2
            .=.+510              ; Reserve 512 bytes (256 words) for Array[]
            .globl   _array
*   25  }
```

(b) Resulting assembly code.

9.3 Structures

We have seen that arrays are data structures grouping many objects having the same type under a single name. Many real situations require organizations of objects having many different properties, but coming under the same banner. As an example, consider a monitoring system in a hospital ward containing up to ten patients. Treating each patient (rather unfeelingly) as a composite object, then a record would contain data such as the hospital number, age, date and an array of physiological readings, such as heart rate, temperature, blood pressure etc. This would be continuously gathered, and perhaps once an hour transferred to a file on magnetic disk for later analysis. These ten objects could be defined in C as:

```
struct med_record          /* Definition for med_record structure */
    {
    unsigned long hosp_numb;
    unsigned char age;
    unsigned long time;
    unsigned char day;
    unsigned char month;
    unsigned short year;
    unsigned short array[256];
    } patient[10];
```

which defines an array of ten composite objects called patient[0]...patient[9], having the structure defined by the tag med_record. Inside this structure are seven objects of various kinds, there can even be other structures. Structures in C are analagous to records in Pascal.

Any member of a structure can be accessed within the scope of the definition by using the Dot (structure element) operator; for example:

```
patient[6].month = 3;
```

makes the object month inside the structure named patient[6] = three.

The tag med_record is optional, and is the name of the structure template. Objects can be given this template any time later, for example dog_1 and dog_2 may be defined as:

```
struct med_record dog_1, dog_2;
```

Thus only a template (which does not cause storage to be allocated) can be declared, and definitions can occur at any following point, for example within other functions.

Taking an example closer to the theme of this book, consider a compound peripheral interface such as the 6821 PIA [4]. We have already seen how this can be interfaced to a MPU in Figs 1.9 and 3.14; here we look at the internal register structure as described in Fig. 9.5. There are six programmer-accessible 8-bit registers living in an address space of four bytes, as determined by the state of the Register Select bits RS0 RS1.

266 *More Naked C*

Sharing a slot are the Data Direction and Data I/O registers. Which of the pair is actually connected to the data bus when addressed is determined by the state of bit-2 of the associated Control register. Each of the eight I/O bits may be set to in or out, as defined by the corresponding bits in the Data Direction register; for instance if ddr_a is 00001111b, then Data register_A has its upper half set as input and lower half as output. Once a Direction register has been set up, then its slot can be changed to the I/O port, by setting the appropriate Control register's bit-2 high.

Each of the six component parts of a PIA can be defined as a pointer, in the way described in the last section, and treated in the normal way. However, if there are several PIAs in the system, then a template for this device as a single compound object can be made and used for each physical port of this kind.

Lines C1 – C9 of Table 9.6(a) declare a *template* describing the PIA as a structure of pointers, thus each register is characterized as an address. Two PIAs are defined based on this declaration, port0 and port1 in lines C12 and C13. Some

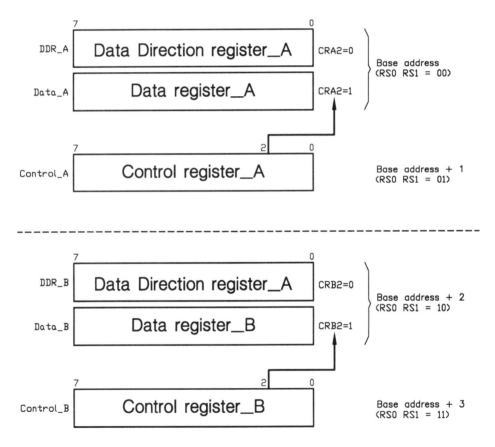

Figure 9.5: Register structure of a 6821 PIA.

of the registers are qualified as pointer-to volatile, as bits read from the outside world will change independently of the software. I have declared these structures to be const and stored in absolute memory, that is static. This means that the structure elements, which here are constant addresses, will be stored in ROM along with the program (assembly lines 3 and 5 in Table 9.6(b)) and any attempt to change these pointers will be flagged by the compiler as an error. Such structures are initialized in the same way as a comparable array. Notice in lines C12 and C13 how the casts are the same as in the template definition.

Functions can take structures as parameters and return them. In both cases the structure name alone is sufficient; for example in line C15 the passed parameter is port0, and this causes *copies* of all six elements to be pushed into the stack prior

Table 9.6: The PIA as a structure of pointers (*continued next page*).

```
C1:   struct PIA                            /* Template for PIA             */
C2:   {
C3:     unsigned volatile char *data_a;     /* I/O port A                   */
C4:     unsigned char *ddr_a;               /* Data Direction register A    */
C5:     unsigned volatile char *control_a;  /* Control register A           */
C6:     unsigned volatile char *data_b;     /* I/O port B                   */
C7:     unsigned char *ddr_b;               /* Data Direction register B    */
C8:     unsigned volatile char *control_b;  /* Control register B           */
C9:   };

C10: main()
C11:   {
C12:     static const struct PIA port0 = {(unsigned volatile char*)0x8000,
                 (unsigned char*)0x8000, (unsigned volatile char*)0x8001,
                 (unsigned volatile char )0x8002, (unsigned char*)0x8002,
                 (unsigned volatile char*)0x8003};

C13:     static const struct PIA port1 = {(unsigned volatile char*)0x8020,
                 (unsigned char*)0x8020, (unsigned volatile char*)0x8021,
                 (unsigned volatile char*)0x8022, (unsigned char*)0x8022,
                 (unsigned volatile char*)0x8023};

C14:     void initialize(struct PIA);/* Declare a function accepting a structure*/

C15:     initialize(port0);
C16:     initialize(port1);

C17: /* Main body sends out of port1's B reg the sum of port0 & port1's A reg*/

C18:     *(port1.data_b) = *(port0.data_a) + *(port1.data_a);
C19:   }

C20: /* Function sets up a PIA as a simple input A and output B port       */

C21: void initialize(struct PIA port)
C22:   {
C23:     *(port.control_a) = 0;
C24:     *(port.ddr_a) = 0;
C25:     *(port.control_a) = 04;
C26:     *(port.control_b) = 0;
C27:     *(port.ddr_b) = 0xFF;
C28:     *(port.control_b) = 04;
C29:   }
```

(a) C source code.

Table 9.6: The PIA as a structure of pointers (*continued next page*).

```
 1:            .text
 2:            .even
 3:  L5_port0: .long   32768,32768,32769,32770,32770,32771 ; Struct PIA port0
 4:            .even
 5:  L51_port1: .long  32800,32800,32801,32802,32802,32803 ; Struct PIA port1
*    1   struct PIA                         /* Template for PIA              */
*    2   {
*    3     unsigned volatile char *data_a;  /* I/O port A                    */
*    4     unsigned char *ddr_a;            /* Data Direction register A*/
*    5     unsigned volatile char *control_a; /* Control register A          */
*    6     unsigned volatile char *data_b;  /* I/O port B                    */
*    7     unsigned char *ddr_b;            /* Data Direction register B*/
*    8     unsigned volatile char *control_b; /* Control register B          */
*    9   };

*   10   main()
*   11   {
*   12     static const struct PIA port0 = {(unsigned volatile char*)0x8000,
                (unsigned char*)0x8000, (unsigned volatile char*)0x8001,
                (unsigned volatile char*)0x8002, (unsigned char*)0x8002,
                (unsigned volatile char*)0x8003};
*   13     static const struct PIA port1 = {(unsigned volatile char*)0x8020,
                (unsigned char*)0x8020, (unsigned volatile char*)0x8021,
                (unsigned volatile char*)0x8022, (unsigned char*)0x8022,
                (unsigned volatile char*)0x8023};
*   14     void initialize(struct PIA);
*   15     initialize(port0);
 6:            .even
 7:  _main:    adda.l   #-24,sp          ; Prepare to push 24 bytes
 8:            move.l   #L5_port0,-(sp)  ; i.e. the six pointers of port0
 9:            move.l   #24,d0           ; out onto the System stack
10:            jsr      a~pushstr        ; Using this library subroutine
11:            jsr      _initialize
12:            lea      24(sp),sp        ; Restore the Stack Pointer
*   16     initialize(port1);
13:            adda.l   #-24,sp          ; Repeat above to pass struct PIA
14:            move.l   #L51_port1,-(sp) ; port1 to initialize()
15:            move.l   #24,d0
16:            jsr      a~pushstr
17:            jsr      _initialize
18:            lea      24(sp),sp
*   17   /* Main body sends out of port1's B reg the sum of port0 & 1's A reg */
*   18     *(port1.data_b) = *(port0.data_a) + *(port1.data_a);
19:            movea.l  L51_port1+12,a1  ; Point A1 to port1's data_b reg
20:            movea.l  L5_port0,a2      ; Point A2 to port0's data_a reg
21:            moveq.l  #0,d7
22:            move.b   (a2),d7          ; Get port0's data_a byte
23:            movea.l  L51_port1,a2     ; Point A2 to port1's data_a reg
24:            moveq.l  #0,d6
25:            move.b   (a2),d6          ; Get port1's data_a byte
26:            add.l    d6,d7            ; Add them
27:            move.b   d7,(a1)          ; and send to port1's data_b reg
28:            rts
*   19   }
*   20   /* Function sets up a PIA as a simple input A and output B port     */
```

Table 9.6: (*continued*) The PIA as a structure of pointers.

```
*    21   void initialize(struct PIA port)
*    22      {
29:               .even
30:_initialize:  link    a6,#0
*    23        *(port.control_a)=0;
31:              movea.l   16(a6),a1       ; Get control_a passed on stack at A6+16
32:              clr.b     (a1)            ; and clear it
*    24        *(port.ddr_a)=0;
33:              movea.l   12(a6),a1       ; Get ddr_a passed on stack at A6+12
34:              clr.b     (a1)            ; and clear it
*    25        *(port.control_a)=04;
35:              movea.l   16(a6),a1       ; Get control_a again
36:              move.b    #4,(a1)         ; Make it 00000100b
*    26        *(port.control_b)=0;
37:              move.l    28(a6),a1       ; get control_b passed on stack at A6+28
38:              clr.b     (a1)            ; Clear it
*    27        *(port.ddr_b)=0xFF;
39:              movea.l   24(a6),a1       ; Get ddr_b passed on stack at A6+24
40:              move.b    #-1,(a1)        ; Make it FFh (i.e. -1)
*    28        *(port.control_b)=04;
41:              movea.l   28(a6),a1       ; Get control_b again
42:              move.b    #4,(a1)         ; Make it 00000100b
43:              unlk      a6
44:              rts
45:              .globl    _main
46:              .globl    _initialize
47:              .globl    a~pushstr
*    29      }
```

(b) Resulting assembler code.

to the Jump (assembly lines 7–10).

Pass by copy is the same technique as used for ordinary single objects, and, as such, the elements themselves cannot be altered by the function. In the situation depicted in Table 9.6, the structure elements are pointers, so although we cannot alter their copies in function initialize(), we can alter the pointed-to variable (i.e. PIA registers) through them. Thus, line C25 means that the contents of structure type PIA named port element control_a is assigned to zero, that is * port.control_a = 0;. As port is an element by element copy of structure type PIA named port0 (if called up from line C15), the contents of port0's control_a register are affected.

Strangely, structure objects are passed by copy, whereas the equivalent process with arrays causes a pointer to the array to be passed. This latter is much more efficient, as only one object (the pointer) has to be pushed on to the stack prior to the Jump to Subroutine, irrespective of the size of the array. Structures are generally smaller than arrays are likely to be, and presumably for this reason the less efficient pass by value copy technique is used. In Table 9.6(b) this is done by moving the System Stack Pointer down 24 bytes (6 × 4-byte pointers), pushing the base address of the structure on to the System stack, the byte size in **D0.L**, and using the machine library subroutine (see Section 9.4) a~pushstr to do the moving, lines 7–12 and 13–18.

Just like a simple object, a structure's address can be passed instead. This is the more efficient method of passing a structure to a function. Thus to pass the medical record of patient[6] to a function store(), which will store it on disk, we could use the calling statement:

store(&patient[6]);

where patient[6] is the name of the structure and &patient[6] its address.

The sizeof operator will give the size of the whole structure. This may be greater than the total of the individual elements, as some machines enforce storage boundaries, which effectively pads out elements with holes. For example in the 68000 family, non-byte objects normally begin at even addresses. An example of this is shown in the use of the .EVEN assembler directive in lines 2 and 4 of Table 9.6. The & operator (i.e. address-of) can also be used to generate a pointer to any element in a structure; thus &patient[6].hosp_number is the address of the unsigned long object hosp_number lurking inside structure patient[6], the latter being a composite object structured as declared by the med_record template.

If we pass a pointer as a parameter to a function, then it must be declared in the function declaration and heading as being a pointer to a particular object; thus for the store() function we would have:

void store(struct med_record * ptr)

which says that ptr is a pointer to a structure type med_record, as passed to store(). Another example is seen in line C14 of Table 9.7, where we are passing a pointer to a structure of type PIA (in line C21).

Given that a function has received a pointer declared to be to a structure, how is it to get at the individual elements? Well, the contents of the pointer to a structure are deemed to be the same as the structure's name. Thus in:

x = (* ptr).hosp_number;

x will take on the value of the first element of a structure type med_record, passed to a function using a pointer. Thus (*ptr) is the equivalent of med_record, assuming that ptr is a pointer to that structure. The parentheses are needed as the structure member operator . has a higher precedence than the indirection * operator. The use of pointers to a structure is so common that C has a special Structure Pointer operator ->, which replaces the (*). pair arrangement thus:

x = ptr -> hosp_number;

To compare the two methods of passing structures to functions, I have repeated Table 9.6(a) in Table 9.7, but using pointers to structures. This time the function

Table 9.7: Sending pointers to structures to a function.

```
C1:   struct PIA                              /* Template for PIA              */
C2:   {
C3:       unsigned volatile char *data_a;     /* I/O port_A                    */
C4:       unsigned char *ddr_a;               /* Data Direction register_A     */
C5:       unsigned volatile char *control_a;  /* Control register_A            */
C6:       unsigned volatile char *data_b;     /* I/O port_B                    */
C7:       unsigned char *ddr_b;               /* Data Direction register_B     */
C8:       unsigned volatile char *control_b;  /* Control register_B            */
C9:   };

C10: main()
C11: {
C12:     static const struct PIA port0 = {(unsigned volatile char*)0x8000,
                 (unsigned char*)0x8000, (unsigned volatile char*)0x8001,
                 (unsigned volatile char*)0x8002, (unsigned char*)0x8002,
                 (unsigned volatile char*)0x8003};

C13:     static const struct PIA port1 = {(unsigned volatile char*)0x8020,
                 (unsigned char*)0x8020, (unsigned volatile char*)0x8021,
                 (unsigned volatile char*)0x8022, (unsigned char*)0x8022,
                 (unsigned volatile char*)0x8023};

C14:     void initialize(struct PIA *);/* Declare ftn accepting a ptr to struct */

C15:     initialize(&port0);
C16:     initialize(&port1);

C17: /* Main body sends out of port1's B reg the sum of port0 & port1's A reg */

C18:     *(port1.data_b) = *(port0.data_a) + *(port1.data_a);
C19: }

C20: /* Function sets up a PIA as a simple input A and output B port          */

C21: void initialize(struct PIA * ptr_2_port)
C22: {
C23:     *ptr_2_port -> control_a = 0;
C24:     *ptr_2_port -> ddr_a = 0;
C25:     *ptr_2_port -> control_a = 04;
C26:     *ptr_2_port -> control_b = 0;
C27:     *ptr_2_port -> ddr_b = 0xFF;
C28:     *ptr_2_port -> control_b = 04;
C29: }
```

prototype in line C14 declares that the passed parameter is a pointer to a structure of type PIA. In line C21, I have named this pointer ptr_2_port, and in lines C23–C28 the -> operator has been used to access the structure elements. Notice how the addresses of the two structures `port0` and `port1` are passed to `initialize()` in lines C15 and C16.

Using the same compiler to process the C source of Table 9.7 gives 178 bytes as compared to 324 bytes resulting from Table 9.6(a). No longer do we need to call up a library function to pass a copy of the structure in its entirety. A similar advantage accrues when a function returns a pointer to a structure, as compared to an actual structure.

If we use pointers to structures, then we can map the structure anywhere we want within the microprocessor's memory map, rather than placing it where the compiler wants to [5, 6]. Thus, for our example of a PIA, we could define a structure comprising six `char` objects (not, as before, pointers to objects) and assign the pointer to this structure at the base address of the actual physical PIA. For example:

&port0 = (struct PIA *)0x8000; _Casts 0x8000 as a pointer to a structure of template_ `PIA`

states that the address of `port0`, previously used to name a structure of type `PIA`, is to be 8000h. As in previous pointer assignments, we must use a cast to convert the constant to the appropriate type, which in this case is pointer-to `struct PIA`.

This gives us a major headache, as the Data and Data Direction registers share the same address, so our six structure members cannot all have unique addresses. The way around this problem is to use a **union**. A union is declared and initialized in the same way as a structure, but all union members occupy the same place in memory. Consider the union template called `share`, in lines C1 – C5 of Table 9.8(a). Here the union has two members, `ddr` and `data`, both `char` (byte)-sized. Lines C6 – C12 declare the structure `PIA` which has four `char`-sized members, two of which are unions of type `share` called `a` and `b`. As a union appears as one location only, this satisfies the physical criteria that a PIA occupies only four bytes of memory.

An element in a union can be accessed by using the Dot operator in the same manner as a structure; for example `b.ddr` is the `ddr` object in a union called `b`. Where a union is buried within a structure, then the Dot operator can be used twice, thus `port0.b.ddr` is the object `ddr` inside union `b` inside `struct port0`. In Table 9.8, we are passing a pointer to a structure of type `PIA` to the function, so the equivalent (in line C30) is `pntr_2_port -> b.ddr`. We see from Table 8.4 that both the `->` and `.` operators have the same precedence, and associate right to left, so parentheses are not required.

In Table 9.8, I have not defined the structure as being `static` or `const`, as opposed to Tables 9.6 and 9.7. This leads to the structure addresses being stored in the frame (assembly lines 4 and 5) at run time rather than being in absolute memory at load time (`static`). The qualifier `const` would not change this, but would produce a warning if the program tried to meddle with these addresses. Compiling this source produced 130 bytes of machine code.

Although the procedure outlined in Table 9.8 seems best, there can be problems. The resultant assembly code has located the four elements at `Base` (line 26), `Base+1` (line 24), Base+2 (line 32) and Base+3 (line 34), that is at sequential addresses. However, many compilers will pad out elements to begin at even addresses. Indeed the circuit of Fig. 3.14 shows the PIA elements located at four sequential even addresses (eight bytes), as address line a_0 is not provided by the 68000 MPU. Most compilers permit various alternative storage configurations for structures, and with collusion with the hardware engineer, a suitable scheme can be devised. Nevertheless, the awareness of hardware circuitry intruding on software matters leads to portability problems if the circuitry or/and processor is changed.

Table 9.8: Unions (*continued next page*).

```
C1:  union share          /* Template for shared DDR and Data registers    */
C2:  {
C3:  unsigned char ddr;
C4:  unsigned volatile char data;
C5:  };
C6:  struct PIA           /* Template for PIA                              */
C7:  {
C8:  union share a;       /* Shared registers A side                       */
C9:  unsigned volatile char control_a;
C10: union share b;       /* Shared registers B side                       */
C11: unsigned volatile char control_b;
C12: };
C13: main()
C14: {
C15: struct PIA *pntr_2_port0 = (struct PIA *)0x8000;/* port0's base @ 8000h*/
C16: struct PIA *pntr_2_port1 = (struct PIA *)0x8020;/* port1's base @ 8020h*/

C17: void initialize(struct PIA *);/*Decl ftn taking ptr to struct type PIA */

C18: initialize(pntr_2_port0);
C19: initialize(pntr_2_port1);

C20: /* Main body sends out of port1's B reg the sum of port0 & port1's A reg */
C21: pntr_2_port0->b.data = pntr_2_port0->a.data + pntr_2_port1->a.data;
C22: }
C23: /* Function sets up a PIA as a simple input A and output B port        */

C24: void initialize(struct PIA * pntr_2_port)
C25: {
C26: pntr_2_port->control_a = 0;
C27: pntr_2_port->a.ddr = 0;
C28: pntr_2_port->control_a = 04;
C29: pntr_2_port->control_b = 0;
C30: pntr_2_port->b.ddr = 0xFF;
C31: pntr_2_port->control_b = 04;
C32: }
```

(a) C source code.

Table 9.8: Unions (*continued next page*).

```
*1  union share              /* Template for PIA                                 */
*2  {
*3      unsigned char ddr;
*4      unsigned volatile char data;
*5  };
*6  struct PIA               /* Template for PIA                                 */
*7  {
*8      union share a;       /* Shared registers A side                          */
*9      unsigned volatile char control_a;
*10     union share b;       /* Shared registers B side                          */
*11     unsigned volatile char control_b;
*12 };
*13 main()
*14 {
1:              .text
2:              .even
3:  _main:      link    a6,#-12
*15     struct PIA *pntr_2_port0 = (struct PIA *)0x8000; /* port0's base @ 8000h */
4:              move.l  #8000h,-4(a6); Pointer to port0 stored at A6-4
*16     struct PIA *pntr_2_port1 = (struct PIA *)0x8020; /* port1's base @ 8020h */
5:              move.l  #8020h,-8(a6); Pointer to port1 stored at A6-8
*17     void initialize(struct PIA *);
*18     initialize(pntr_2_port0);
6:              move.l  -4(a6),(sp)   ; Push out port0's address on stack
7:              jsr     _initialize   ; to pass to function initialize()
*19     initialize(pntr_2_port1);
8:              move.l  -8(a6),(sp)   ; Repeat for port1's address
9:              jsr     _initialize
*20 /* Main body sends out of port1's B reg the sum of port0 & port1's A reg     */
*21     pntr_2_port0->b.data = pntr_2_port0->a.data + pntr_2_port1->a.data;
10:             move.l  -4(a6),a1     ; Point A1 to port0
11:             move.l  -4(a6),a2     ; Also A2
12:             moveq.l #0,d7
13:             move.b  (a2),d7       ; Get port0's data_a byte at 8000h
14:             move.l  -8(a6),a2     ; Point A2 to port1
15:             moveq.l #0,d6
16:             move.b  (a2),d6       ; Get port1's data_a byte at 8020h
17:             add.l   d6,d7         ; Add them
18:             move.b  d7,2(a1)      ; Send result to port0's data_b register
19:             unlk    a6
20:             rts
*22 }
```

Table 9.8: (*continued*) Unions.

```
*23 /* Function sets up a PIA as a simple input A and output B port    */
*24  void initialize(struct PIA * pntr_2_port)
*25  {
21:              .even
22: _initialize: link     a6,#0
*26 pntr_2_port->control_a = 0;
23:              move.l   8(a6),a1  ; Point A1 to port base address passed in A6+8
24:              clr.b    1(a1)     ; Clear base+1, that is Control reg A
*27 pntr_2_port->a.ddr = 0;
25:              move.l   8(a6),a1  ; Again (bad minimization!)
26:              clr.b    (a1)      ; This time clear base; that is DDR A
*28 pntr_2_port->control_a = 04;
27:              move.l   8(a6),a1  ; Yet again!
28:              move.b   #4,1(a1)  ; Make Control reg A = 00000100b
*29 pntr_2_port->control_b = 0;
29:              move.l   8(a6),a1  ; and again!
30:              clr.b    3(a1)     ; Base+3 is control reg B
*30 pntr_2_port->b.ddr = 0xFF;
31:              move.l   8(a6),a1  ; And yet again!
32:              move.b   #-1,2(a1) ; Base+2 is DDR B, make it 11111111b
*31 pntr_2_port->control_b = 04;
33:              move.l   8(a6),a1  ; and again!
34:              move.b   #4,3(a1)  ; Make Control B = 00000100b
35:              unlk     a6
36:              rts
37:              .globl   _main
38:              .globl   _initialize
*32  }
```

(b) Resulting assembler code.

One final note on pointers to structures. If arithmetic is attempted on such objects, then one is taken to be the size of the structure. For example:

`pointer = pntr_2_port0 + 1;`

would give a value to `pointer` of four more than `pntr_2_port0` (assuming no holes in the structure). This could be exploited if a system has a multitude of, say, PIAs stored sequentially, which could then be treated as an array of structures. Of course, `pointer` would have to be defined as a pointer-to structure type PIA, before being used in this way.

9.4 Headers and Libraries

We have already seen that for clarity at assembly-level, it is better to name constant objects at the head of the program module. Thus, as an example, the locations of the counter and interrupt flag are named as COUNTER and INT_FLAG in lines 11 and 12 of Table 6.2(b). The .DEFINE (some assemblers use EQU) directive is used to replace any susequent occurrences of these identifiers by the constants 9000h and 9080h respectively (e.g. lines 22 and 23).

As well as clarifying the source code, grouping all such definitions as a **header**, makes changing the program to reflect hardware alterations easier. Thus if INT_FLAG was subsequently moved to 9002h, then only the header definition line need be changed, and the program reassembled. In a large program, changing perhaps 20 references to 9080h is, at the very least, tedious and error prone.

Where a large modular program is being developed, the likely complex header can be written as a separate file and included at the top of each module using the .INCLUDE (or equivalent) directive. For example:

```
.include "hardware.h"
```

Header files are conventionally given a .h suffix.

Although the .DEFINE directive is normally used as a straight text replacement mechanism, some assemblers permit more sophisticated processing. For example line 7 of Table 5.2 used .DEFINE to evaluate an expression mathematically, which was then used to substitute for the delay parameter's name N.

The C language extends the principle of headers by using a preprocessing stage to evaluate directives, which in all cases are identified with a leading # character. Conceptually the preprocessor is a separate program fronting the compiler proper and sometimes, physically, is as shown in Fig. 7.5.

Some typical substitutions are:

```
#define  TRUE     1
#define  FALSE    0
#define  ERROR   -1
#define  FOREVER_DO  for(;;;)
#define  I/O_PORT (char*)0x8000
#define  PYE     22/7
```

Conventionally, the token which is to be replaced is capitalized. It must be separated from both #define and the replacement text by at least one space or tab. The replacement text is everything from this point to the end of line. Some compilers insist that the # character begins the line (no spaces) and that there is no whitespace between the # and define. Table 9.9 repeats 9.5(a) using a header for each module. Notice how the addresses, including casts, are named.

The #define directive can do more than simple text and mathematics substitutions, it can be used to define macros with arguments, rather like in Table 7.6. Consider the definition:

```
#define MAX(X,Y)  ((X)>(Y) ? (X) : (Y))
```

Although MAX(X,Y) looks like a function call, any reference to MAX later expands into in-line code, for instance:

```
temperature = MAX(t1,t2);
```

Table 9.9: Using #define for text replacement.

```
#define    FOREVER_DO       while(1)
#define    ANALOG_PORTX     (char*)0x6000
#define    ANALOG_PORTY     (short*)0x6002

short Array[256];                        /* Global array of 256 words            */
/* The background routine is defined here                                        */
void display(void)
   {
   register unsigned char x_coord;       /* The x co-ordinate                    */
   char *const dac_x = ANALOG_PORTX;     /* 8-bit X-axis D/A converter           */
   short *const dac_y = ANALOG_PORTY;    /* 12-bit Y-axis D/A converter          */
   FOREVER_DO                            /* Do forever                           */
      {
      *dac_y = Array[x_coord];           /* Get array[x] to Y plates             */
      *dac_x = x_coord++;                /* Send X co-ordinate to X plates       */
      }
   }

/* The foreground interrupt service routine is defined here                      */

#define    COUNT_PORT       (short*)0x9000
#define    INTERRUPT_FLAG   (char*)0x9080

void update(void)
   {
   static unsigned short last_time;      /* The last counter reading             */
   static unsigned char update_i;        /* The array update index               */
   short * const counter = COUNT_PORT;   /* The counter is at 9000/1h            */
   char * const int_flag = INTERRUPT_FLAG; /* The external interrupt flag        */
   *int_flag = 0;                        /* Reset external interrupt flag        */
   Array[update_i++] = *counter-last_time; /* Difference is new array value      */
   last_time = *counter;                 /* Last reading is updated              */
   }
```

Here X will be replaced by t1 and Y by t2 giving the equivalent:

```
temperature = ((t1)>(t2) ? (t1) : (t2));
```

Notice how the macro definition was carefully parenthesized to avoid problems with complex parameter substitutions. For example:

```
#define SQR(X)   X*X
--------------------
y = SQR(1+7);
```

will result in $y = 1+7*1+7$; which is 9, rather than 64! The solution is to define SQR(X) as:

```
#define SQR(X)   (X)*(X)
```

which results in $y = (1 + 7) * (1 + 7)$; which is 64, as desired.

When defining macros, there must be no space between the macro name and the opening (, otherwise simple substitution will occur; thus:

```
#define  SQR (X)  (X)*(X)
```

causes $y = \text{SQR}(z);$ to become $y = (z)(z) * (z);$!

Care should be taken in defining very complex macros, especially using the ++ and -- operators, as their expansion with compound operators can be difficult to predict. Any macro or text substitution can be undefined subsequently by using the #undef directive.

Most of the remaining preprocessor directives involve conditional compilation [7]. Consider the following code fragment:

```
#ifndef  MPU
#error  Microprocessor type not defined
#endif

#if  MPU == 68K
     typedef short   WORD;
     typedef int     LONG_WORD;

#elseif  MPU == 6809
     typedef  int    WORD;
     typedef  long   LONG_WORD;

#else
#error Unknown microprocessor type
#endif
```

There are quite a number of new keywords used in this example. The purpose is to introduce two new types of C objects, namely WORD and LONG_WORD, rather than use char, int etc. The C operator typedef allows the writer to use synonyms for object types of any complexity. For example the FILE type available to hosted C compilers to open, close, write to or read from a named file on disk is a synonym for a complex structure type.

Now to make the types WORD and LONG_WORD portable, the underlying base type must be chosen according to the target processor. Thus for example, int is a 16-bit word in most 6809 and 8086 compilers, and usually 32 bits for 68000 and 80386 target compilers. Our example defines the WORD and LONG_WORD types differently according to the state of the variable MPU, which has been set by the operator prior to using the compiler. For example in the MSDOS operating system:

```
SET MPU = 68K
```

in the startup autoexec.bat file will do the needful. Alternatively putting a #define MPU = 68K in the first line of the header would have the same effect.

The #ifndef of the first line says that if MPU is not defined, print an error message (#error). The #endif that follows, closes this sequence down. The #if, #elseif and #else directives that follow, have their obvious meanings, and delimit one of three actions depending on the state of MPU.

The #include directive is used to read in a specified file at the point at which it occurs. In C there are two versions, for example:

```
#include "hardware.h"
#include <hardware.h>
```

In the case of the former, the preprocessor assumes that the file hardware.h is in the same directory as itself. In the latter, various other specified directories are searched as well, usually a special header subdirectory. The details are compiler dependent. Usually the quotes version is used for your own private include files, whereas the angle bracket form specifies standard library header files. Of course, files other than headers may be included, such as other C source programs.

All C compilers come with a set of **libraries**, which give the writer facilities to do complex mathematics, input and output routines, file handling, graphics etc. These libraries consist of a number of functions (see Table 9.10) in object code form, together with a dictionary. Such libraries are added to the linker's command line as shown in Fig. 7.5; however, the linker does not treat a library object file in the same way as a normal program object file. Rather than adding all the object code in a library to the existing code, only functions which are referred to and declared **extern** by the user's modules are extracted. Thus functions are selectively added.

Most compilers come with a librarian utility. This allows the programmer to make up a library of his/her own favorite functions or, more dangerously, alter the commercial ones. The linker scans libraries in the order they are named in its command line; thus it is possible to replace unsatisfactory commercial functions by home-brew ones.

Old C did not specify a standard library, although many of the more common functions became a de facto part of the language. The ANSI standard does specify a de jure common core library [8], but most compilers have additional libraries to deal with operating system-specific functions, graphics, communications etc.

In general the standard libraries are only relevant in a hosted environment. In a free-standing situation, such as met in embedded microprocessor targets, many library functions are either irrelevant or require modification.

Most compilers that are not operating-system specific use libraries at several levels. The lowest of these is the machine library, which holds basic subroutines which the assembly-level source code can use without the writer at the high-level being aware of their existence. Thus, for example, an integer multiplication in a 6809 MPU target requires a 16×16 operation, although the processor itself has only an integral 8×8 MUL instruction. It is likely that the C-originated assembly code will include a JSR to the requisite subroutine held in the machine-level library. An example of this is given in line 10 of Table 9.6, where the subroutine a~pushstr is used by the compiler to implement the passing of a structure to a function (see also Table 14.6, line 115).

The next up in the hierarchy of libraries provides low-level support routines used by the user callable libraries, and includes all the operating-system interface routines. For example, they may contain subroutines to obtain a character from a terminal (typically called inch for input character) and to output a single character (typically outch for output character). The actual code here depends on the hardware. In a non-hosted environment, the writer will alter such routines to suit the system.

The user-callable libraries contain all the functions which may be explicitly called from the C program. These are the ANSII standard libraries and the various high-level options, such as graphics. Such libraries make use of the low-level support library when interacting with the environment.

Variations, include optional integer libraries (suitable for embedded applications where the normal floating-point functions may not be required) and libraries coded to make use of mathematics co-processors.

Given that libraries comprise a number of functions external to the user's program, such functions that are to be called must be declared extern and prototyped in the normal way. To avoid this chore, compilers come with a number of standard header files which may be #included as appropriate at the head of the user program. Table 9.10 shows the header file math.h provided with the Cosmic/Intermetrics cross 6809 C compiler V3.31, as an example. This declares most of the standard ANSII maths library. As can be seen, the majority of maths functions take double float arguments and return a double float value.

This header file is designed to be used by several related compilers. If the variable _PROTO has been defined, then any text of the form __(a) will be replaced by just a:

```
#ifdef _PROTO
    #define __(a) a
#else
    #define __(a) ()
#endif
```

For example, on this basis the first True line will be converted by the preprocessor to:

```
double acos (double x);
```

which is the normal ANSII C function prototype. However, if _PROTO is not defined we will get:

```
double acos (());
```

which is suitable for an old C-style compiler, which does not support prototyping. Notice how the internal variable __MATH__ is defined at the top of the header. This lets subsequent headers know that the math.h header is present.

Finally, the ANSII committee have authorized the #pragma directive, as a pragmatic way of introducing compiler dependent directives, which may be anything the compiler writer wishes. An example of this from the same compiler is:

#pragma space [] @ dir

which instructs the compiler to store (i.e. space) all non-`auto` data objects (designated `[]`) in direct memory. That is, use the Direct address mode for `static` and `extern` data objects instead of the default Extended Direct addressing mode. Obviously this is very target specific, and considerations of this nature are the subject of the next chapter.

Table 9.10: A typical `math.h` library header (with added comments).

```
/*   MATHEMATICAL FUNCTIONS HEADER
 *   copyright (c) 1988 by COSMIC
 *   copyright (c) 1984, 1988 by Whitesmiths, Ltd.
 */

#ifndef __MATH__
#define __MATH__  1

/*   set up prototyping                                                   */
#ifndef __
#ifdef _PROTO
#define __(a)    a
#else
#define __(a)    ()
#endif
#endif

/*   function declarations                                                */
double acos   __((double x)); /* Computes the radian angle, cos  of which is x  */
double asin   __((double x)); /* Computes the radian angle, sine of which is x  */
double atan   __((double x)); /* Computes the radian angle, tan  of which is x  */
double atan2  __((double y, double x));
/* Computes the radian angle of y/x. If y is -ve the result is -ve.
   If x is -ve the magnitude of the result is greater than pi/2           */
double ceil   __((double x)); /* Computes the smallest integer >=to x     */
double cos    __((double x)); /* Computes the cosine of x radians, range [0,pi] */
double cosh   __((double x)); /* Computes the hyperbolic cosine of x      */
double exp    __((double x)); /* Computes the exponential of x            */
double fabs   __((double x)); /* Obtains the absolute value of x          */
double floor  __((double x)); /* Computes the largest integer <= x        */
double fmod   __((double x, double y));/* Computes the floating-pt remainder of x/y */
double log    __((double x)); /* Computes the natural logarithm of x      */
double log10  __((double x)); /* Computes the common  logarithm of x      */
double modf   __((double value, double *pd));
/* Extracts the integral and fractional parts                             */
double pow    __((double x, double y));   /* Raises x to the power of y   */
double sin    __((double x)); /* Computes the sine of x rads, range [-pi/2,pi/2] */
double sinh   __((double x)); /* Computes the hyperbolic sine of x        */
double sqrt   __((double x)); /* Computes the sqr root of x; if x -ve returns 0 */
double tan    __((double x)); /* Computes the tan of x rads, range [-pi/2,pi/2] */
double tanh   __((double x)); /* Computes the hyperbolic tangent of x     */
int    abs    __((int i));    /* Obtains the integer absolute value of i  */

#endif
```

References

[1] Kernighan, B.W. and Ritchie, D.M.; *The C Programming Language*, 2nd. ed., Prentice-Hall, 1988, Section A7.3.2.

[2] Jaeschke, R.; Recursion, Variable Classes and Scope, *DEC Prof.*, **3**, no. 4, 1984, pp. 84–93.

[3] Jeaschke, R.; Pointers to Functions, *Programmer's Journal*, **3**, no. 2, 1985, pp. 20–21.

[4] Cahill, S.J.; *Digital and Microprocessor Engineering*, 2nd. ed., Ellis Horwood/Simon and Schuster, 1993, Section 5.3.

[5] Jouvelot, P.; De L'Assembleur aux Languages Structures: Le Language 'C'; *Micro Systems* (France), no. 42, June 1984, pp. 102–112.

[6] Banahan, M.; *The C Book*, Addison-Wesley, 1988, Section 8.2.2.

[7] Banahan, M.; *The C Book*, Addison-Wesley, 1988, Chapter 7.

[8] Banahan, M.; *The C Book*, Addison-Wesley, 1988, Chapter 9.

CHAPTER 10

ROMable C

In the last two chapters we have seen that it is possible to take a source program written in C and compile to assembly level. This assembly code can then be linked and converted into a machine code file, ready for loading, as described in Chapter 7. In that chapter, we observed that the environment of a hosted computer is very different to that of a naked system. In the former situation, each operator request for a program run causes the relevant machine-code file to be loaded into computer RAM (usually from disk) and execution to commence. In a **naked system** the program is normally permanently resident in ROM. Thus the initializing loading stage is eliminated. A compiler producing code which can be run in ROM is known as a ROMable compiler.

At the very least, a ROMable compiler must provide the means to put program code and constant data in one section of memory (i.e. ROM), and variable data in another (i.e. RAM). However, there remain several other problems to overcome before a high-level sourced program can successfully run in a naked system. Typical of these are the means to set up the System stack, Reset and Interrupt vector tables, link in hand-assembled routines and implement exception service routines. Handling MPU-specific tasks, such as setting interrupt mask bits in the **Status/CCR register**, and portability issues raise their heads.

Most of these hardware-related activities are handled by the operating system, but in a free-standing environment the programmer must provide such services as are required by the executing software. In this chapter we examine this aspect of software design in more detail.

10.1 Mixing Assembly Code and Starting Up

As far as a microprocessor is concerned, life begins after a Reset. It is the responsibility of the design engineer to ensure that the various fixed restart vectors are in their predetermined place prior to this event. This ensures that the MPU can make it to the start of the executable program. There are several other chores that must be performed before the processing proper can commence. Typically, bits must be twiddled in the Status/Code Condition register, such as the Interrupt Mask and State flags. Dynamic exception vectors must be loaded, stacks set up and Page/Segment registers initialized. C is a powerful language, but its power does not extend down into specific machine registers.

As C cannot make itself a congenial environment, we must do this for it by writing a **startup routine** in a native assembly code and use the linker to join the two together in holy matrimony [1].

Ignoring interrupts, which we will cover in the next section, we have three tasks to perform:

1. Set up the System Stack Pointer to the top of System stack.

2. Ensure that the Restart vector is in the appropriate ROM location.

3. Go to the C program.

Table 10.1 shows a possible implementation. Three source listings are given. The startup routine proper simply puts a suitable address into the System Stack

Table 10.1: Elementary startup for a 6809-based system (*continued next page*).

```
        .processor m6809
;*********************************************************************
;* Startup code for non-interrupt system                             *
;* Assumes RAM up to 07FFh                                           *
;*********************************************************************
        .external _main     ; _main is outside this file
_Start: lds        #0800h   ; Point Stack Pointer to top of RAM
        jsr        _main    ; Go to C code
        bra        _Start   ; Should it return then repeat
        .public    _Start   ; Make this routine known to linker
        .end
```

(a) Startup source executable code, `startup.s`.

```
        .processor m6809
;*********************************************************************
;* Vector table, Reset vector only                                   *
;*********************************************************************
        .external _Start    ; Start is outside this file
        .word      [6]      ; Miss out the interrupt vectors
RESET:  .word      _Start   ; Put restart address here
        .end
```

(b) The source vector table, `vector.s`.

```
main()
    {
    static int i;
    while(1)   {i++;}
    }
```

(c) A dummy C function, `fred.c`.

Table 10.1: (*continued*) Elementary startup for a 6809-based system.

```
1                    ; Compilateur C pour MC6809 (COSMIC-France)
2                           .list      +
3                           .processor m6809
4                           .psect     _bss
5  0001    00 00  L3_i:     .byte      0,0

6   ;**************************************************************
7   ;* Startup code for non-interrupt system                      *
8   ;* Assumes RAM up to 07FFh                                    *
9   ;**************************************************************
10                          .psect     _text
11                          .external  _main      ; _main is outside this file
12 E000 10CE0800 _Start:    lds        #0800h     ; Point Stack Pointer to top of RAM
13 E004 BDE009              jsr        _main      ; _main is outside this file
14 E007 20F7                bra        _Start     ; Should it return then repeat
15                          .public    _Start     ; Make this known to the linker

16          ;   1   main()
17          ;   2       {
18          ;   3           static int i;
19          ;   4           while(1)     {i++;}
20 E009 7C0002   _main:     inc        L3_i+1
21 E00C 2603                jbne       L4
22 E00E 7C0001              inc        L3_i
23 E011 20F6       L4:      jbr        _main
24          ;   5           }
25                          .public    _main

26  ;*********************************************************************
27  ;* Vector table, Reset vector only                                   *
28  ;*********************************************************************
29                          .external _Start     ; Start is outside this file
30 FFF2                     .word     [6]        ; Miss out the interrupt vectors
31 FFFE E000      RESET:    .word     _Start     ; Put restart address here
32                          .end
```

(d) Resulting code.

Pointer and jumps to the subroutine/function created by the C compiler. Traditionally this is named main(). At assembler level the Whitesmiths group compilers transform function names with a leading underscore, hence _main. This is not universally the case; for example a leading period or lagging underscore are common. However named, normally the main C function is written as an endless loop, and thus there will not be a return. In the illustrative situation depicted in Table 10.1(c), this is a (rather useless) counter function, continually incrementing the variable i.

The vector table in the 6809 MPU lives in a different region of memory from the program text, and for this reason is written in Table 10.1 as a separate module. At link time, it will be put into the appropriate location.

The resulting linked assembly code was produced using the Cosmic/Intermetrics 6809 C cross-compiler version 3.31, with the linker set thus:

```
lnk09 +data - b1 +text -b0xE000 startup.o fred.o +text -b0xFFF2 vector.o
```

The startup.s and vector.s files are assembled to their relocatable object versions startup.o and vector.o. The compiler then converts fred.c to fred.o and links startup.o to this code followed by vector.o. startup begins at E000h (+text -b0xE000) and vector at FFF2h (+text -b0xFFF2), where we are assuming ROM between E000h and FFFFh (e.g. a 2764 EPROM). The code in Table 10.1(d) shows everything in its proper place.

If desired, the Vector module could be written in C and compiled to vector.o before linking. A possible C routine with the same role as Table 10.1(b) is given in Table 10.2. This is an example of an array of pointers to functions, where vector[n] is a const pointer to function n (the name of a function is its address, thus main is the pointer to function main()) [2]. Only the Reset vector is shown; to expand the function to include interrupt vectors just replace the null pointers by the root name of the appropriate handler function (see Table 10.7).

ANSII C permits a pointer of any kind to be assigned the constant zero, as is done on line C2 of Table 10.2, that is a void or null pointer. No legitimate data should be held at this address (i.e. 0000h) [3]. For this reason, the

Table 10.2: Using arrays of pointers to functions to construct a vector table.

```
extern Start();
(* const vector[])() = {0,0,0,0,0,0,Start};
```

(a) C source code.

```
1                      ; Compilateur C pour MC6809 (COSMIC-France)
2                                      .list    +
3                                      .processor m6809
4                                      .psect   _text
5                      ;   1   extern Start();
6                      ;   2   (* const vector[])() = {0,0,0,0,0,0,Start};
7    FFF2 0000         _vector: .byte       [2]
8    FFF4 0000                  .byte       [2]
9    FFF6 0000                  .byte       [2]
10   FFF8 0000                  .byte       [2]
11   FFFA 0000                  .byte       [2]
12   FFFC 0000                  .byte       [2]
13   FFFE E000                  .word       _Start
14                              .public     _vector
15                              .external   _Start
16                              .end
```

(b) Resulting code after the link.

linker script above, which assumed memory between 0000h and 07FFh (e.g. a 6116 RAM), started the data bias at 0001h (+data -b1) rather than the more obvious 0000h bias.

Starting up a 68000 MPU-based system can be done in the same way as for the 6809, with a separate text bias for the Vector and Startup routines, typically 00000h and 00400h respectively. However, as the program usually directly follows the Vector table, a composite Vector/Startup module may be created and linked in at zero. As shown in Table 10.3, the User Stack Pointer is setup and the state changed to User before entering the main C routine (see also Table 14.10).

Table 10.3: A simple Startup/Vector routine for a 68000-based system.

```
~~1WSL 3.0 as68k    Thu Apr 13 14:45:17 1989

1                             .extern _main      * _main is outside this file
2 00000 0000a000     SSP: .long   0xA000         * Initial setting of SSP
3 00004 00000416     PC:  .long   _Start         * Restart PC value
4 00008                        .  =.+0x3F8       * Skip up to 0400h

5 00400 207c 00001000 _Start: movea.l #0x1000,a0 * Fix to set-up User Stack Ptr
6 00406 4e60                   move    a0,usp    * Privileged instruction
7 00408 027c dfff              andi.w  #0xDFFF,sr* Bit 13 changes to User state
8 0040c 4eb9 00000416          jsr     _main     * Go to C routine
9 00412 6000 fff8              bra     _Start    * Repeat if returns
10                             .end
```

Tables 10.1 and 10.3 are simple examples of incorporating assembly routines with C code. They are elementary because no data is explicitly passed between them. It would have been quite easy to pass the value, say, i = 6 to main() in Table 10.1, but we would have to know how the C compiler handles such variables, as each has its own house rules. In fact, that particular compiler would have expected i to be passed in Accumulator_D rather than through the System stack (see Table 10.6). Thus in any particular compiler, a knowledge of its operation is needed in order to mesh the two successfully.

Before giving an example, why use a mixture of the two languages, except for startup? It is an accepted rule of thumb that a program will spend some 90% of its time in around 10% of the code [4]. Where time is of the essence, replacing this code by equivalent assembly-based subroutines will be beneficial. Another candidate for assembly code is the creation of library routines (see also Table 10.16). As these will be used by many different projects, time spent in refining such code can be justified in some cases.

Our example here involves the creation of a library subroutine to return the unsigned short int square root of an unsigned short int parameter. The function is to mimic the C function:

```
unsigned short sqroot(unsigned short);
```

for the Whitesmiths 68000 C compiler version 3.2.

The relevant house rules for this compiler are:

1. Integral and pointer parameters are extended to four bytes and pushed onto the System stack least significant byte first. Where there is more than one parameter, then the compiler works along the list from right to left.

2. Registers **D3–D5** and **A3–A7** are guaranteed unaltered by the function on return.

3. Integral and pointer parameters are returned in **D7.L**.

There are of course also rules for floating-point and structure objects.

The algorithm implemented by Table 10.4 uses Newton's numerical method [5]. This states that if we guess an initial value for $\sqrt{x}$, usually $\frac{1}{2}x$, then:

$$\text{NEW\_ESTIMATE} = \frac{1}{2}(\text{OLD\_ESTIMATE} - x/\text{OLD\_ESTIMATE})$$

and if we keep going round the loop, the estimate will converge to the desired value. In our listing, I have exited whenever NEW_ESTIMATE = OLD_ESTIMATE or when the number of interations reaches 20. The latter is necessary, as numerical techniques often produce an oscillating outcome (for example $x = 65535$ produces an estimate alternating between 255 and 256), or even do not converge. Without an unconditional out, such functions may go into an unscripted endless loop.

In Table 10.4, all variables are held in registers, so no frame is created and **SP** is used as the reference to obtain the passed variable x (MOVE.W 4(SP),D7). Furthermore, none of the preserved registers are used, therefore they do not require saving and retrieving. The answer is returned in **D7** as required.

Calling up the function from a C program is done in exactly the same way as any function actually written in C, for example:

```
x = sqr_root(27U);
```

where the suffix U indicates an unsigned type constant. Of course sqr_root() will be external to the C program, so an `extern` declaration must be made in the normal way before sqr_root() is called, thus:

```
extern unsigned int sqr_root(unsigned short);
```

One of the disadvantages of using any high-level language is the loss of the ability to use any special feature of the underlying processor. For example, it may be necessary to lock out any interrupt occurring during a specific part of the code. How could we handle a 6809-based system with the requirement to stop at a specific point and use the SYNC instruction (see page 164) to continue when an interrupt subsequently occurs? Of course, we could write the code as part of an assembly subroutine and link it in as previously shown, but this is not very efficient for short sequences.

Table 10.4: A C-compatible assembler function evaluating the square root of an **unsigned int**.

```
        .processor  m68000

; ************************************************************************
; *  Calculates the square root to the nearest lower integer             *
; *  using Newton's method where an original estimate of n/2 is made     *
; *  and successive estimates are = (old_estimate + n/old_estimate)/2    *
; *  Exit either after 20 iterations or when new and old estimates       *
; *  are the same                                                        *
; *  EXAMPLE : Return for n = 18 is 4                                    *
; *  ENTRY   : short unsigned int is passed on the Stack at SP+4/5       *
; *  EXIT    : Return in D7.W as an unsigned short int, max 256          *
; *  EXIT    : D0/D1/D2 and CCR altered                                  *
; ************************************************************************
;
_sqr_root: move.w 4(sp),d7  ; Copy n to D7.W
           cmp.w  #1,d7     ; n = 0 or 1?
           bhi    CONTINUE  ; IF higher than continue
           bra    EXIT      ; ELSE exit with answer = n
CONTINUE:  lsr.w  #1,d7     ; Create first estimate by dividing by 2
           move.w #19,d0    ; 19+1 iterations count in D6.W
; After initialization repetitively build up new estimate in D2.L
LOOP:      move.w d7,d1     ; Copy estimate into D1.W
           clr.l  d2        ; Copy n into D2 as a 32-bit clone
           move.w 4(sp),d2  ; for the division following
           divu   d7,d2     ; [D2.W] = n/old_estimate
           move.w d2,d7     ; Move it to D7.W
           add.w  d1,d7     ; [D7.W] = old_estimate + n/old_estimate
           lsr.w  #1,d7     ; Divide by 2 to give the new estimate
           cmp.w  d1,d7     ; Compare new with old estimates
           dbeq   d0,LOOP   ; IF equal exit ELSE dec loop count; IF not -1 repeat
EXIT:      rts              ; ELSE exit with answer in D7.W
           .public _sqr_root; Make known to the outside world
           .end
```

Many C compilers permit the insertion of assembly source lines interleaved in the C source code. Although this is a common feature, it is not standard, and thus is very implementation dependent. Where it is available, the keyword asm is usually involved. For example, the Aztec C compilers use #asm and #asmend to sandwich such code. The Microtec equivalent uses a #pragma asm – #pragma endasm sandwich. Our illustration in Table 10.5 uses the Whitesmiths group built-in function _asm() for this purpose. Here I have forced a LDS #0800h assembler line in at the beginning of the C code. This obviates the need for the Startup module, but the Vector module must still be linked in. _asm() can take several lines of assembly code as its argument between double quotes, and use \n and \t to indicate New Line and Horizontal Tab respectively.

Table 10.5: Using in-line assembly code to set up the System stack.

```
main()
  {
  static int i;
  _asm("lds #0800h      ; Point Stack Pointer to top of RAM");
  while(1)    {i++;}
  }
```

(a) C source.

```
; Compilateur C pour MC6809 (COSMIC-France)
        .list    +
        .processor m6809
        .psect  _bss
L3_i:   .byte   [2]
;   1   main()
;   2           {
        .psect  _text
;   3           static int i;
;   4           _asm("lds #0800h     ; Point Stack Pointer to top of RAM");
_main:  lds     #0800h      ; Point Stack Pointer to top of RAM
;   5           while(1)    {i++;}
        inc     L3_i+1
        jbne    L4
        inc     L3_i
L4:     jbr     _main
;   6           }
        .public _main
        .end
```

(b) Resulting assembly code.

The Microtec `asm()` can optionally use an assembly command to return data to a C object. For example:

```
switch = asm(unsigned char, "move.b  9000h,d0");
```

which assigns the value read from $9000h$ to an unsigned byte C variable.

Despite its flexibility, assembler windows should be used sparingly, as it seriously compromises the portability of such code (see Section 10.4).

It is possible to call a function whose absolute location is known from a C program, but which cannot be accessed in relocatable object form by the linker. This is likely to occur when the target system has a resident operating system/monitor, and the C user program wishes to use those external resources. Another situation which requires this facility, is where a preprogrammed mathematics package is resident, for example the 6839 floating-point ROM.

As an example, assume that a certain ROM-based 6809-monitor has a subroutine called OUTCH (OUTput CHaracter) located at $F830h$. This sends out a single character, passed to it in Accumulator_B, to the terminal. We wish to make use of this subroutine in implementing a C function which sends a character ch to the terminal whenever called.

Now we noted on page 286 (see also Table 10.2), that in ANSII C the name of a function is a pointer to that function, that is its address. Thus, it might be thought that the statement (0xF830)(ch); would pass ch and jump to the subroutine at F830h. However, 0xF830 is an integer constant so we must first cast it to type pointer-to a void function taking a single char parameter, that is (void(*)(char))0xF830. This complex cast reads from inside out: pointer to function (*)/ taking a char (char)/ returning void. The whole is enclosed by the cast's parenthesis and qualifies the constant 0xF830. Note how the complex type reads from inside out first right then left. This is the normal way of constructing compound types.

In Table 10.6 I have used a header to replace the name OUTCH by this casting. It would be normal to use a header to define the resources available in such a co-resident ROM. Thus the statement:

(OUTCH)('\n');

translates in Table 10.6 to:

LDB #10
JSR 0F830h

as desired.

Table 10.6 defines a function known as void new_line(void) which is designed to send a Line Feed (ASCII code 10) to the terminal. This simply in turn sends out '\n' to OUTCH. The character '\n' is C'ese for New Line (or Line Feed). As an alternative, an absolute value may be cast to char and passed to OUTCH; thus in this case OUTCH((char)10) is a direct equivalent to line C4, but rather less readable. Other useful C escape sequences (or tokens) for non-printable characters are \t for

Table 10.6: Calling a resident function at a known address.

```
1                   ; Compilateur C pour MC6809 (COSMIC-France)
2                                   .list    +
3                                   .processor m6809
4                                   .psect   _text
5                   ;   1  #define OUTCH    (void(*)(char))0xf830
6                   ;   2  void new_line(void)
7                   ;   3  {
8                   ;   4      (OUTCH)('\n');
9    E000  C60A     _new_line:      ldb      #10
10   E002  BDF830                   jbsr     0f830h
11                  ;   5  }
12   E005  39                       rts
13                                  .public _new_line
14                                  .end
```

Horizontal Tab (ASCII code for HT is 9), \v for Vertical Tab (VT = 11), \b for Back Space (BS = 08), \r for Carriage Return (CR = 13), \f for Form Feed (FF = 12), \a for Audible alert (BEL = 7).

Two points to notice concerning the coding. As previously mentioned, the Cosmic/Intermetrics 6809 V3.3 compiler passes its first integral type parameter in its Accumulator_D rather than on the System stack. With a byte (char) parameter only, the right half of **D** is used, that is Accumulator_B. If OUTCH did not expect its parameter in this register, then a line of assembly code would be needed to match the C function parameter passing convention to that of the monitor subroutine. For instance, TFR B,A if OUTCH expected its parameter in Accumulator_A. Also, any registers which the C-function house rules say should be preserved should be saved before calling up the alien subroutine. This compiler does not make any assumptions concerning the return state of the 6809's registers.

The function (OUTCH)() should not be declared in new_line(), as the use of a fixed address in the function call is an equivalent procedure. Neither should it be declared **extern**, as it will not be linked in.

10.2 Exception Handling

Interrupts and their software cousins are handled using the techniques discussed in the previous section. In order to process an interrupt correctly, the software must arrange for:

1. The service routine start address to be in its correct vector location.

2. Any registers not preserved by the compiler's function house rules to be saved and retrieved.

3. The service function to be terminated by a Return From Interrupt operation (e.g. RTI, RTE, IRET) rather than a Return From Subroutine (e.g. RTS, PULS PC, RET).

Consider the program of Table 9.5. There are two functions here, the background main function called display() and the interrupt service function update(). Function update() is not explicitly entered, or indeed known, by background function display(); they communicate through global object Array[], which is known to both of them.

We look first at the 6809 processor and assume the use of $\overline{\text{IRQ}}$ to switch context. As the entire processor state is automatically saved, all our interrupt handler (IRQ_handler in line 11 of Table 10.7(a)) has to do is jump to the subroutine _update, and on return do a RTI. The address of IRQ_handler is placed in the IRQ vector in Table 10.7(b). Thus, when an $\overline{\text{IRQ}}$ interrupt occurs, the processor will save its state and go via the IRQ vector (FFF8:9h) to the stub IRQ handler in the startup routine. This simply jumps to the appropriate C function and terminates with a RTI. Notice that this startup routine clears the **I** mask bit in the **CCR**

Table 10.7: 6809 startup for the system of Table 9.5.

```
1                            .processor m6809
2                ;***********************************************************
3                ;* Startup code for background display() and IRQ entered update() *
4                ;* Assumes RAM up to 07FFh                                        *
5                ;***********************************************************
6                            .external _display, _update ; Both routines are outside
7  E000 10CE0800      _Start: lds    #0800h          ; Stack Pointer to top of RAM
8  E004 1CEF                  andcc  #11101111b      ; CLI
9  E006 BDE00F                jsr    _display        ; Go to background C code
10 E009 20F5                  bra    _Start          ; If it returns, then repeat
11 E00B BDE035  IRQ_handler:  jsr    _update         ; Go do function update()
12 E00E 3B                    rti                    ; Exit IRQ handler
13                            .public _Start, IRQ_handler; Make known to linker
14                            .end
```

(a) Startup showing the IRQ handler routine.

```
1                            .processor m6809
2                ;***********************************************************
3                ;* Vector table, IRQ and vector only                        *
4                ;***********************************************************
5                            .external _Start, IRQ_handler
6  FFF2                       .word   [3]            ; Miss out SWI2, SWI3 and FIRQ
7  FFF8 E00B             IRQ: .word   IRQ_handler    ; Put IRQ handler address here
8  FFFA                       .word   [2]            ; Miss out SWI & NMI
9  FFFE E000           RESET: .word   _Start         ; Put restart address here
10                            .end
```

(b) Vector table showing the IRQ handler address.

(line 8), which allows the MPU to respond to $\overline{\text{IRQ}}$ requests. The I mask has been automatically cleared after a Reset.

The situation would be a little more complex if $\overline{\text{FIRQ}}$ were used to initiate the exception. In this situation, only the **PC** and **CCR** are automatically saved. Thus the handler must use a Push/Pull pair to sandwich the JSR, in order to preserve the state. This is the situation for all 68000-based interrupts and the Push/Pull sandwich is clearly seen at INT2_handler in lines 15 and 17 of Table 10.8. The house rules of this compiler (Whitesmiths V3.2) are such that registers **D3** to **D5** and **A3** to **A7** are preserved in any C function, so only the remaining registers are saved by the handler. The three interrupt mask bits are set to 001 in line 12 to enable level-2 interrupts (they were set to 111 when the MPU was Reset).

Both Tables 10.7 and 10.8 are linked in with the C code in exactly the same manner as for the corresponding Tables 10.1 and 10.3. Software interrupts and exceptions are handled in the same way as hardware interrupts. Where interrupt vectors are stored in RAM rather than the normal ROM, the startup routine must dynamically load the address before enabling the interrupt mask.

294 ROMable C

Table 10.8: 68000 startup for the system of Table 9.5.

```
~~1WSL 3.0 as68k   Wed Apr 19 15:45:50 1989
1  **************************************************************************
2  * Startup for background display and INT2 entered update()               *
3  **************************************************************************
4                         .extern _display, _update * Both routines are outside
5  00000 0000a000         SSP: .long    0xA000     * Initial setting of SSP
6  00004 00000426         PC:  .long    _display   * Restart PC value
7  00008                       .=.+96              * Go to Level2 int vector
8  00068 00000416         INT2: .long INT2_handler * Addr. of INT2 handler here
9  0006c                       .=.+916             * Move up to 400h

10 00400 207c 00001000 _Start: movea.l #0x1000,a0* Make to set-up User Stack
11 00406 4e60                  move    a0,usp
12 00408 46f8 0100             move.w  0x0100,sr * User state, Int mask = 001
13 0040c 4eb9 00000426 ENTER:  jsr     _display  * Go to background C routine
14 00412 6000 fff8             bra     ENTER     * Repeat if returns

15 00416 48e7 e3f0 INT2_handler:movem.l d0-d2/d6/d7/a0-a2,-(sp)
                                                  * Save relevant regs
16 0041a 4eb9 00000466         jsr     _update   * Go to INT2 service routine
17 00420 4cdf 0fc7             movem.l (sp)+,d0-d2/d6/d7/a0-a2
                                                  * Restore original state
18 00424 4e73                  rte
19                             .end
```

In a realistic system, the startup is likely to be more complex than these examples show. For example, any programmable I/O devices should be configured before enabling interrupts. If the exception service routine communicates through global variables and these are presumed to have an initial value, then this too should be done in the startup module. This will be described in the following section.

The double-hop response to an interrupt slows down the MPU's response to a request. There are two ways around this problem. The first involves writing the interrupt service routine (ISR) entirely in assembly language; thus the handler becomes the whole routine. If the ISR is of any size, it is likely that it will be in a separate file or library, and will be added in through the linker.

Some compilers allow the programmer to specify a C function as an interrupt service routine. In such cases the generated assembly code includes an entry sequence that saves all used registers that the compiler's house rules state must be preserved. On exit these are returned and a RTI/RTE generated at exit. Like assembly windows, these are extensions to the ANSII standard and are highly compiler specific. As an example, the Mictrotec Research Paragon C cross to 68000/68020 V3 compiler requires such functions to be sandwiched by the $INTERRUPT directive, for example:

```
#define $INTERRUPT
< definition of function ifred() >
```

#undef $INTERRUPT

Function ifred() will then be coded as an interrupt service routine rather than a subroutine. As many functions as required may be sandwiched.

To illustrate the effect of $INTERRUPT, consider the Real-Time Clock program of Table 8.6. Compiling this as an interrupt service function with the Paragon compiler, gives the code in Table 10.9. Notice how the registers are saved and restored at the beginning and end of the routine, and the terminating RTE. In this situation, the address of clock(), that is .clock, should be placed in the appropriate vector, rather than that of an intermediate handler.

The Whitesmiths group compilers versions 3.3 and up, use the prefix @port to specify interrupt service functions, see Tables 14.6 and 14.12. Thus:

```
@port void clock(void)
{body}
```

would give us the Real-Time Clock interrupt service function for these compilers.

Using interrupts in high-level code is fraught with difficulties. Unlike assembly code, an interrupt will be serviced in the middle of a high-level instruction. If, for example, we had a global int variable i which was shared between background and foreground routines, then an interrupt in the middle of an instruction i++ may well produce intriguing results, for instance:

```
i++
        inc     L3_i+1      ; Increment lower byte
<<<- - - - - - - - - - - - - Interrupt - - - - - - - - - - - - - >>>
        bne     L4          ; IF not zero THEN continue
        inc     L3_i        ; ELSE increment upper byte
    L4:
```

Here we have assumed a 6809 compiler with a 16-bit int. To increment this object, the lower byte has been incremented first, and only if this rolls over to zero is the upper byte incremented (Table 10.1, lines 20–22). If i was initially 00FFh, then the first INC produced i = 0000. If an interrupt now occurs, and the ISR used, say, i to update array[i], then array[0] will be altered instead of array[256]! Clearly a compiler that used the sequence:

```
LDD     L3_i
ADDD    #1
STD     L3_i
```

would be better; however, long 4-byte integers will still be prone to disjoint global problems like this.

C compilers for 8-bit processors normally use absolute memory locations to hold floating-point numbers, rather than internal registers, and this non-recursive mode makes floating-point arithmetic particularly prone to this problem. Even 16/32-bit devices, which can handle all sizes of integers in one indivisible machine instruction, require multiple floating-point operations, unless using a mathematical

Table 10.9: clock() configured as an interrupt function.

```
Microtec Research ASM68008 V6.2a  Page 1 Thu Apr 20 11:16:34 1989
Line Address
  1                              *    Paragon  MCC68K Compiler   Version 3.1
  2                                   OPT   NOABSPCADD,E,CASE
  3                                   list10_9   IDNT
  4                                   SECTION    9,,C
  5                                   XDEF    .clock
  6    00000400 48E7 C0C0     .clock: MOVEM.L  D0/D1/A0/A1,-(SP)
  7    00000404 202F 0014             MOVE.L   20(SP),D0
  8                              * 1     unsigned char Seconds,Minutes,Hours;
  9                              * 2     #define    $INTERRUPT
 10                              * 3     void clock(void)
 11                              * 4     {
 12                              * 5       if(++Seconds>59)
 13    00000408 207C 0000 E000          MOVE.L   #.Seconds,A0
 14    0000040E 5210                    ADDQ.B   #1,(A0)
 15    00000410 1010                    MOVE.B   (A0),D0
 16    00000412 0C00 003B               CMPI.B   #59,D0
 17    00000416 6332                    BLS.S    _L2
 18                              * 6       {
 19                              * 7         Seconds=0;
 20    00000418 4239 0000 E000          CLR.B    .Seconds
 21                              * 8         if(++Minutes>59)
 22    0000041E 207C 0000 E002          MOVE.L   #.Minutes,A0
 23    00000424 5210                    ADDQ.B   #1,(A0)
 24    00000426 1010                    MOVE.B   (A0),D0
 25    00000428 0C00 003B               CMPI.B   #59,D0
 26    0000042C 631C                    BLS.S    _L3
 27                              * 9         {
 28                              * 10          Minutes=0;
 29    0000042E 4239 0000 E002          CLR.B    .Minutes
 30                              * 11          if(++Hours>23)
 31    00000434 207C 0000 E004          MOVE.L   #.Hours,A0
 32    0000043A 5210                    ADDQ.B   #1,(A0)
 33    0000043C 1010                    MOVE.B   (A0),D0
 34    0000043E 0C00 0017               CMPI.B   #23,D0
 35    00000442 6306                    BLS.S    _L4
 36                              * 12          {
 37                              * 13            Hours=0;
 38    00000444 4239 0000 E004          CLR.B    .Hours
 39                              * 14          }
 40                                    _L4:
 41                              * 15        }
 42                                    _L3:
 43                              * 16      }
 44                                    _L2:
 45                              * 17      return;
 46    0000044A 4CDF 0303               MOVEM.L  (SP)+,D0/D1/A0/A1
 47    0000044E 4E73                    RTE
 48                                    SECTION 14,,D
 49                                    XDEF    .Seconds
 50    0000E000 00            .Seconds: DCB.B   1,0
 51    0000E001 00                      DCB.B   1,0
 52                                    XDEF    .Minutes
 53    0000E002 00            .Minutes: DCB.B   1,0
 54    0000E003 00                      DCB.B   1,0
 55                                    XDEF    .Hours
 56    0000E004 00            .Hours:  DCB.B    1,0
 57    0000E005 00                     DCB.B    1,0
 58                                    END
```

co-processor. Thus, in general it is inadvisable to use floating-point global variables which can be altered by interrupt service routines. Similar considerations apply to any global compound-element structure and multiple-byte integers for 8-bit MPUs. Of course it is always possible to mask out interrupts during sensitive processing.

Interrupt problems occurring due to disjoint operations are particularly pernicious because they appear very rarely and apparently at random. As they are not reproducible to order, it is virtually impossible to track them down!

If global variables have to be shared, the normal advice is to ensure that only the highest order of interrupt service function making use of the variable actually does the changing. Here the background function is treated as level 0. Thus in our Real-Time Clock, the interrupt function clock() is permitted to change the global variables Seconds, Minutes and Hours, with the background and any lower priority interrupts only *reading* these variables. Higher priority interrupt functions should not make any reference to these variables.

This procedure is not foolproof. Consider a background function turning off the central heating pump each morning at 9 am, that is 09:00:00. It turns the pump off and on by pulsing a toggling flip flop. It is now 09:59:59. The program reads Hours as 09. Getting interested, it is about to read Minutes when an interrupt occurs and alters the time to 10:00:00. On return, Minutes and Seconds are then read as 00:00, and the processor thinks it is 9 am, toggles the flip flop and turns the pump on again! This may happen perhaps once a year, but when it does, the switching will continue at 180° from the proper sequence. The cure is to mask out the interrupt when the time is being read, or to read it several times in quick succession — and not to use a toggling flip flop as the pump interface!

10.3 Initializing Variables

Targeting C to a ROM-based system presents problems concerning the data portions of the program. This is where variables go when they are static and/or global (extern). Recall from Section 8.2:

1. auto variables can be initialized in their definition, but the resulting code is identical to a definition subsequently followed by an assignment. As shown in Table 8.3(b), these fixed values are moved into memory each time the local area in which their scope applies is entered, that is run-time setup. Uninitialized variables have an indeterminate value until assigned.

2. static and global variables (static or otherwise) can be initialized in their definition. The resulting code leads to a compile-time set-up, where the constants are placed in memory by the loader, see Table 8.3(a). When the program starts, it assumes that these values are already in situ, put there by some outside agency (the loader). On subsequent executions, any altered variables will not regain their original values, unless a load precedes the run. Uninitialized static/global variables are given an explicit zero value, as for example in Table 10.9.

3. `static` or global objects declared `const`, are placed by the compiler in the text area of memory. In an embedded system, this will be in ROM, and is useful for look-up tables and string constants. Such objects are always present, with their initial values placed there by the *one and only* load into the EPROM programmer. Table 9.2 shows an example of this situation.

Category-2 above constitutes a problem in a ROM-based system, as there is no load prior to each run, and therefore RAM-based `static` and global variables will not have their pre-initialized values. The state of RAM on power-up is indeterminate. The obvious way around this is to adapt the style of the C source, so that no assumptions are made regarding the initial values of such variables. Although most algorithms are amenable to this approach, there are pitfalls to trap the unwary.

The standard problem here is the use of libraries. Since this is code written by someone else, you can never quite be sure how initialized data is handled. In practice, library routines likely to be used by embedded systems, will probably not use pre-initialized `static`/global data, but beware of file and I/O routines.

Although pre-initialized variables can usually be avoided, the safest approach is to use the startup routine to initialize the data program sections in RAM. To do this, the compiler must arrange for an image of initialized data to be present in ROM (usually following the text area). On startup, this is copied byte by byte to the correct place in RAM before going to `main()`. The actual details of how this is done vary considerably from compiler to compiler. To give the reader an overview, we will briefly look at three products, the Aztec C68K/ROM V3.30c, the Microtec Paragon MCC68K V3.1, and Cosmic/Intermetrics V3.3 compilers.

The C language specifies that all `static` and/or global variables are assumed to be zero unless explicitly pre-initialized in their definition. In an embedded system this can simply be implemented by clearing the RAM chip(s) in their entirety at startup. It would be more efficient if only the appropriate number of bytes were cleared, although in a small system this startup burden is likely to be of little consequence.

Most compilers place non-zero explicitly pre-initialized and default zero variables in different but related data program sections. In our compiler examples the non-zero pre-initialized variables go into `DSEG`, `Section 13` and `.PSECT _data`. Uninitialized (or explicitly zeroed) `static`/global variables go into `BSS`, `Section 14` and `.PSECT _bss` respectively (BSS is an archaic expression Block Start Symbol, originally used to denote a block of memory common between various programs). The three compilers put the program into `TSEG`, `Section 9` and `.PSECT _text` respectively. Table 10.9 shows the Paragon code using `Section 9` for program code (line 4) and `Section 14` for the three uninitialized global variables represented by the labels `.Seconds`, `.Minutes`, `.Hours` in line 48. Table 10.5(b) shows the use of the `_bss` segment for the uninitialized `static` variable `i`.

To assist writing the startup routine, the programmer needs to know, for example, how long the two data sections are and where they start. Most linkers create certain reserved public symbols giving this information. In the case of our three example compilers these are:

Paragon:
?RAM_START   Where the data section begins.
?ROM_START   Where the image begins in ROM.
?ROM_SIZE   How many bytes the image is.

Aztec:
__H0_org & __H0_end   Code segment start and end+1.
__H1_org & __H1_end   Initialized data segment start and end+1.
__H2_org & __H2_end   Uninitialized data segment start and end+1.

Cosmic:
__text__   Code segment end+1.
__data__   Initialized data segment end+1.
__bss__   Uninitialized data segment end+1.

The linker allows the programmer to set the starting address of each section separately. If the BSS/Section 14/_bss sections are not biased in this way, then they normally follow directly on from the corresponding DSEG/Section 13/_data section.

Finally how do the compilers produce an image of the pre-initialized data in ROM? The Aztec compiler does this automatically with the image following on directly from the TSEG portion; that is starting at __H0_end. Its length is __H1_end − __H1_org. Using this information, a possible startup for this Aztec compiler is shown in Table 10.10. This is written as an extension to Table 10.3, but using the Aztec's assembler syntax (standard Motorola). Operation is self-evident from the comments; however, note that if a segment does not exist, then the org and end labels are made the same by the linker, so a zero difference signals non-existence.

The Paragon product provides two library routines, which help this copy process. These are .initdata and memclr(). The former is designed to be called directly from the startup routine, for example jsr .initdata, and takes no parameters directly. The latter is normally used from the C program, requiring a pointer to the first byte and an int count.

The linker must be informed through its command file (see Table 7.10) that an image of Section 13 is required, by using the directive initdata. Thus in line 19 of Table 10.11, we tell the linker to generate an image of Section 13 starting at 6000h (unfortunately there is no symbol denoting the end of Section 9, the text). In the startup, jsr .initdata will then use the linker-generated symbols automatically to do the copying. Line 10 informs the linker that Section 14 (uninitialized variables) is to follow Section 13. In doing this, RAM can easily be cleared from ?RAM_START+?ROM_SIZE upwards.

In the Cosmic/Intermetrics compilers, the linker is followed by a hexer utility, which generates machine code in the requested format for the EPROM programmer (see Table 7.5). Each program section can be shifted to a new start point by the hexer; however, as the text remains unaltered, the program still assumes its data is at the linker's (original) data bias. Thus, to produce an image of the data section following on from the text we have:

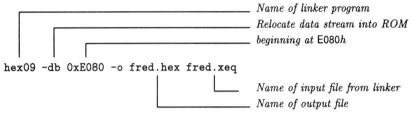

which says produce the (Intel coded) machine code file with the data bias (-db) reset to E080h. The output file is called fred.hex and the input (from the linker) is fred.xeq. The net result of this process is to create a copy of the initialized

Table 10.10: A startup for the Aztec compiler initializing statics/globals.

```
; **********************************************************************
; * Startup for Aztec 68K v3.30c                                        *
; * Copies initial values of statics/globals into RAM                   *
; **********************************************************************
          public   __H0_org, __H0_end
          public   __H1_org, __H1_end
          public   __H2_org, __H2_end
          cseg              ; Code segment
SSP:      dc.l     $A000    ; Initial setting of SSP
PC:       dc.l     _Start   ; Restore PC value
          ds.l     $3F8     ; Skip up to 0400h

_Start:   movea.l #$1000,a0  ; Prepare to set up User Stack Pointer
          move    a0,usp     ; Privileged instruction

; calculate length of initialized data segment in D0.L
          move.l  #__H1_end,d0 ; End of DSEG, +1
          sub.l   #__H1_org,d0 ; Start of DSEG
          beq     ZERO_OUT     ; Don't copy if none
          lsr.l   #1,d0        ; Convert to word length
; Now copy initialized data image in ROM in RAM
          movea.l #__H0_end,a0 ; Image is at end of TSEG
          movea.l #__H1_org,a1 ; Point to start of DSEG
LOOP1:    move.w  (a0)+,(a1)+  ; Copy each word
          dbf     d0,LOOP1     ; D0 holding count

; Calculate length of unitialized data segment in D0.L
ZERO_OUT: move.l  #__H2_end,d0 ; End of BSS, +1
          sub.l   #__H2_org,d0 ; - start of BSS
          beq     CONTINUE     ; Don't zero out if none
          lsr.l   #1,d0        ; Convert to word length
; Now zero unitialized section in RAM
          movea.l #__H2_org,a0 ; A0 points to base of BSS
LOOP2:    clr.w   (a0)+        ; Clear each word
          dbf     d0,LOOP2     ; Until reaches -1

CONTINUE: andi    #$DFFF,sr    ; Clear bit 13 to change to user state
          jsr     _main        ; Go to C routine
          bra     _Start       ; Repeat if returns
          end
```

Table 10.11: A typical lod68k command file to produce an image of initialized data in ROM for use in the startup code.

```
********************************************************************************
* This is a prototype command file for the Microtec linker                     *
* Puts a copy of initialized data in ROM for the startup                       *
********************************************************************************
* Section 0 is for the entry code, e.g. vector table, in ROM                   *
* Section 9 is for the program in ROM usually                                  *
* Section 13 holds initialized local static and global variables, in RAM       *
* Section 14 is for other vars, e.g. Global and uninitialized statics          *
* Put initialized static/globals after uninitialized same                      *
order 13,14              * Put Section 14 after 13                             *
sect 0=0                 * Vector table starting at 00000                      *
sect  9=0400h            * Program starts at 0400h                             *
sect  13=0E000h          * Any data is at E000h up (RAM)                       *
absolute 0,9,13          * Put only these ROM sections in .hex file            *
* Copy section in ROM at zzzzzh for initialized local static                   *
* data produced in Section 13 in RAM, if relevant                              *
* In entry program subroutine .initdata will copy it back                      *
* always at runtime into RAM                                                   *
initdata 13,6000h
list d,s,t,x,c           * Public symbols in object module                     *
*; Local symbol table to object module; Lists it; and public                   *
* symbol table; Produces a cross-reference listing                             *
load startup             * Start up assembler routine                          *
load fred                * Then the compiled user program                      *
load 68000.lib\mcc68kab.lib  * and  absolute library                           *
end                                                                            *
********************************************************************************
```

data in ROM, beginning at E080h but leaving the actual data area unchanged. An example is given in Table 10.12(b).

The Cosmic/Intermetrics 6809 compiler does not produce start_of labels (e.g. Text segment start), but including all programs sections in the startup routine, as shown in Table 10.12(a), defines these local symbols according to the biases set in the linker. Thus if the linker's data bias is 1, then Start_data is 0001h. This routine is similar to that of Table 10.10, but with differing symbols.

Cosmic/Intermetrics provide a utility toprom with their compilers version 3.32 and up, to modify their linker output to create this image automatically. The starting address in RAM and end address of this image in ROM are also embedded into the start of this ROM record, and are used by their provided startup routine. This works in the same way as outlined above, but with less hassle.

As can be seen in Table 10.12, this compiler supports the use of the 6809's direct page (or zero page) address mode (see page 35) as a non-ANSII extension. Any static or extern data object can be placed into the assembler's _zpage program section by preceding it by the directive @dir. Thus, altering line C3 in Table 10.5 to:

@dir static int i;

will change the .psect _bss to .psect _zpage and the two following INC commands will use Direct rather than Extended addressing, as shown in Table 10.13

Table 10.12: A startup for the Cosmic compiler, initializing statics/globals and setting up the **DPR** for zero page.

```
            .processor m6809
;************************************************************
;* Startup routine for Cosmic 6809 V3.3 supporting zero page  *
;* and copying initial values of statics/globals into RAM     *
;************************************************************
            .external _main, __text_, __data_, __bss_
Start_data: .psect   _data         ; Define beginning of data section
Start_bss:  .psect   _bss          ; Define beginning of bss section
Start_zero: .psect   _zpage        ; Define beginning of zero page section

            .psect   _text
_Start:     lds      #0800h        ; Point Stack Pointer to top of RAM
; Now clear bss region
            ldx      #Start_bss    ; Point to beginning of BSS
LOOP1:      cmpx     #__bss_       ; End yet?
            beq      INIT_DATA     ; IF yes THEN move on
            clr      0,x+          ; ELSE clear byte and advance pointer
            bra      LOOP1
; Now setup data region
INIT_DATA:  ldx      #Start_data   ; Point to beginning of data
            ldy      #0E080h       ; Point to beginning of image
LOOP2:      cmpx     #__data_      ; End yet?
            beq      ZERO_PAGE     ; IF yes THEN move on
            lda      0,x+          ; ELSE get byte
            sta      0,y+          ; and move it
            bra      LOOP2
; Set up DPR to point to zero page
ZERO_PAGE:  ldd      #Start_zero   ; Start of zero page
            tfr      a,dp          ; Top byte to Direct Page register
            jsr      _main         ; Go to C code
            bra      _Start        ; Should it return then repeat

            .public  _Start        ; Make this routine known to linker
            .end
```

(a) Assumes an image of the data lies at E080h on up.

:20E0000034463362AE5E8C000C2F074F5F8E0000200F8E0001EC5E58495849308BEC02AE3A

:05E020008432C435C08C

:20 E080 00 0000000100000001000000020000000600000018000000780000002D0000013B0 51

:14 E0A0 00 00009D80000589800005A0A0026115001C8CFC00 E0

:00E000011F

(b) Copying the data segment of a modified Table 9.2 into ROM at E080h upwards.

(see also Table 14.8). All such objects in the file can be placed in the zero page by using the ANSII directive #pragma:

#pragma space[]@dir

but it must be remembered that a page in the 6809's space is only 256 bytes long.

The bias for this page can be set in the linker; for instance:

ln09 +zpage -b0x8000

Table 10.13: Zero-page storage with the Cosmic 6809 compiler.

```
1                         ; Compilateur C pour MC6809 (COSMIC-France)
2                                   .list      +
3                                   .processor m6809
4                                   .psect     _zpage
5                         L3_i:     .byte      [2]
6                         ;    1    main()
7                         ;    2    {
8                                   .psect     _text
9                         ;    3         @dir static int i;
10                        ;    4         while(1)      {i++;}
11   E000  0C01           L1:       inc        L3_i+1
12   E002  2602                     jbne       L4
13   E004  0C00                     inc        L3_i
14   E006  20F8           L4:       jbr        L1
15                        ;    5    }
16                                  .public    _main
17                                  .end
```

sets it to 8000h. This will be the value of Start_zero in Table 10.12. Bringing this down to Accumulator_D and then doing a TFR A,DP sets up the **Direct Page register** to the upper address byte (80h in this example) as required.

I have assumed in Table 10.12(a) that the initial state of the zero page does not matter. If it does, then all 256 bytes can be cleared or an image copied from ROM.

10.4 Portability

To the microprocessor engineer, portability is one of the major attractions of a high-level language. Thus a company upgrading a 6502-based product line to, say, the 68000 family, can continue to use the bulk of the original software, without a substantial change. In reality, the migration of software between differing systems, at the lowest to the highest level, is fraught with difficulties to the unwary [6].

As an example of low-level problems that can occur, most of the newer families of MPU are software downwards compatible. Thus the 80386 MPU has an 8086 emulation mode and the 68020 MPU is object code compatible to the 68000. Consider the CLR <memory> instruction in the 68000/8 MPU. This is implemented as a classical read–modify–write operation, although the data read is irrelevant (see page 25). This means that the address of <memory> is put out on the address bus twice. A devious hardware engineer may deliberately make use of the resulting double address decoder pulse, by using CLR, say, to increment a counter twice. At some time later, probably after this ingenious engineer has left, the company decides to upgrade to a 68020-based microcomputer. They have been assured the 68000 code will directly run under 68020 control. So it does, or does it? Motorola have speeded up CLR on the 68020 MPU and subsequent family members, by dispensing with the initial useless Read cycle, ergo a counter incrementing at half its proper rate! Abstruse bugs like this are difficult and very expensive to unearth, but abound where software is migrated between systems.

At the higher level, one solution to the portability problem is to define a virtual machine (i.e. having a hypothetical structure) together with a UNiversal Computer-Oriented Language (UNCOL) [7]. Each physical machine would have a translator from UNCOL to its particular machine code. With such a scheme, a high-level language would only require the one machine-independent compiler to UNCOL.

Unfortunately no UNCOL exists in practice, although several half-hearted attempts towards this goal have been made. At one time, A-natural [8] was in vogue as a kind of standard assembly language, but its close relationship to the 808x-MPU family led to its eventual demise. Some software engineers consider C as an UNCOL. Certainly its origins as a high-level assembly language used to port the operating system UNIX to various hardware hosts [9] would seem to fit it into that role. Amongst its other virtues, the relative lack of dialects, now enforced by the ANSII standard, makes C one of the most portable of the higher languages. But even here, 100% portability is a pipe-dream, and the term transportable is a more apt description.

Considerations of portability depend on the type and scope of the software. This can roughly be categorized as follows:

1: Operating System independent
A program in this category will run in the same way, irrespective of its cocooning operating system. Thus, for example, the program given in Table 10.14(a) should execute equally well on a Hewlett Packard Apollo work station (680x0-based) under UNIX and on an IBM PC (80x86-based) under MSDOS.

2: Operating System specific
Programs which take advantage of special features of some operating system, and can therefore only run on hardware supporting that operating system.

3: System and machine specific
A further restriction on category-2, but also relies on a specific hardware feature. Hardly portable at all!

4: Unhosted
Typical of embedded microprocessor circuits. Cannot rely on ANSII-standard I/O functions. Both super portable and not portable at all!

Old C had only a de facto standard library, as defined by Kernighan and Ritchie [10]. Compiler writers were free to provide any library functions they felt like. This of course made porting software a nightmare, unless the same compiler was available across the target range. The ANSII standard now provides an essential core of standardized library functions, which must be available no matter what the eventual target is [11]. Thus, in principle, using C compilers conforming to this standard should make category-1 portability easy to achieve.

Two of these standard functions are used in Table 10.14. `scanf()` is a formatted Read function, taking input from the standard input channel `stdin` (usually the keyboard), according to a list of format tokens [12]. Thus:

`scanf("%u",&n);`

means go to `stdin` and get an unsigned decimal integer (`%u`), which will be put away at the address of n (i.e. assigned to n). Other formats tokens are `%d`, `%ld`, `%x`, `%f` etc., for Decimal integer, Long Decimal integer, heXadecimal integer and decimal Floating-point.

`printf()` is the formatted write to standard output function counterpart (`stdout` is normally the VDU screen or printer), which sends messages with variable values replacing embedded format tokens [12]. Thus:

`printf("The sum of all integers up to %u is : %lu\n",num,sum);`

prints the message in quotes, with the format token `%u` replaced by the decimal value of num at that point in the program, and the long decimal value of sum likewise. Notice the use of `\n` to give a new line. Table 10.14(b) shows a run-time example.

Table 10.14: A portable C program using ANSII library I/O routines.

```
#include <stdio.h>
main()
    {
    unsigned short num,i;
    unsigned long sum;
    printf("Enter number \n");
    scanf("%d", &num);
    for(sum=0,i=num; i>1; i--)
        {sum += i;}
    printf("The sum of all integers up to %d is : %ld\n", num,sum);
    }
```

(a) C source code.

```
Enter number
35
The sum of all integers up to 35 is :   629
```

(b) Typical run.

Compilers come with a set of header files, giving amongst other things, prototypes of all the library functions. Some of these are `<stdio.h>` for the standard input/output functions and `<stdlib.h>` for utility functions. Table 9.10 shows a typical `<math.h>` mathematics function header.

Unfortunately, even with the ANSII standard, many details are left as implementation dependent. For example, the size of ints (typically 16 or 32 bits), whether an unqualified char is signed or unsigned, the direction of truncation for / (divide) and the sign of the result for % (remainder) are machine-dependent for negative operands. File handling, for example rules for naming, and various system-related constants, such as the End of File constant (EOF is usually -1), are operating-system specific.

Most implementation and operating-system foibles tend to be obscure and difficult to track down. As a simple example of the former, consider the code fragment:

```
int i;
for (i=0; i<32768; i++)        {do this;}
```

This will work perfectly well in an implementation which maps int on to a 32-bit word, but this will be done forever on a 16-bit implementation with its largest value of +32767 (7FFFh).

To reduce the possibility of this kind of problem, system-dependent variables should be gathered together into a header file, which can easily be altered if the software is transposed. Also standard types can be defined. Thus:

```
SIGNED_32 i;
for (i=0; i<32768, i++)        {do this;}
```

where the header contains the typedef

```
typedef long SIGNED_32
```

for a 16-bit implementation and

```
typedef int SIGNED_32
```

for a 32-bit int size.

Programs generated by a compiler with extended libraries and/or using special operating-system specific features, are category-2 portable. A considerable rewrite will be necessary to port such software to different machines, especially in the latter situation. Severe problems arise where some special host hardware feature is utilized. Graphics-oriented software frequently comes into this category-3, and the concept of portability to another operating system/host is then virtually meaningless.

Porting embedded microprocessor C code presents the engineer with its own set of particular problems. Provided that the source does not make use of non-ANSII features, the bulk of raw code will translate to any target. C compilers are available for the majority of CPUs, from mainframe down to microcontroller. Some examples resulting from our sum-of-integers source are presented without comment in Table 10.15.

Most remarks made previously also apply to this category-4 portability. In particular the hardware-oriented nature of free-standing systems leads to code which makes assumptions concerning the structure in memory of data. For example, byte ordering in some processors places the most significant bits of a word in the lower byte address (the so called Little-Endians); others do the opposite (the Big-Endians). Thus breaking up a 16-bit int word into two chars named byte1 and byte2 in this manner:

Table 10.15: Compiling the same source with a spectrum of CPUs (*continued next page*).

```
;:ts=8
;main(n)
;unsigned char n;
        public  _main
_main:  link    a6,#.2
        movem.l .3,-(sp)
;{
;static unsigned short sum;
        bss     .4,2
;for(sum=0;n>0;n--)
        clr.w   .4
        bra     .8
;   {sum+=n;}
.7      move.l  #0,d0
        move.b  11(a6),d0
        add.w   d0,.4
.5      sub.b   #1,11(a6)
.8      tst.b   11(a6)
        bhi     .7
;return(sum);
.6      move.l  #0,d0
        move.w  .4,d0
.9      movem.l (sp)+,.3
        unlk    a6
        rts
;}
.2      equ     0
.3      reg
        dseg
        end
```

(a) Aztec 68000 MPU V3.30c.

```
;main(n)
;unsigned char n;
_main:  push    bp
        mov     bp,sp
        mov     cx,word ptr 4[bp]
;{
;static unsigned short sum;
;for (sum=0;n>0;n--)
        mov     word ptr [026c],0
        jmp     L2
;   {sum+=n;}
L1:     add     word ptr [026c],c
        dec     cx
L2:     or      cx,cx
        jne     L1
;return(sum);
        mov     ax,word ptr [026c]
;}
        pop     bp
        ret
        .public _main
        .end
```

(b) Zortech 8086 MPU V3.0 with debugger V1.02.

```
;:ts=8
;main(n)
;unsigned char n;
        public   main_
main_:  jsr     .csav#
        fcb     .3
        fdb     .2
;{
;static unsigned short sum;
        bss     .4,2
;for (sum=0;n>0;n--)
        stx     .4
        stx     .4+1
        jmp     .6
.5      clc
        lda     #255
        ldy     #11
        adc     (4),Y
        sta     (4),Y
        lda     #255
        adc     #0
.6      ldy     #11
        lda     (4),Y
        sta     24
        stx     25
        txa
        cmp     24
        sbc     25
        jcs     .7
;   {sum+=n;}
        lda     (4),Y
        sta     24
        stx     25
        clc
        lda     .4
        adc     24
        sta     .4
        lda     .4+1
        adc     25
        sta     .4+1
        jmp     .5
;return(sum);
.7      lda     .4
        sta     8
        lda     .4+1x
        sta     9
        rts
;}
.2      equ     0
.3      equ     0
        public .begin
        dseg
        cseg
        end
```

(c) Aztec 6502 MPU V3.20c.

Table 10.15: Compiling the same source with a spectrum of CPUs (*continued next page*).

```
         NAME    summ(18)                              NLIST   D
         RSEG    CODE(0)                               LIST    E,L
         RSEG    UDATA(0)            ; Version 1.5 Compiler 860818 P Code Gen 860715
         PUBLIC  main(150,191)       ; Source: summ Prog: Date: 25-MAY-1989 12:50:15
         EXTERN  ?CL6801_1_15_L07                      NAME    summ
         RSEG    CODE                                  DSEG
* 1.     main(n)                                       DEFS    00002H
         P6801                                         EXTPUB  GLOB_
* 2.     unsigned char n;            GLOB_:            STKLN   100H
* 3.     {                                             EXTRN   ENTZ2_
main:    PSHB                                          XSEG
         PSHA                        CONST_:           CSEG
* 4.     static unsigned short sum;  M_summ:           JP      ENTZ2_
* 5.     for (sum=0;n>0;n--)         ; 1.  main(n)
         CLRB                                          EXTPUB  MAIN_
         CLRA                        ; 2.  unsigned char n;
         STD     ?0000               ; 3.  {
?0002:   TSX                         ; 4.  static unsigned short sum;
         LDAB    1,X                 ; 5.  for (sum=0;n>0;n--)
         PSHB                        ; 6.       {sum+=n;}
         CLRB                        ; 7.  return(sum);
         TSX                         MAIN_:            PUSH    HL
         SUBB    0,X                 ;     line 3
         INS                         ;     line 5
         BCC     ?0001                                 XOR     A
* 6.          {sum+=n;}                                LD      (GLOB_+0FFFFH),A
?0003:   TSX                         Q_2:              LD      HL,00000H
         LDAB    1,X                                   ADD     HL,SP
         CLRA                                          LD      A,(HL)
         PSHB                                          AND     A
         PSHA                                          JR      Z,Q_1 ;
         LDD     #?0000                                JR      Q_3 ;
         PULX                        Q_4:              LD      HL,00000H
         PSHB                                          ADD     HL,SP
         PSHA                                          LD      A,(HL)
         PSHX                                          DEC     A
         PULA                                          LD      (HL),A
         PULB                                          JR      Q_2 ;
         PULX                        Q_3:;     line 6
         ADDD    0,X                                   LD      HL,00000H
         STD     0,X                                   ADD     HL,SP
         TSX                                           LD      C,(HL)
         PSHB                                          LD      A,(GLOB_+0FFFFH)
         PSHA                                          ADD     A,C
         PSHX                                          LD      (GLOB_+0FFFFH),A
         PULA                                          JR      Q_4 ;
         PULB                        Q_1:;     line 7
         PULX                                          LD      A,(GLOB_+0FFFFH)
         ADDD    #1                                    LD      L,A
         PSHB                                          LD      H,00000H
         PSHA                                          POP     DE
         PSHX                                          RET
         PULA                        ; 8.  }
         PULB                                          DSEG
         PULX                                          ORG     GLOB_
         LDAB    0,X                                   XSEG
         DEC     0,X                                   ORG     CONST_
* 7.     return(sum);                                  END
         BRA     ?0002               ; Code Bytes: 49 (4) Constant Bytes: 0
?0001:   LDD     ?0000               ; Data Bytes: 2
* 8.     }                           ; Constant Bytes: 0
?0005:   PULX
         RTS                         (e) Microtec Z80 MPU V1.5.
         RSEG    UDATA
?0000:   RMB     2
         END

(d) IAR 6801 MCU V1.15/MD2.
```

Table 10.15: (*continued*) Compiling the same source with a spectrum of CPUs.

```
Transputer DECODE (V1.2) of t_sum.bin
ID T8 "occam 2 V2.1"
"CC_transputer V2.0"
SC 0
TOTALCODE 148 0
STATIC 2
      1  main(n)

CODESYMB "main" 00000030
               71  00030  ldl     1
               30  00031  ldnl    0
            20 20  00032  ldnl    MODNUM
            BF 60  00034  ajw     -1
               D0  00036  stl     0
      2  unsigned char n;
      3  {
      4  static unsigned short sum;
      5  for(sum=0;n>0;n--)
               40  00037  ldc     0
               70  00038  ldl     0
               E1  00039  stnl    1
               13  0003A  ldlp    3
               F1  0003B  lb
               40  0003C  ldc     0
               F9  0003D  gt
            A0 21  0003E  cj      00050
      6  {sum+=n;}
               70  00040  ldl     0
               31  00041  ldnl    1
               13  00042  ldlp    3
               F1  00043  lb
               F5  00044  add
               70  00045  ldl     0
               E1  00046  stnl    1
               13  00047  ldlp    3
               F1  00048  lb
               41  00049  ldc     1
               F4  0004A  diff
               13  0004B  ldlp    3
            FB 23  0004C  sb
            0A 61  0004E  j       0003A
      7  return(sum);
               70  00050  ldl     0
               31  00051  ldnl    1
               B1  00052  ajw     1
            F0 22  00053  ret
      8  }
```

(f) Parallel C INMOS Transputer T825 V2.0.

```
; Compilateur C pour MC68HC16 (COSMIC-France)
        .psect    _bss
        .even
L3_sum: .byte     [2]
        .psect    _text
        .even
_main:  pshm      x,d
        tsx
        .set      OFST=0
        clrw      L3_sum
L1:               ; line 5, offset 7
        ldab      OFST+3,x
        beq       L11
        clra
        addd      L3_sum
        std       L3_sum
        dec       OFST+3,x
        bra       L1
L11:              ; line 6, offset 27
        ldd       L3_sum
        ldx       0,x
        ais       #4
        rts
        .public   _main
        .end
```

(g) COSMIC/Intermetrics V.3.32 6816 MCU.

```
  1  smain(n)
  2  unsigned char n;
smain:    .entry    smain,^m,r2,r3.
          subl2     #4,sp
          movab     $DATA,r3
  3  {
  4  static unsigned short sum;
  5  for(sum=0;n>0;n--)
          clrw      (r3)
          moval     4(ap),r2
          movzbl    (r2),r0
          beql      sym.2
          nop
  6  {sum+=n;}
sym.1:    movzwl    (r3),r1
          movzbl    (r2),r0
          addl2     r1,r0
          cvtlw     r0,(r3)
          decb      (r2)
          movzbl    (r2),r0
          bneq      sym.1
  7  return(sum);
sym.2:    movzwl    (r3),r0
          ret
  8  }
```

(h) DEC V.2.3-024 VAX 750 minicomputer.

```
          NAME      test(16)
          RSEG      CODE(0)
          RSEG      DATA(0)
          PUBLIC    main
          EXTERN    ?CL6811_3_00_L07
          RSEG      CODE
          P68H11
  1  main(n)
  2  unsigned char n;
  3  {
main:              PSHB
                   PSHA
  4  static unsigned short
  5  for(sum=0;n>0;n--)
                   CLRB
                   CLRA
                   STD       ?0000
?0002:             TSX
                   LDAB      1,X
                   CMPB      #0
                   BLS       ?0001
  6  {sum+=n;}
?0003:             TSX
                   LDAB      1,X
                   CLRA
                   ADDD      ?0000
                   STD       ?0000
                   DEC       1,X
  7  return(sum);
                   BRA       ?0002
?0001:             LDD       ?0000
  8  }
                   PULX
                   RTS
                   RSEG      DATA
?0000:             FCB       0,0
                   END
```

(i) IAR 6811 MCU V3.00E.

```
byte1 = word/256; byte2 = word;   and   byte1 = word>>8 ; byte2 = word;
```

will only be equivalent for the latter case. Particular problems arise in reconciling targets with segmented address spaces and special I/O instructions (e.g. the 80x86 family) to code targeted to processors with a linear address space and memory-mapped I/O (e.g. the 680x0 family).

In practice, most portability problems occur in handling I/O and files. Operating systems are designed to act as an insulating layer between applications programs (software that you write) and such considerations. Most small and medium-sized embedded systems are self-standing, or at most a ROM-based monitor may be resident.

Without this decoupling, it is likely that the designer will have to write the startup/support code and library routines to handle, where applicable, interrupts, fault response, memory management and device protocol. A good deal of this is processor dependent, and so must be coded at assembler level, which by definition is non-portable.

A larger embedded system may be able to support the overhead of a resident commercial operating system. The majority of standard operating systems are not suitable for this category of system, supposing as they do a fairly standard computer environment. More relevant real-time systems software can be purchased, but hardly add to the portability score. Sometimes a single-board computer may be available which mimics a standard computer architecture, such as an IBM PC. This can then be used in certain circumstances with a standard operating system, such as MSDOS. A ROM version of the system software is available where a magnetic disk bulk storage unit is not required.

An embedded configuration is characterized by a rich variety of I/O devices, such as lamps, 7-segment and alphanumeric LCD displays, switches, keypads, analog to digital converters and many more exotic examples. Using standard I/O library routines, such as in Table 10.14, is hardly practicable in these situations. Instead special device drivers must be developed. These can be written in C, but care must be taken, as machine and architectural considerations intrude at this level. Standard ANSII and other library routines which do not access I/O can be utilized in the normal way.

Where peripherals resembling standard computer terminals will be attached to the system, then the ANSII I/O routines can be used in the usual way. These routines, such as printf() and scanf() as well as file input/output make use of the base routines putchar() and getchar(). Thus if putchar() and getchar() are written to suit the target hardware, then the higher-order input/output library routines can be used in the normal way.

As an example, consider a self-standing circuit based on an embedded 68000-MPU. This system runs under an operating system monitor which communicates with a terminal through a bidirectional serial link through a UART. The monitor has two subroutines to send and receive single characters along this link. Subroutine OUTCH is located at 7F1Eh and sends out one 8-bit character located in the bottom byte of **D0**. Subroutine INCH at 7F00h waits until a character is received and returns

Table 10.16: Tailoring the ANSII I/O functions to suit an embedded target.

```
int putchar(unsigned char c)
{
_asm("clr.l    d0\n");
_asm("move.b   7(sp),d0    * Get c out of stack widened to int \n");
_asm("jsr      0x7F1E      * OUTCH \n");
_asm("clr.l    d7\n");
_asm("move.b   d0,d7       * Return(c); \n");
}
```

(a) The putchar() function in maximot.h.

```
int getchar(void)
{
_asm("clr.l    d7\n");
_asm("jsr      0x7F00    * INCH \n");
_asm("move.b   d0,d7     * Return it \n");
}
```

(b) The getchar() function in maximot.h.

```
#include <maximot.h>
#include <stdio.h>
main()
{
printf("Hello world");
}
```

(c) A silly main() function to print out a string.

```
*    1  #include <maximot.h>
*    2  #include <stdio.h>
L5:                          ; This is the string "Hello world",0
        .byte     72,101,108,108,111,32,119,111
        .byte     114,108,100,0
*    3  main()
*    4  {
        .even
_main:  link      a6,#-4     ; Open a frame to send a pointer to the string
*    5  printf("Hello world");
        move.l    #L5,(sp)   ; Push out the pointer to the string
        jsr       _printf    ; Goto the printf() function declared in <stdio.h>
        unlk      a6         ; Close down the frame
        rts                  ; and return to the Startup routine

        .globl    _main
        .globl    _printf
```

(d) The resulting source code, with printf() extracted from the library.

with it in the lower byte of **D0**.

The definition of the C function putchar() is:

- Accepts an unsigned character as its single parameter

- Returns this character as an int

giving the declaration unsigned char putchar()(int c), where c is the character to be sent out. The definition of this function is given in Table 10.16(a), and simply extracts c from the System stack (seven bytes up from **SP**), jumps to subroutine OUTCH and then widens and copies it to **D7.L**, the normal return register for this compiler (Cosmic V3.32).

The definition of getchar() is:

- Does not take any input parameter

- Returns the received character, widened to an int

- If there is a problem getting this character, then a special End Of File (EOF) is returned. In this compiler EOF is -1 (FFFFFFFFh)

giving the declaration int getchar(void). The definition of this function is given in Table 10.16(b). As the monitor function INCH does not return an error condition, the EOF protocol is not implemented (a more sophisticated INCH subroutine would detect problems such as parity violation or overrun).

Finally to illustrate the concept, a main function printing out a simple string is shown in Table 10.16(c). The two tailored functions are included as a header file <maximot.h>; alternatively they could be incorporated in a library. The library function printf() is declared in the ANSII standard header file <stdio.h> (for STanDard Input/Output). The actual use of printf() is commented in the listing, and is straightforward in this simple example. The actual machine code produced by this example (with an integral-only version of printf()) was 2950 bytes. Although this may seem extravagant, printf() is an extremely versatile and flexible function. If all we required of printf() was to output fixed strings, the ANSII library function puts() (for PUT String) would give a much more economical solution. Similarly gets() is a more limited input library function. A string in C is defined as an array of character codes terminated with 00h (see line 5 in Table 10.16(d)). Of course both gets() and puts() use the base functions getchar() and putchar().

References

[1] Lawrence, P and Mauch, K.; *Real Time Microcomputer Systems Design*, McGraw-Hill, 1987, Section 7.6.

[2] Banahan, M.; *The C Book*, Addison-Wesley, 1988, Section 5.6.

[3] Kernighan, B.W. and Ritchie, D.M.; *The C Programming Language*, Prentice-Hall, 2nd. ed., 1988, Section 5.4.

[4] Doyle, J.; C – An Alternative to Assembly Programming, *Microprocessors and Microsystems*, **9**, no. 3, April 1985, pp. 124–132.

[5] Crenshaw, J.W.; Square Roots are Simple?, *Embedded Systems Programming*, **4**, no. 1, Nov. 1991, pp. 30–52.

[6] Dettmer, R.; A Movable Feast: The TDF Route to Portable Software, *IEE Review*, **39**, no. 2, March 18th, 1993, pp. 79–82.

[7] Goor, A.J. van de; *Computer Architecture and Design*, Addison Wesley, 1989, Section 2.5.3.

[8] Reid, L. and McKinly, A.P.; Whitesmiths C Compiler, *BYTE*, **8**, no. 1, Jan. 1983, pp. 330–344.

[9] Johnston, S.C. and Ritchie, D.M.; Portability of C Programs and the UNIX System, *The Bell System Technical Journal*, **57**, no. 6, part 2, 1973, pp. 2021–2048.

[10] Kernighan, B.W. and Ritchie, D.M.; *The C Programming Language*, Prentice-Hall, 1978, Chapter 7.

[11] Kernighan, B.W. and Ritchie, D.M.; *The C Programming Language*, Prentice-Hall, 2nd. ed., 1988, Chapter 7 and Appendix B.

[12] Barclay, K.A.; *ANSI C Problem Solving*, Prentice-Hall, 1990, Appendix F.

Part 3

Project in C

In this part we follow through an embedded microprocessor-based product from inception through hardware and software design to the testing and debugging of a functional prototype. Both C and assembly-level software implementations are considered to compare the two techniques. In a similar vein a 6809 and 68008-based system are designed in order both to emphasize the portability of a high-level language and to illustrate the feasibility of C as the language of choice in an 8-bit target, as well as the more traditional 16/32 bit product.

As well as an exercise in programming in C, the project is used as a vehicle to examine some of the products that are available to aid in the investigation of software veracity and also the interaction between hardware and software.

CHAPTER 11

Preliminaries

Trend monitoring is a common instrumentation requirement. The aneroid recording barometer, providing hard copy of typically a month's atmospheric pressure, is an everyday example. The techniques used to acquire and display such data in any particular situation depend on the signal characteristics.

Short-duration non-repetitive events are typically captured and displayed on a storage oscilloscope. Until relatively recently, electrostatic storage cathode ray tubes (CRTs) were used to combine both memory and display functions. Most current equipment uses semiconductor RAM for storage, in conjunction with a CRT display. Digital storage oscilloscopes have the advantage that acquired data can be subsequently read out to a computer for analysis and onto an X-T recorder for hard copy. Single-shot events lasting from 10^{-8} to 10^8 seconds can readily be accommodated.

Repetitive events require a slightly different approach. Normally it is necessary to view the signal for the last few cycles. For events lasting from several seconds upwards, a chart recorder gives both storage and display; for example the aneroid barometer. The maximum pen writing speed (slew rate) of typically 500 cm/s places a lower limit on the event cycle duration. Even where the writing speed is adequate, fast events can lead to records physically occupying a great deal of space.

Where fast sub-millisecond repetition cycle signals are to be observed, the oscilloscope is the instrument of choice. The timebase is triggered at some unique point on the waveform and adjusted to display the duration of interest. Although the cycle time can be very short indeed, any changes must be long term in order to be observed.

The standard oscilloscope display relies on the persistence of human vision, which 'sees' images repeated at a rate of 50 per second or more as flicker free; the principle of television and the cinema. Slow timebase rates (where the scan duration is measured in seconds) give a moving-spot display, as the luminance lifetime of standard CRT phosphor is typically 10 ms. Long-persistence phosphors are available, and the classic example of an application of this technique is radar. Figure 11.1 shows a simulated trace of the electrical activity of the human heart using a long-persistence CRT. The nominal period of this electrocardiogram (ECG or EKG) signal is between 20 and 180 beats/minute. Using a timebase of 200 ms/cm, gives a potential 2-second record length.

316 *Preliminaries*

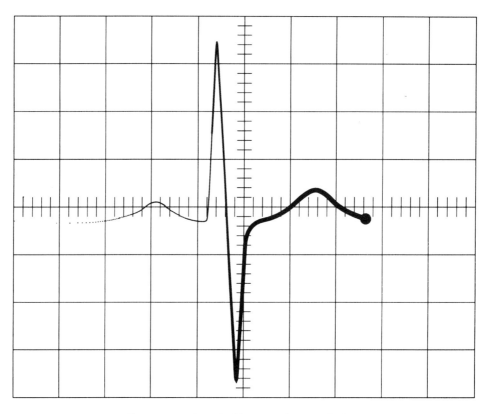

Figure 11.1: A typical long-persistence display.

The main problem here, besides the variable brightness trace, is the fixed nature of the phosphor's luminance lifetime; thus the CRT must be selected with the application in mind. A digital solution, in conjunction with a standard CRT, provides a much more flexible solution. Using a MPU to control the acquisition, storage and display of the data, means that additional features, such as freeze, back spacing and signal processing, can also be accomplished. Furthermore, once the data is in situ it can be used in ways not related to the display function.

In essence, we need to continuously sample the signal at a suitably slow rate, while concurrently scanning and displaying several seconds' worth of past data at a faster rate suitable for the human eye. A typical sequence, showing the resulting scrolling trace, is shown in Fig. 11.2. This diagram shows file snapshots taken at $\frac{1}{4}$ window intervals. A window here is defined as the time past shown on the display. The most historical data is shown to the left of the display, and this results in the trace scrolling to the left as new data is acquired. In implementing this process for our project, we will have created a time-compressed memory. Unlike Fig. 11.1, this technique does not rely on the phosphor luminance lifetime.

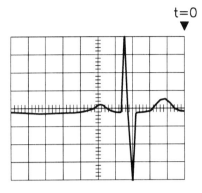
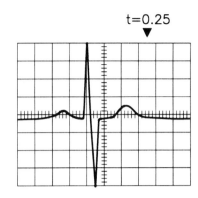
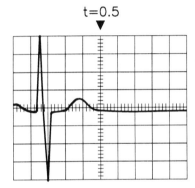
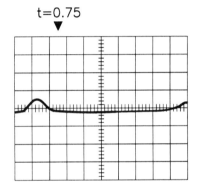
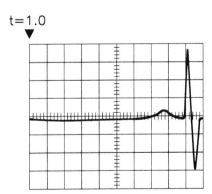

Figure 11.2: Characteristic scrolling display of a time-compressed memory.

11.1 Specification

The customer specification is the rock on which the enterprise is built. As such, it should be treated with the same respect afforded to the foundation of any building.

The product request will normally originate either from the customer, or as a projected need from marketing personnel. Unless the objective is the exact replacement of a product already on the market, for example a central heating controller, such a request is likely to be couched in the language of the application rather than in technical terms. There will be obvious boundary constraints of both a financial and technical nature, but other concerns may well involve complying, say, with legal rulings, such as medical safety requirements.

In essence, the design team must tease as much information as possible from the originator; take away the request and return with a set of proposals. This will involve consideration of the following questions:

- What is it to do?

- Is it possible?

- How is it to be done?

- Can the request be modified?

The outcome of these deliberations is communicated to the customer, and after several iterations a concrete specification will emerge, provided that the project is thought viable. It is important that the specification be decided at this point, not least to avoid the phenomena of 'creeping featurism'. The document will be used as the basis of a suitably detailed implementation, culminating in a working prototype. It may even be used as a legal document, should litigation occur!

With this discussion in mind, let us begin the process with the specification on which our project is based. The customer has asked us to construct a portable ECG/EKG monitor with the following outline specification:

1: Input
Three-lead ECG/EKG signal with integral amplifier having a bandwidth of 0.14 Hz to 50 Hz.

2: Output
100 mm (4″) width standard CRT, displaying a nominal two seconds worth of data.

3: Data accuracy
±0.5% of full scale.

4: Display resolution
Better than 0.5 mm.

5: Facilities
Freeze on demand. Sampling variation of −50% to +100% around nominal.

A prototype circuit is to be built, to demonstrate the feasibility of the proposal and to win customer approval. It will allow simple field trials to be undertaken. The prototype will use commercial power supplies and an oscilloscope as a display. These standard components can be bought in or designed in-house according to production design considerations.

If we make an initial decision to choose a MPU-based implementation as our starting point (see also Section 11.2), then we can do a primary feasibility study on paper.

With an upper frequency of 50 Hz, Shannon's sampling theorem tells us that a minimum of 100 samples will be required per second, say, 128 as a round (binary) number. For a 2-second record length, this requires 256 data points.

At 128 samples per second, the sample period is 7.8 ms. As a first strategy, we could have one complete scan across the CRT in this time, and thus a new sample would be taken each screen scan (probably during flyback). Allowing 1 ms for flyback, the 256 samples would be displayed in 6.8 ms; giving 26.6 μs per dot on the screen. In this time, the microprocessor would need to increment the X co-ordinate, get the next sample from the array, and send out the X and Y co-ordinates. This is probably getting close to the limit of what a general-purpose MPU is capable of, especially if a simpler microcomputer unit (see Section 11.2) is chosen. As the scanning rate here is 128 per second, we can afford to consider two new samples during each full screen scan. At 64 scans per second, this is still well above the flicker rate, but the trace will jerk two dots left after each scan (see Fig. 11.2). We now have $7.8 \times 2 = 15.4$ ms (two sample periods) for the scan plus flyback; that is 14.4 ms per scan. Dividing by 256 gives 56 μs per screen dot, which is a satisfactory compromise.

In summary, for a time-compressed memory complying with the customer specification, as shown in Fig. 11.3, we have:

1. Sampling rate   128 samples/second

2. Memory capacity   256 samples (for a 2-second frame)

3. Scan rate   64 frames/second

4. Flyback time   1 ms nominal

5. Time between steps   56 μs, at two samples per scan

The average adult has a resting heartrate of 72 beats per minute (0.83 Hz), with a variation between 40 and 180 beats per minute over all conditions. Although the frequency range is essentially contained in the range 0.14 Hz to 50 Hz [1], most of the energy lies below 20 Hz. Thus the 128 Hz sampling rate will give at least six samples per cycle, which is just adequate for reasonable visual representation. Increasing the sampling rate to, say, 512 Hz, would require a 1024-word data store and consequently a 10-bit digital to analog converter. Furthermore, eight samples per scan would be needed to keep the dot rate on the screen the same. Remember that the dot rate is the time used by the MPU to get and send out the new X and Y values to the CRT amplifiers.

Of course designing a prototype and subsequent modifications is only the beginning of the process. Setting up a production line is expensive, and the alternative of subcontracting all or part of this activity is one of the major design decisions that will be taken at this point. With the assumption of in-house manufacture, which is only really feasible for large scale production, the next stage is the construction of several preproduction prototypes. In making a few units, as if for sale, the production team will be verifying that the system can be economically built on an assembly line. Electronic devices are relatively standard, but mechanical components, such as printed-circuit boards, switches, connectors, case and artwork are somewhat variable. Decisions must be made regarding methods of construction, second-sourcing of components, stock levels and even down to whether to use surface mount or sockets for the integrated circuits. Just as important, but often overlooked, is how and when to test components, subsystems and the final product.

The production literature covers assembly details and wiring patterns. In some cases programs for computer-aided manufacture (CAM) facilities will be covered under this heading. Included in this category is the testing documentation. This may be either a tester's manual or software for automatic test equipment (ATE).

Post-production documentation covers service manuals and of course the user's handbook. The quality of this material will often add considerably to the customer's satisfaction, which hopefully will eventually increase the reputation of the manufacturer and eventually increase sales.

11.2 System Design

As shown in Fig. 11.4, there are several critical steps between agreeing a specification and actually getting down to the minutia of hardware/software design [2]. One global decision involves the selection of the system transducers, since these form the interface between the electronics and the real world. These will be chosen on the basis of an analysis of the parameters involved, together with their measurement and interconversion to an analogous electrical quality. In our case, standard ECG/EKG configured pads (see Fig. 11.3) sense the bioelectrical potentials and a 100 mm CRT acts as the output device.

The choice of transducer is not unduly influenced by the technology which will be used for the central processing electronics. However, their selection at this time, coupled with a system task analysis, will permit a speculative block diagram to be made of the system. This system formulation is shown for our specific project in Fig. 11.3.

At this point some thought can be given to the technology of the central electronics. At the functional level, this will involve a partition between the analog and digital processes. For example, should an input signal be filtered before the A/D conversion (analog filter) or after (a digital software filter)?

At the digital end of things, the choice essentially lies between random logic

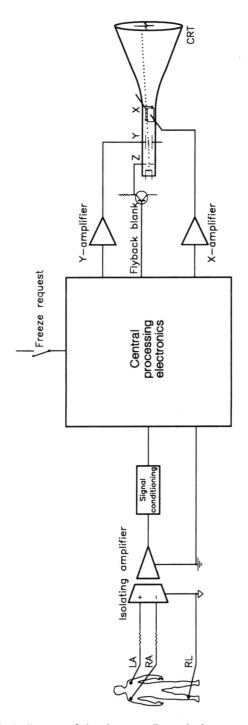

Figure 11.3: Block diagram of the electrocardiograph time compressed memory.

(hard-wired digital combinational and sequential circuitry) and programmable logic (microprocessor-based software-directed hardware). Conventional logic is often best for small systems with few functions, which are unlikely to require expansion. Indeed the present project was based on a random-logic time-compressed memory predecessor. In larger mass-produced products, this type of logic may appear in the guise of programmable arrays, semi and fully custom-designed integrated circuits.

Microprocessors work sequentially doing one thing at a time, while random logic can process in parallel. Thus, where nanosecond speed is important, conventional logic is indicated (but note that analog electronics is even faster). It is possible to run many microprocessor chips in parallel, the transputer being the seminal example. The conventional approach uses mixed logic with a microprocessor in a supervisory role controlling the action of supporting random logic and analog circuitry.

In keeping with the objective of this book, we choose a microprocessor-based implementation. In such cases, the processing tasks must be partitioned between hardware and software. As an example, consider an extension to our specification, where the time between ECG/EKG peaks is continually measured, and is to be displayed on a separate alphanumeric readout. Now we have a choice between using an expensive intelligent display, which incorporates an integral ASCII decoder [3], or a cheaper dumb display, where the segment patterns are picked out by software. The former will cost more on a unit basis, but the latter will require money before the product is launched, to design the software-driver package. This of course is a fairly trivial example, but in general hardware is available off the shelf and therefore has a low initial design cost and takes some load off the central processor. At this level, software is rarely obtainable off the shelf and requires initial investment in a (fairly highly paid) software engineer, but is usually more flexible than a hardware-only solution. In some cases, technical considerations rule out one or other approach. Thus in our example, it is likely that the processor will not have time both to display the waveform and to pick out the peak (a more difficult task than it seems) in software. Hence, external peak-picking hardware is indicated, as shown in Fig. 6.1. Of course, this hardware could be another MPU running a peak-detection software routine! Thus, when technically feasible, a software-oriented solution is indicated for large production runs, where the initial investment is amortized by a lower unit cost.

With a provisional task allocation between hardware and software, what choices has the designer available in implementing the hardware? There are three main approaches to the problem.

In situations where the ratio of design cost to production numbers is poor, a system implementation should be considered. This entails using a commercial microcomputer, such as an IBM PC, as the processor. Such instruments are normally sold with keyboard, VDU, magnetic and random access memory, which are sufficient for most tasks. Ruggedized rack mounting industrial and portable versions are also available. Generally, the hardware engineer will be concerned only to customize the system, by designing specialist supporting hardware and interface circuitry. The software engineer will create a software package based on this

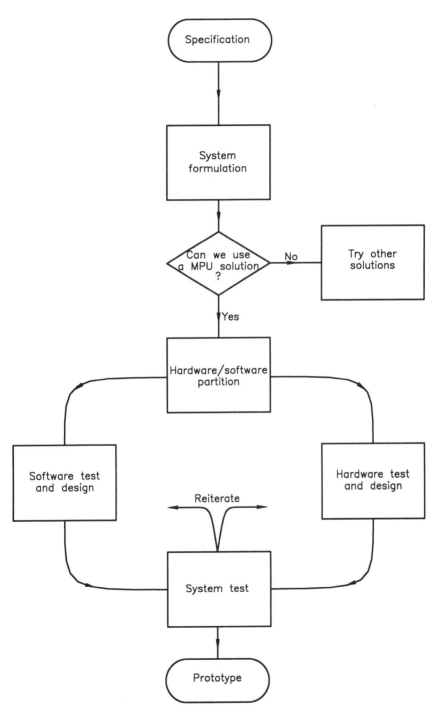

Figure 11.4: A broad outline of system development.

microcomputer, which will drive the hardware. The microcomputer will support commercially available development packages, such as editors, assemblers, compilers and debuggers, which facilitates the software design process at low cost.

Tailoring a general purpose machine to a semi-dedicated role requires a relatively low investment up-front and low production expenses. Furthermore, documentation and the provision of service facilities are eased, as a pre-existing commercial product is used. Technically this type of implementation is bulky, but where facilities such as a disk drive and VDU are needed, the size and unit cost are not necessarily greater than a custom-designed equivalent. Sometimes the customer may already possess the microcomputer; the vendor simply selling the hardware plug-in interface and software package. This can be an attractive proposition for the end user.

Thus a system-level implementation is indicated when low-to-medium production runs are in prospect and the system complexity is high; for example computerized laboratory equipment. For one-offs this approach is the only economic proposition, provided that such a system will satisfy the technical boundary constraints. For instance, it would be obviously ridiculous (but technically feasible) to employ this technique for a washing-machine controller.

At the middle range of complexity, a system may be constructed using a bought-in single-board computer (SBC). Sometimes several modules are used (eg. MPU, memory, interface), and these are plugged into a mother-board carrying a bus structure. If necessary these may be augmented with in-house designed cards to complete the configuration.

Although the cost of these bought-in cards is many times that of the material cost of the self-produced equivalent, they are likely to be competitive in production runs of up to around a thousand. Like system-implemented configurations, they considerably reduce the up-front hardware expenses and do not require elaborate production and test facilities. By shortening the design time, the product can be marketed earlier, and subsequently the economics improved by substitution of in-house boards. Whilst more expensive than a system based on a commercial microcomputer, a board-level implementation gives greater flexibility to configure the hardware to the specific product needs. Furthermore, in a multiple-card configuration, at least some of the standard modules can be used for more than one product (e.g. a memory card) thus gaining the cost benefits of bulk buying. Reference [4] gives an example of this approach.

Neither system or board-level implementations provide an economical means of production for volumes much in excess of a thousand, with the exception of high complexity-value products. In many cases, technical demands, such as size and speed, preclude these techniques even for small production runs. In such situations, a fundamental chip-level design, as outlined in Fig. 11.5, is indicated.

Chip-level involves implementation at the integrated circuit (or even silicon) level. In this situation, the circuit is fabricated from scratch, giving a configuration dedicated to the specific application. Given that the designers require an intimate knowledge of, for example, a spectrum of integrated circuits, CAD techniques for PCB and ASIC design and software techniques, and must keep an eye on the

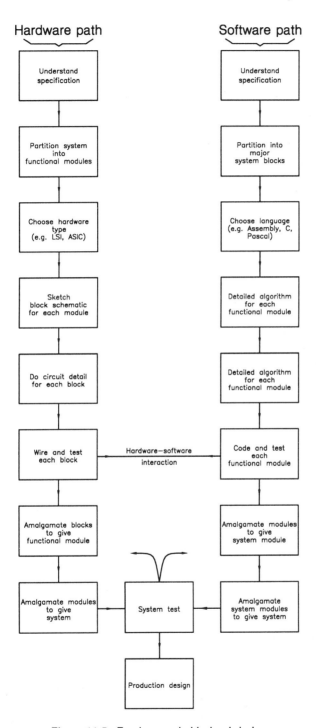

Figure 11.5: Fundamental chip-level design.

326 *Preliminaries*

eventual production process, the cost of developing, testing and production of such systems is large. Furthermore, the up-front expenses are high and may cause cash flow problems. However, the materials cost is low, and this approach is often the best technical solution to the problem.

Software implementation for most dedicated MPU-based systems, irrespective of the hardware implementation category, is developed in an analogous way to chip-level design. This is emphasized in Fig. 11.5 by the two parallel tracks. Other than monitors and operating systems, there is little in the way of off-the-shelf firmware packages. There is, however, a considerable body of high-level routines published, which can be adapted and compiled down to the chosen target. Nevertheless, in general, software design is an expensive proposition, and is difficult to amortize in small production runs.

Standard microprocessor-based circuitry uses a MPU chip, together with individual ICs for memory and I/O interface. Address decoding and other support logic (glue logic) is likely to be implemented using programmable logic devices to reduce the chip count. Higher production runs may economically utilize single-chip microcontroller units (MCUs). MCUs are single devices with the MPU, RAM, ROM, I/O interface and address decoder all integrated together [5]. As an example, the

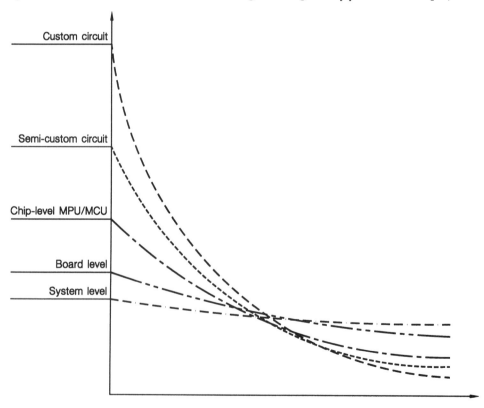

Figure 11.6: A cost versus production comparison.

68HC11E9 MCU has an 8-channel A/D converter, two serial and up to four parallel ports, timers, 256-byte RAM, 512-byte EEPROM, 12 kbyte of ROM/EPROM and a watch-dog timer on the one chip! The processing power of MCUs is generally less than their MPU equivalents, but newer devices such as the 68HC16 and the 68300 series (software compatible with the 68020 MPU) are somewhat more powerful (and expensive!).

Compilers for the more common MCUs (e.g. 6805, 6811, 6816, 8051) are available, but frequently have restrictions due to the limited architectures of the core processor. Table 10.15(a), (f) and (g) are examples of 6801, 6816 and 6811 MCU compiled output respectively. Note especially the length of the 6801's code, it being one of the older devices which did not have a programming structure designed with a high-level language implementation in mind.

It is possible to integrate the fabrication pattern of some of the simpler MPUs onto your own custom IC, such as the 6805 device. These are held in the library of the appropriate CAD package. Provided that memory and other patterns are available, in principle a custom MPU-based system can be integrated onto one silicon chip. However, this approach is only applicable to very large scale production runs and requires considerable skill. Custom hard-disk controllers are typical of products which use this technique.

Figure 11.6 gives an idealized picture of the interaction of the various technologies with regard to production levels. These relationships make no presumption as regards the technological advantages of the different approaches.

References

[1] Riggs, T., et al.; Spectral Analysis of the Normal Electrocardiogram in Children and Adults, *J. Electrocardiology*, **12**, no. 4, 1979, pp. 377–379.

[2] Wilcox, A.D.; *68000 Microcomputer Systems*, Prentice-Hall, 1987, Part 1.

[3] Cahill, S.J.; *The Single-Chip Microcomputer*, Prentice-Hall, 1987, Appendix 3.

[4] Blasewitz, R.M. and Stern, F.; *Microcomputer Systems*, Hayden, 1987, Section 9.5.

[5] Cahill, S.J.; *The Single-Chip Microcomputer*, Prentice-Hall, 1987.

CHAPTER 12

The Analog World

Most real-world parameters are analog in nature. Some examples are temperature, pressure and light intensity. An analog parameter is a continuum — limited in practice between an upper and lower level. Thus a dry-bulb thermometer can be read to whatever resolution is necessary, between, say, $-10\,°C$ and $+180\,°C$. Below this, the mercury disappears into its bulb, and above this the top of the tube is blown off! Theoretically, the quantum nature of matter sets a lower limit to the continuous nature of things, but in practice noise levels and the limited accuracy of the device generating the signal sets a realistic upper limit to resolution.

Digital circuitry deals with patterns of symbols, which represent amplitudes. Depending on the number and type of digits making up the pattern, only a finite total of values are possible. Most of us tend to use denary (decimal) digits to represent our numbers, while computers prefer binary. Patterns made up of eight binary digits can represent up to 2^8 (256) discrete values, whilst 16 bits can resolve down to $\frac{1}{2^{16}}$ ($\frac{1}{65536}$) of full scale (see Table 12.1).

Physically, bits are represented in hardware as two values of a signal parameter. Most commonly this is voltage, and typical ranges are $0-0.8\,V$ and $2.0-5.0\,V$ for logic 0/logic 1 respectively. Other values, such as $\pm 12\,V$, and parameters such as frequency, for instance $1.2/2.4\,kHz$, are in regular use. Digital signals are in fact analog signals with analog characteristics, such as finite transition times and noise. Although digital systems designers must be cognizant of these signal properties, they are normally regarded as secondary effects, caused by the intrusion of an imperfect world!

Given that we wish to do our processing using digital techniques, in our case a microprocessor, conversion to and from analog signals is necessary. The aim of this chapter is to overview this process, with an eye to our specific project. As well as the A/D and D/A converters themselves, we will look at some of the consequences of the digitization of analog signals.

12.1 Signals

Our project specifically targeted the adult ECG/EKG signal as the system input. The physical origins of this signal and the use of electrodes as the sensor are outside the scope of this text; the interested reader is directed to references [1, 2] for this information. The most common configuration measures the potential difference across the chest, the LA (left arm) and RA (right arm) leads in Fig. 11.3. The RL (right leg) is used as the reference point. As the RA–LA potential is rarely more than a few millivolts, the following amplifier must provide differential gain to the order of 1000 (60 dB). However, there is more to this stage than just gain. A good ECG/EKG amplifier must have the following properties, in addition to the usual requirements of linearity, slew rate and frequency response:

1. A common-mode rejection ratio (differential gain/common-mode gain) of at least 80 dB. Common-mode signals arise from interference from external sources, typically mains hum. Such extraneous signals are important, as they are usually much greater than the signal of interest. Signals appearing on both leads should not affect a differential amplifier's output, but in practice there will be some feedthrough.

2. Suppression of baseline drift. Large, essentially d.c., voltages can appear across the electrodes, due to electrolytic action at the skin interface. These are not constant, but change slowly with time. Straight amplification by 1000 would cause the amplifier to saturate.

3. As a safety requirement, leakage current between electrodes, that is through the body, to be less than $10\,\mu A$ [2]. Because of this, an isolating amplifier is recommended. This uses a front end with optical or transformer coupling to the normally-powered main gain block. This front end can either be battery powered or supplied through an isolating power supply (sometimes integral with the amplifier).

4. Protection against pacemaker spikes and, if applicable, defribrillator surges of around 25 kV!

Because of safety considerations [2], I have resisted the temptation of describing an ECG/EKG amplifier here. A range of commercially available isolating amplifiers is available for biomedical applications, a typical example being the Burr Brown ISO100P. Either disposable or reusable silver/silver chloride electrodes are available from any medical supply house [3]. If necessary shave and clean the skin with surgical spirit before application. However, all the hardware and software to be presented can be fully and safely tested in comfort using a sinusoidal or function generator.

Auxiliary circuits, such as filters, level shifters and sample and hold circuits, lumped together in Fig. 11.3 as the signal conditioning process, are discussed at the appropriate point later.

Quantizing a signal obviously distorts the original information. In essence, quantizing is the comparison of the analog quantity with a fixed number of levels. The nearest level is then the value taken in expressing the original in its digital equivalent. Thus in Fig. 12.1, an input voltage of 0.4285 full scale is 0.0536 above

330 *The Analog World*

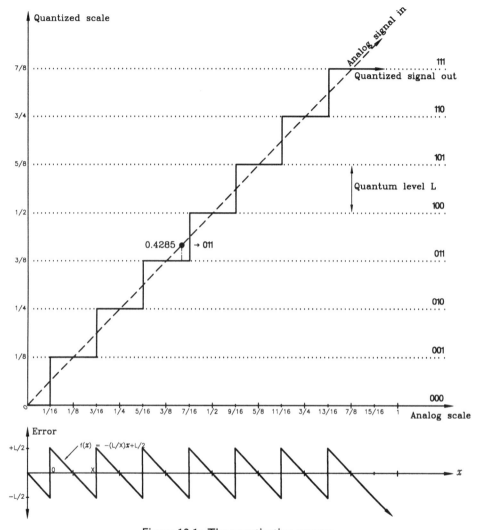

Figure 12.1: The quantization process.

quantum level 3 and 0.0714 below level 4. Its quantized value will then be taken as level 3 and coded as 011*b* in a 3-bit system.

The residual error of −0.0536 will remain as quantizing noise, and can never be eradicated (see Fig. 12.2(d)). The distribution of quantization error is given at the bottom of Fig. 12.1, and is affected only by the number of levels. This can simply be calculated by evaluating the average of the error function squared. The square root of this is then the root mean square (r.m.s.) of the noise.

$$\mathcal{F}(x) = -\frac{L}{X}x + \frac{L}{2}$$

The mean square is:

$$\frac{1}{X}\int_0^X \mathcal{F}(x)^2\,dx = \frac{1}{X}\int_0^X \left[\frac{L^2}{X^2}x^2 - \frac{L^2}{X}x + \frac{L^2}{4}\right]dx$$

$$= \frac{1}{X}\left|\frac{L^2}{3X^2}x^3 - \frac{L^2}{2X}x^2 + \frac{L^2}{4}x\right|_0^X = \frac{L^2}{12}$$

Giving a r.m.s. noise value of $\frac{L}{\sqrt{12}} = \frac{L}{2\sqrt{3}}$

A fundamental measure of a system's merit is the signal to noise ratio. Taking the signal to be a sinusoidal wave of peak to peak amplitude $2^n L$ (see Fig. 12.2), we have an r.m.s. signal of $\frac{\left(\frac{2^n L}{2}\right)}{\sqrt{2}}$, that is $\frac{\text{peak}}{\sqrt{2}}$. Thus for a binary system with n binary bits, we have a signal to noise ratio of:

$$\frac{\left(\frac{2^n L}{2\sqrt{2}}\right)}{\left(\frac{L}{\sqrt{12}}\right)} = \frac{2^n \sqrt{12}}{2\sqrt{2}} = 1.22 \times 2^n$$

In decibels we have:

$$\text{S/N} = 20\log 1.22 \times 2^n = 6.02n + 1.77\,\text{dB}$$

The dynamic range of a quantized system is given by the ratio of its full scale $(2^n L)$ to its resolution, L. This is just 2^n, or in dB, $20\log 2^n = 20n\log 2 = 6.02n$. The percentage resolution given in Table 12.1 is of course just another way of expressing the same thing.

Table 12.1: Quantization parameters.

Binary bits	Quantum levels (2^n)	% resolution	Resolution Dynamic range	S/N ratio (dB)
4	16	16.25	24.1 dB	26.9 dB
8	256	0.391	48.2 dB	49.9 dB
10	1024	0.097	60.2 dB	61.9 dB
12	4096	0.024	72.2 dB	74.0 dB
16	65,536	0.0015	96.3 dB	98.1 dB
20	1,048,576	0.00009	120.4 dB	122.2 dB

The exponential nature of these quality parameters with respect to the number of binary-word bits is clearly seen in Table 12.1. However, the implementation complexity and thus price also follows this relationship. For example, a 20-bit conversion of 1 V full scale would have to deal with quantum levels less than 1 μV apart. Compact disks use 16-bit technology for high quality music. Pulse-code modulated telephonic links use eight bits, but the quantum levels are unequally spaced, being closer at the lower amplitude levels. This reduces quantization hiss where conversations are held in hushed tones! Linear 8-bit conversions are suitable for most general purposes, having a resolution of better than $\pm\frac{1}{4}\%$. Actually video looks quite acceptable at a 4-bit resolution, and music can just be heard using a single bit (i.e. positive or negative)!!

The analog world treats time as a continuum, whereas digital systems sample signals at discrete intervals. The sampling theorem [4] states that provided this interval does not exceed half that of the highest signal frequency, then no information is lost. The reason for this theoretical twice highest frequency sampling limit, called the Nyquist rate, can be seen by examining the spectrum of a train of amplitude modulated pulses. Ideal impulses (pulses with zero width and unit area) are characterized in the frequency domain as a series of equal-amplitude harmonics at the repetition rate, extending to infinity [5]. Real pulses have a similar spectrum but the harmonic amplitudes fall with increasing frequency.

If we modulate this pulse train by a baseband signal $A\sin\omega_f t$, then in the frequency domain this is equivalent to multiplying the harmonic spectrum (the pulse) by $A\sin\omega_f t$, giving sum and different components thus:

$$A\sin\omega_f t \times B\sin\omega_h t = \frac{AB}{2}(\sin(\omega_h + \omega_f)t + \sin(\omega_h - \omega_f)t)$$

More complex baseband signals can be considered to be a band-limited (f_m) collection of individual sinusoids, and on the basis of this analysis each pulse harmonic will sport an upper (sum) and lower (difference) sideband. We can see from the geometry of Fig. 12.2(b) that the harmonics (multiples of the sampling rate) must be spaced at least $2 \times f_m$ apart, if the sidebands are not to overlap.

A low-pass filter can be used, as shown in Fig. 12.2(d), to recover the baseband from the pulse train. Realizable filters will pass some of the harmonic bands, albeit in an attenuated form. A close examination of the frequency domain of Fig. 12.2(d) shows a vestige of the first lower sideband appearing in the pass band. However, most of the distortion in the reconstituted analog signal is due to the quantizing error resulting from the crude 3-bit digitization. Such a system will have a S/N ratio of around 20 dB.

In order to reduce the demands of the recovery filter, a sampling frequency somewhat above the Nyquist limit is normally used. This introduces a guard band between sidebands. For example the pulse code telephone network has an analog input bandlimited to 3.4 kHz, but is sampled at 8 kHz. Similarly the audio compact disk (CD) uses a sampling rate of 44.1 kHz, for an upper music frequency of 20 kHz. This means that with a 16-bit sample and 70 minute play period, a CD must store around 3000 Mbits!

A more graphic illustration of the effects of sampling at below the Nyquist rate is shown in Fig. 12.3. Here the sampling rate is only 0.75 of the baseband frequency. When the samples are reconstituted by filtering the resulting pulse train, the outcome, shown in Fig. 12.3(b), bears no simple relationship to the original. This spurious signal is known as an alias.

Returning now to our project, we have established that the sampling rate will be 128 per second. As the baseband of interest is limited to 20 Hz, this seems to give us a Nyquist margin of around 300%. However, the ECG/EKG signal does have

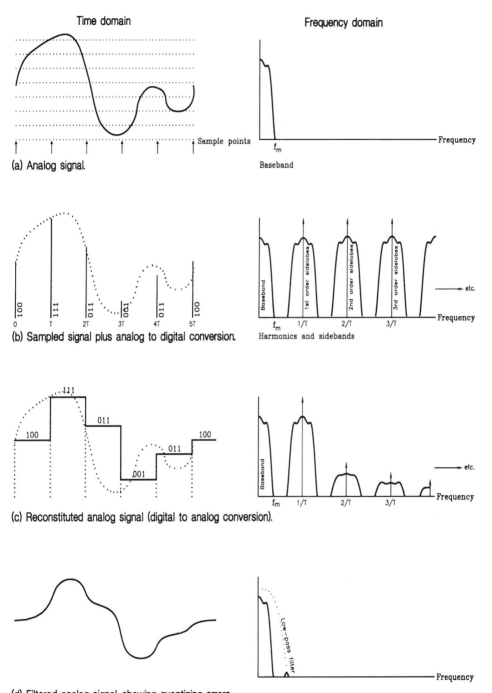

Figure 12.2: The analog–digital process.

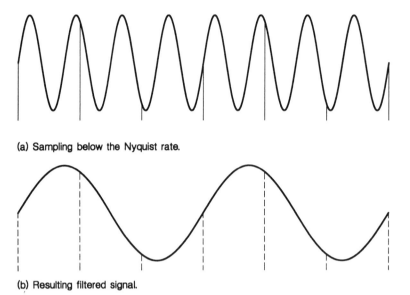

(a) Sampling below the Nyquist rate.

(b) Resulting filtered signal.

Figure 12.3: Illustrating aliasing.

components beyond 1 kHz; and noise, both from external and internal (e.g. muscle noise) sources, will have a spectrum extending well above the Nyquist limit. Thus, an anti-aliasing filter will be required as part of the front end signal conditioning process.

Many filter designs exist. That shown in Fig. 12.4 is a 4th-order Butterworth low-pass filter using the multiple-feedback configuration. The overall gain in the passband is designed to be unity, with the −3 dB frequency at 24 Hz. Design equations and other relevant data is given in reference [6]. In practical situations, component tolerances will cause wide deviations from this figure; this is especially true of the large capacitors used for low frequencies. Reference [6] gives a tuning procedure using $R_1 R_3$, where more precise results are required. The transfer characteristic of Fig. 12.4(b) is a real transfer of an untuned circuit, using 0.1% resistors and ±1% capacitors. Actual preferred values, as shown bracketed, were used. From this characteristic, the gain at the Nyquist frequency of 64 Hz is −32 dB down from the passband.

12.2 Digital to Analog Conversion

Digital to Analog (D/A) conversion can be defined as the production of an analog signal whose amplitude is proportional to the quantitative magnitude of the input digital word. Natural binary code is weighted in ascending order of powers of two. Thus a 4-bit binary number can be written as:

$$A = (b_3 \times 2^3) + (b_2 \times 2^2) + (b_1 \times 2^1) + (b_0 \times 2^0)$$

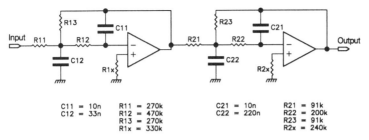

(a) The circuit diagram.

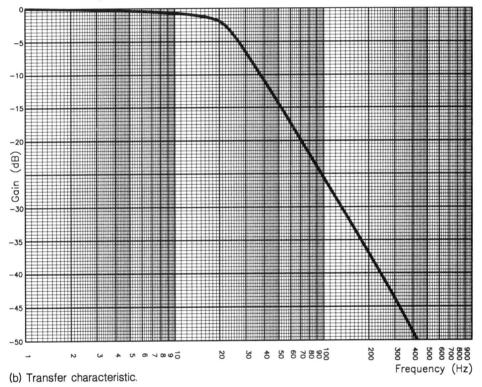

(b) Transfer characteristic.

Figure 12.4: A 4th-order anti-aliasing filter.

where b_n is the nth binary digit, either 0 or 1. In general:

$$A = \sum_{k=0}^{M-1} b_k \times 2^k$$

for an M-digit word.

With this definition in mind, the use of a suitably weighted resistor network is suggested. Four resistors, each switched to V or earth by one digital bit, feeding the virtual earth of an operational amplifier's summing junction, will give a composite

336 The Analog World

analog output, which is a function of the digital pattern and the resistor values. Thus using an 8 kΩ resistor driven by b_0, 4 kΩ driven by b_1, 2 kΩ by b_2 and 1 kΩ by b_3 will feed currents in ascending orders of two into the summing junction.

In a practical situation, the use of weighted resistors, leads to severe accuracy problems. These are the result of the wide range of resistance values; for example, in a 12-bit system, if b_0 switches in 1 kΩ then b_{11} will have to switch in a 2.048 MΩ resistor. As well as the problem of matching the ratio of all these resistors, the precision analog switches have to carry an equally wide spread of currents.

One way around the matching problem is to use a ladder network, such as shown in Fig. 12.5(a). Looking left at node A we 'see' a resistance 2R. At node B we have $R + 2R \parallel 2R = 2R$. By symmetry, the resistance looking left is always 2R. To a precision reference voltage V_{ref} at node D, the ladder appears to be this 2R resistance in parallel with the 2R resistor to switch b_3. If the output is short circuited to ground, then irrespective of the switch position a current $I_{ref} = \frac{V_{ref}}{R}$ will flow into node D from V_{ref}, of which 50% will flow down through the switch.

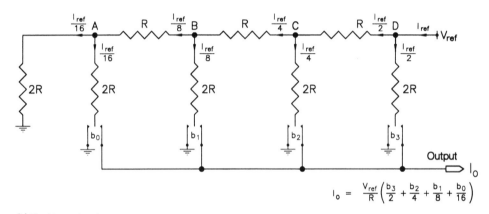

(a) Resistor network.

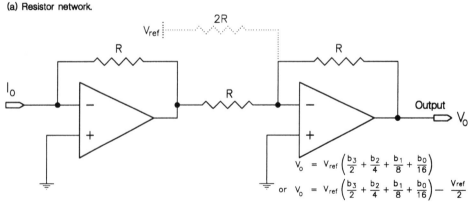

(b) Current to voltage converter with optional bipolar output.

Figure 12.5: The R-2R current D/A converter.

The current $\frac{I_{ref}}{2}$ arriving at node C similarly splits into two parts, with $\frac{I_{ref}}{4}$ being switched by b_2. By continuing this line of argument, it can be seen that each switch leftwards controls half of its neighbour's current. The output current is then:

$$I_O = \frac{I_{ref}}{16} \sum_{k=0}^{3} b_k \times 2^k \qquad \text{where } b_k = 1 \text{ or } 0$$

which is the desired transformation ratio.

The network can be extended ad infinitum by replacing the termination resistor by the requisite ladder sections. Irrespective of the number of bits, the absolute value of resistor is irrelevant; only the 2:1 ratio matters. This is important where the ladder is implemented on a monolithic silicon integrated circuit, where accurate absolute values are impossible to fabricate, but ratios are accurately implemented.

The switches shown in the diagram are of course electronic and controlled by the digital bit pattern. In long words, where the ratio of switchable current is very large, alternative ladders are available which have the property of equal currents but different voltages [7]. Capacitor-based ladders can also be used.

The output quantity of this network is current. This can conveniently be converted to voltage by feeding this into the inverting input of a feedback operational amplifier, as shown to the left of Fig. 12.5(b). This appears as a virtual earth to the network, and gives an output voltage of $-I_O R$. A further stage inverts this negative-going voltage, giving $V_O = I_O R$.

This final stage may be optionally used to subtract half full-scale voltage, to give a bipolar output. Examining Fig. 12.6(b) shows that the most significant bit (b_3) acts as a sign bit with 1 for positive and 0 for negative. The remaining magnitude digits are identical to the equivalent 2's complement code. Thus a computer working in 2's complement code need only invert the sign bit before outputting to the D/A converter, to give a true bipolar output. This modified 2's complement code is known as offset binary code.

Real D/A converters have characteristics which differ from the ideal transfer relationship; as shown in Fig. 12.7 [8]. As an example, consider an 8-bit D/A converter going from $01111111b$ to $10000000b$. In all, eight switches must change. Unless these are exactly matched, and their ladder legs too, it is inevitable that a blip on the characteristic will occur. If this mismatch is more than one bit (less than 0.4% in this example) then it is possible that the trend (larger binary codes give larger analog outputs) will be violated. In such cases, the converter is non-monotonic. Besides these non-linear errors, the converter may exhibit a constant offset and gain error. Unlike the non-linear errors, both these linear errors may be trimmed out using the operational amplifier buffers.

Manufacturers specify their error figures in different ways. For example the AD7528's relative accuracy is measured as the maximum deviation of any code from the ideal, with offset and gain errors eliminated. Depending on the version, this is given as ± 1 bit and $\pm \frac{1}{2}$ bit. Differential non-linearity is the maximum difference between the ideal 1-bit change expected between any two adjacent codes

338 The Analog World

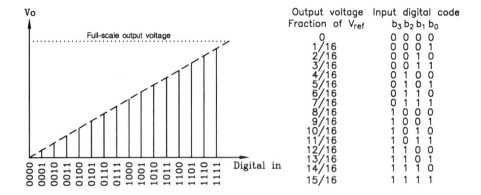

(a) Natural unipolar code conversion.

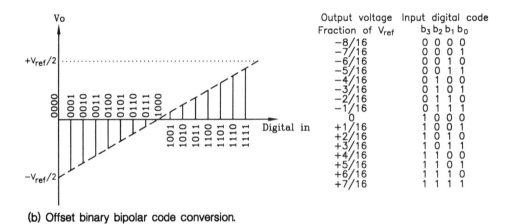

(b) Offset binary bipolar code conversion.

Figure 12.6: Conversion relationships for the network of Fig. 12.5.

and the actual measured value. This is given as ±1 bit, and therefore the converter is guaranteed monotonic (just!). Gain error is the worst case full-scale error due to offset and gain tolerance. It can be as high as ±6 bits, but is easily trimmed out if need be.

The output port chosen for our project is the Analog Devices AD7528 dual 8-bit D/A converter. This provides for both X and Y analog channels in the one device. The AD7528 is microprocessor-compatible, with two integral 8-bit transparent latches. Besides any necessary operational amplifier network, only an external precision reference voltage is required.

The heart of each converter is a current R-2R ladder network, which is essentially an 8-bit version of Fig. 12.5(a). From Fig. 12.8 we see that an additional R resistor connected to the output node is also provided (at pins 3 and 19), designed to be used as the feedback resistor of the first amplifier stage of Fig. 12.5(b). Its use is

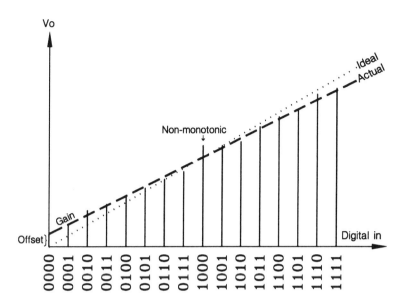

Figure 12.7: A real-world transfer characteristic.

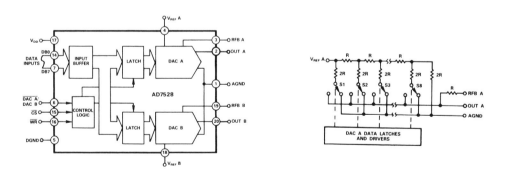

(a) Functional diagram. (b) Simplified functional circuit for DAC A.

Figure 12.8: The AD7528 dual D/A converter. Reprinted with the permission of Analog Devices, Inc.

illustrated in Fig. 12.9.

The original specification in Section 11.1 called for 0.5% resolution and ±0.5% full-scale accuracy. An 8-bit system gives $\frac{1}{256} \approx 0.4\%$ resolution, thus a ±1 bit non-linear accuracy will fall within our target.

From Section 11.1, we have estimated a data rate interval to both D/A converters of 56 μs. The settling time for a step between zero and full current ($00000000b \leftrightarrow 11111111b$) is 400 ns maximum to within a $\frac{1}{2}$ bit of the final value (supply voltage V_{DD} of +5 V and V_{ref} of +10 V). This is around 0.7% of the step period. Of course the amplifier circuitry converting this current to voltage will worsen this figure.

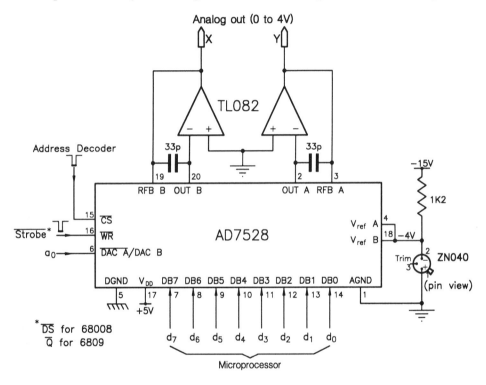

Figure 12.9: Interfacing the AD7528 to a microprocessor.

Both channels are independent, with separate analog sections and 8-bit transparent latches driving the ladder switches. $\overline{DACA}$/DACB (pin 6) directs data from the MPU bus through to the appropriate 8-bit latch whenever both the Chip_Select ($\overline{CS}$) and $\overline{Write}$ ($\overline{WR}$) pins are low. In driving $\overline{DACA}$/DACB from a_0, the AD7528 looks to the MPU as two ordinary 8-bit digital output ports located at adjacent addresses. An address decoder line enables the $\overline{CS}$ input, whilst a strobe of some kind activates $\overline{WR}$ when sending data. In the 6809 MPU, the $\overline{Q}$ inverted clock is normally used for this purpose (see Fig. 1.7), whilst $\overline{DS}$ fulfils this role for the 68008 device. The 68000 MPU would use $\overline{UDS}$ or $\overline{LDS}$ as appropriate, see Fig. 3.10. With a V_{DD} of +5 V, all digital signals are TTL and therefore MPU compatible. The

AD7528 response times are (just!) compatible with a 2 MHz 6809 and no wait-state 8 MHz 68000/68008.

The reference voltage must be supplied externally. In Fig. 12.9, I have used the Plessey ZN040 4.01 V ±1% bandgap voltage reference IC for this purpose. With the amplifier configuration shown, this gives a full-scale voltage output of nominally +4 V. The actual voltage may be trimmed over the range ±5% by connecting pin 2 to the center tap of a 100 kΩ potentiometer across pins 1 and 3. The choice of reference voltage is fairly arbitrary in our case, as both channels are to drive input amplifiers on an oscilloscope. If necessary, a Zener diode will act as a reasonable substitute (anode to top) with 5.6 V having the lowest temperature coefficient. The series resistance of 1.2 kΩ gives a bias current of around 9 mA. The minimum current is given as 150 μA, with a maximum of 75 mA. This would also be suitable for a 5.6 V Zener diode.

V_{ref} can vary over the range ±10 V. By choosing a negative value, the single amplifier/channel configuration used in Fig. 12.9 gives a positive unipolar range. The internal feedback resistor has been used with an external 33 pF polystyrene capacitor in parallel to stabilize the amplifier's high frequency behavior. The TL082 operational amplifiers feature a typical slew rate of 13 V/μs (minimum 8 V/μs). Thus a full-scale swing of 4 V will take around 300 ns. Any operational amplifier can be used here, but note that the general purpose 741 type has a slew rate of only 0.5 V/μs. Analog power supplies of between ±8 V to ±15 V are suitable, and can be used for the reference voltage IC bias.

Ideally the analog ground should be run directly back to the power supply common point, rather than make a direct connection to the noisy digital ground. Where this is done, it is recommended that two back to back signal diodes be connected between them, close to the IC. This reduces the chance of transient voltages injecting noise into the system.

Testing the the D/A converter dynamically is covered in Section 15.2. A simple static test is possible before connecting the digital signals to the system. Firstly meter the V_{ref} inputs and power supplies. Then keeping pins 6, 15 and 16 to digital ground, as well as all eight data lines, check output A is close to analog ground. Bring DB7 to V_{DD}. Now output A should be $\frac{1}{2} V_{ref}$. With all DB lines logic 1, the output voltage will be $\approx V_{ref}$. Repeat with pin 6 at logic 1, for output B. The deselected channel will retain its last value. Connecting all DB lines together to a TTL-compatible square wave generator and monitoring the analog outputs is an alternative test, which also checks delay and slew rate parameters.

Handle carefully to avoid electrostatic discharge damage, and *do not insert into a powered socket*.

12.3 Analog to Digital Conversion

The opposite transformation conversion direction of analog to digital is by far the more complex of the two. One possible scenario involves an array of analog comparators, which simultaneously compare the analog input V_{in} with a series of ascending voltage steps of $\frac{1}{n}V_{ref}$. In Fig. 12.10, these quantum steps are produced by a chain of eight equi-valued resistors, giving $\frac{1}{8}$ full-scale resolution. The output of any comparator is logic 1 if V_{in} exceeds the quantum, otherwise logic 0. Thus eight unique binary patterns are produced, depending on the analog magnitude. A relatively simple logic decoder network converts this unary code to natural binary. A bipolar conversion is easily accomplished by tying the top and bottom of the resistor chain to $+\frac{1}{2}V_{ref}$ and $-\frac{1}{2}V_{ref}$ respectively. These so called flash or parallel converters are expensive, because of the large number of precision circuit components involved, for example 255 high-speed analog comparators for an 8-bit converter. However, this technique is fast, with conversion times of better than 30 ns achievable in commercial products (for example the TRW TDC1007J 8-bit converter [9]).

For most microprocessor applications, the short conversion time is an embarrassment, demanding direct memory access procedures to make full use of this speed. Instead, most MPU-compatible A/D converters use the more prosaic technique of successive approximation. This is the electronic equivalent of the beam balance. Consider an unknown weight placed in one pan of such a balance, and a range of known weights in powers of two available to the operator. A systematic approach to the determination of the unknown quantity is to start with the largest known weight. If this is too light, then it is left in the pan, otherwise it is removed. The same procedure is carried out in descending order until the smallest standard weight has been used. The solution is then the assemblage of weights in the pan, to the resolution of the smallest weight.

One possible 8-bit electronic utilization of this strategy is shown in Fig. 12.11. Here a MPU addresses a digital to analog converter, the equivalent of the beam balance. In software the various bit patterns from $10000000b$ down to $00000001b$ can be added to the subtotal and sent out to the D/A converter. The resulting analog equivalent is compared to the incoming voltage, and the magnitude relationship between them read through the 3-state buffer at bit 7 [10].

A possible coding in **C** is given in Table 12.2. Here constant pointers to the two peripherals are assigned addresses (assumed to be $6000h$ and $6002h$), and two byte-sized variables defined to hold the assemblage of bits and the test pattern. The former is initialized to zero (nothing on the pan) and the latter to $80h$ (largest known weight, $10000000b$).

The `while` loop adds the test weight to the trial pattern, sends it out to the D/A converter and checks the comparator output. If this is logic 1 (i.e. in line 13, the contents of the address `comparator` ANDed with $10000000b$ is non-zero; `*comparator & 0x80`) then the test weight is removed (subtracted); otherwise it is left as part of the aggregrate. Each new test weight is generated by shifting `weight` right once. As `weight` has been declared `unsigned`, this should be a Logic Shift Left operation (but strictly is compiler dependent). After eight passes through the loop, the state of the aggregate is the digital equivalent to V_{in}.

Although the circuit of Fig. 12.11 works, it is fairly slow. Typically an 8-bit conversion will take around $100\,\mu s$ at best. A wide range of stand-alone successive

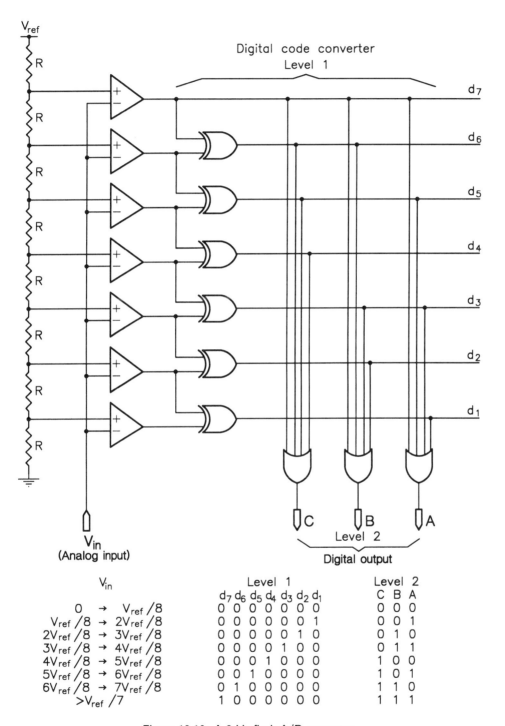

Figure 12.10: A 3-bit flash A/D converter.

344 The Analog World

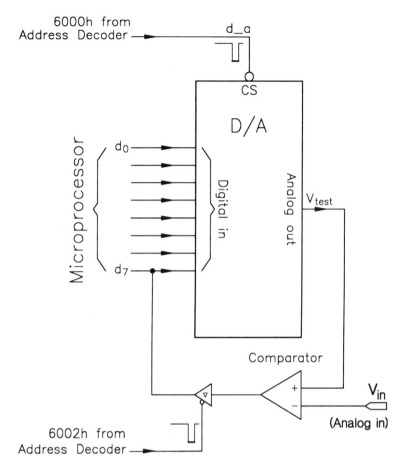

Figure 12.11: A software controlled successive approximation D/A converter.

approximation converters are available to interface directly to MPU buses, taking 10 μs or less to complete an 8-bit conversion. Word sizes up to 16 bits are cataloged.

For our project we have chosen to use the Analog Devices AD7576 8-bit A/D converter, as outlined in Fig. 12.12. The AD7576 is a monolithic device containing all the digital and analog circuitry necessary to implement the successive approximation strategy. In Fig. 12.12, the block labelled SAR is the Successive Approximation Register, holding the bit pattern trial as it is built up (**digital in** Table 12.2). The Control Logic box sequentially sets each flip flop in the SAR, clearing it shortly after, if the comparator (COMP) indicates that the analog output of the D/A converter (DAC) is above the input (A_{in}). The timing of this sequence is a function of the internal Clock Oscillator box, whose frequency is controlled by **CR** components at pin 5. The minimum conversion time is given as 10 μs. An external oscillator may alternatively be used to drive pin 5, and in this situation 2 MHz gives the 10 μs minimum conversion time.

Table 12.2: C driver for Fig. 12.11.

```
#define   d_a_address    (unsigned char *)0x6000
#define   comp_address   (unsigned char *)0x6002

analog_in()
{
unsigned char * const d_a          =    d_a_address;
unsigned char * const comparator   =    comp_address;
register unsigned char digital     =    0;    /* The digital trial          */
register unsigned char weight      =    0x80; /* The walking weight         */

while (weight != 0)
    {
    digital += weight;                 /* Add weight to trial              */
    *d_a = digital;                    /* Send out to d_a converter        */
    if (*comparator & 0x80)            /* IF too big                       */
        {digital -= weight;}           /* THEN remove weight from trial    */
    weight >>= 1;                      /* weight divided by 2              */
    }
return (digital);                      /* Nearest approximation returned   */
}
```

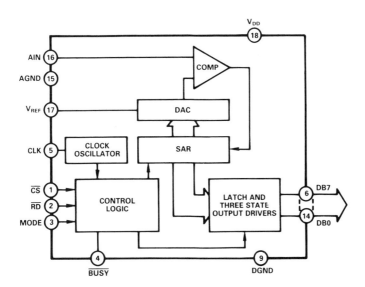

Figure 12.12: Functional diagram of the AD7576 A/D converter. Reprinted with the permission of Analog Devices Inc.

346 *The Analog World*

The AD7576 operates in two modes, depending on the state of the MODE input. If pin 3 is high, then a low-going signal at the $\overline{\text{RD}}$ pin begins the conversion process, provided that the device is enabled ($\overline{\text{CS}} = 0$). $\overline{\text{BUSY}}$ goes low during this process, and returns high when it has been completed. The new data is transferred to the internal latch register on the rising edge of $\overline{\text{BUSY}}$. These latches are interfaced to the data bus via integral 3-state buffers, which are enabled when $\overline{\text{RD}}$ (i.e. Read) is low and the device is enabled. Thus the $\overline{\text{RD}}$ control is a dual purpose Start Convert and Read function, that is in reading data a new conversion is automatically initiated.

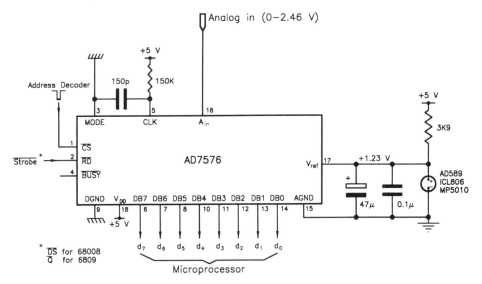

(a) Interface connections.

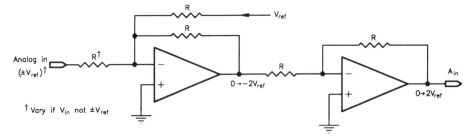

(b) Level shifting a bipolar analog signal to the range 0 to 2 V_{ref}.

Figure 12.13: Interfacing the AD7576 to a microprocessor.

The interface diagram of Fig. 12.13(a) uses the AD7576 in its asynchronous mode (pin 3 low). Here the A/D converter performs continuous conversions. Data in the output latches is always valid, and can be used ($\overline{\text{RD}}$ and $\overline{\text{CS}}$ low) at any time. With the clock CR components shown, the data is never more than 10 μs out of date.

The AD7576 is powered by a single +5 V supply. As this will probably be common with the logic supply, it should be decoupled to analog ground as close to the device as possible, with a recommended 47 µF tantalum capacitor in parallel with a 0.1 µF ceramic capacitor. This supply, and analog ground, should be run directly back to the power supply.

The internal D/A converter requires an external V_{ref} of 1.23 V ±5%. This is provided from an AD589 bandgap reference, and should be decoupled in the same way. With this value of V_{ref}, full scale at $2V_{ref}$ is 2.46 V, giving a nominal 10 mV resolution. The internal D/A converter suffers from the same errors discussed in Section 12.2 The resulting non-linear error (the relative accuracy) is either ±1 or $±\frac{1}{2}$ bit maximum, depending on the device selection type. This is within our specification.

The input analog range is unipolar $0-2V_{ref}$. The simple operational amplifier network shown in Fig. 12.13(b) will convert a bipolar input to the necessary range, by adding a constant bias. This offset may be alternatively incorporated into the anti-aliasing filter. The resulting code is in offset binary form and can, if necessary, be converted to 2's complement form by inverting the MSB (see page 337).

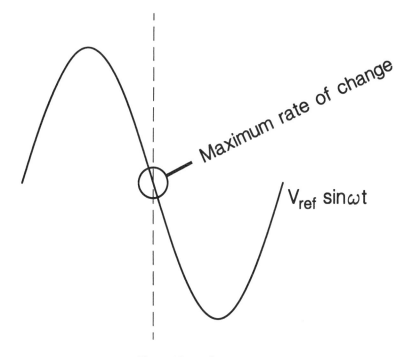

Figure 12.14: Aperture error.

One consideration remains. The analog input is changing during the time the conversion takes place. Accuracy considerations dictate that any change should not exceed one bit during this aperture time. Taking, as a worst-case situation, a sinusoid swinging through the full scale, as shown in Fig. 12.14, then we can determine the rate of change by differentiation:

Rate of change ($\frac{d}{dt}$) is $V_{ref}\omega \cos\omega t$
Maximum when $\cos\omega t = 1$ is $V_{ref}\omega$ volts s^{-1}
Aperture time is 10 μs, therefore:
change in 10 μs (δ) is $10^{-5}V_{ref}\omega$
and thus:

$$\delta \leq 1 \text{ bit}$$
$$10^{-5}V_{ref}\omega \leq \frac{2V_{ref}}{256}$$
$$\omega \leq 781 \text{ radians s}^{-1}$$
$$f \leq 124 \text{ Hz}$$

This is well within our specification, but if, say, a 12-bit conversion was needed within 16 μs, then the upper frequency falls to less than 10 Hz! In such cases a sample and hold (S/H) circuit preceding the A/D converter must be used. This captures the signal, with typically a 40 ns aperture time [11]. The principle of most S/Hs involves a capacitor being charged up during the sample period, and held whilst conversion occurs. As with A/D and D/A converters, S/H circuits are normally obtainable as monolithic integrated circuits.

Although S/H aperture times are low, they may take several μs to stabilize, after which conversion can commence. They tend to droop during hold, as the capacitor looses its charge (typically 20 μV/ms), and suffer from all analog illnesses of the flesh, such as drift, offset and non-linearity. Thus the S/H must be matched to the A/D converter's performance.

References

[1] Friedman, H.H.; *Diagnostic Electrocardiography and Vectorcardiography*, McGraw-Hill, 3rd. ed., 1985.

[2] American National Standards Institute/Association for the Advancement of Medical Instrumentation; *Safe Current Limits for Electromedical Apparatus*, ANSI/AAMI ES1-1985.

[3] American National Standards Institute/Association for the Advancement of Medical Instrumentation; *Pregelled ECG Disposable Electrodes*, ANSI/AAMI ES12-1983.

[4] Shannon, C.E.; Communication in the Presence of Noise, *Proc. IRE*, **37**, Jan. 1949, pp. 10–21.

[5] Julian, M.; *Circuits, Signals and Devices*, J. Wiley/Longman, 1988, Chapter 9.

[6] Graeme, J.G. and Tobey, G.E.; *Operational Amplifiers*, McGraw-Hill, 1971, Chapter 8.

[7] Cahill, S.J.; *The Single Chip Microcomputer*, Prentice-Hall, 1987, Section 6.1.

[8] Clayton, G.B.; *Data Converters*, MacMillan, 1982, Section 6.6.

[9] Allan, R.; Breaking the Data-Conversion Speed Barrier, *Electronics*, **53**, 1980, pp. 109.

[10] Hansen, J.; Creating Software ADCs, *Embedded Systems Programming*, **6**, no. 3, March 1993, pp. 24 – 36.

[11] Gadway, R.; *Sample and Hold, or High-Speed A/D Converters, How do you Decide?*, Burr-Brown Application Note AN-56, May 1973 and *EDN*, Sept. 15, 1972.

CHAPTER 13

The Target Microcomputer

From our discussion, the target microcomputer will have the following facilities:

1. A single-channel 8-bit analog input port.

2. A dual-channel analog output port.

3. A single-bit digital output port for the flyback blank.

4. A single-bit digital input port to read the freeze switch.

All this is in addition to the memory, address decoder and other necessary support circuitry.

In consideration of the requested sampling rate variation of −50% to + 100% around the nominal 128 per second value, both of the following circuits use an oscillator connected to the MPU's interrupt line(s). The interrupt frequency can easily be varied using a potentiometer. Furthermore, a switch connected to this sampling oscillator's Reset acts as a convenient freeze input. No sample rate – no new samples.

The alternative scheme requires a switch port, not only to read the freeze-request switch (see Fig. 11.3), but to read several switches requesting the sampling rate. Although I have not used this technique, the two microcomputers developed in this chapter have 4-bit switch ports provided. This gives a Read-option expansion capability, and is exploited in Chapter 14, where diagnostic software tests are discussed.

The provision of an 8-bit digital port is a little more expensive than the necessary 1-bit output. This is also useful for diagnostic purposes and gives additional scope for expansion.

Microcomputers based on both the 6809 and 68008 MPUs are developed in the next two sections. By using C to target two different MPUs, we will be able to investigate one of the major advantages of a high-level language.

13.1   6809 – Target Hardware

The implementation shown in Fig. 13.1 is based on a 6809 MPU running at 1 MHz. This is set by the 4 MHz crystal/capacitor network Y1, C9, C10. A power-on manual Reset signal of a nominal 100 ms duration is provided by S3, C11, R15. This relies on the Schmitt trigger action of $\overline{\text{RST}}$, which is described in Section 1.1.

Samples are acquired at a rate dictated by the astable network U7, C6, R3, R7. Based on a 555 timer [1], the total period is given by the relationship:

$$t_p = 0.693(R7 + 2R3)C6$$

and can be varied with R3 from nominally 60 – 250 Hz. The 555 is a noisy device, and thus the +5 V supply should be locally well decoupled. By connecting S1 to the 555's Reset, the astable can be halted. Thus no further updates will occur, giving a frozen display. $\overline{\text{NMI}}$ is used as the interrupt input, as its edge-triggered nature obviates the need for an external interrupt flag, such as used in Fig. 6.6. All unused interrupt lines, as well as $\overline{\text{BREQ}}$ and $\overline{\text{MRDY}}$, are tied high through R5. $\overline{\text{HLT}}$ has its own pull-up resistor R4, as this line is frequently used by in-circuit emulators to control the progression of the MPU.

The address map for the system is:

0000 – 07FFh	6116 RAM
2000h	Analog output channel X
2001h	Analog output channel Y
4000h	Analog input channel
8000h	Digital input port
A000h	Digital output port
E000 – E7FFh	2716 EPROM

The address decoder comprises U3 and U4. The 74HCT138 splits the memory map into eight 8 kbyte pages, six of which enable the devices above. All Write-to devices include $\overline{\text{Q}}$ as part of their enabling logic. The digital output port U2 is clocked by the rising edge of this strobe, whilst the dual D/A converter A/D1 is enabled by it. The 6116 RAM uses Q together with $\overline{\text{R/W}}$ as a modified $\overline{\text{Read/Write}}$ control. This is shortened during a Write cycle, as described in Fig. 1.8. $\overline{\text{Output\_Enable}}$ is driven by $\overline{\text{R/W}}$ to ensure that no data is output during the premature ending of a Write cycle. $\overline{\text{OE}}$ of the 2716 EPROM (usually labelled V_{pgm}) is similarly enabled, to prevent accidental writing to a read-only memory. NAND gates U4 provide these auxiliary functions.

Both RAM and EPROM have a 2 kbyte capacity, which is more than adequate for our application. With a 1 MHz clock frequency, any speed selection will be suitable. With a 2 MHz clock, a 300 ns EPROM is required. Although it is possible to purchase such a 2716 (or Texas 2516) it is easier and cheaper to use a 2764 8 kbyte device at this speed (see Fig. 13.3). If desired, an integral battery backup 48Z02 RAM may be directly substituted for the 6116. RAMs with an access time of 150 ns (min) should be used for a 2 MHz processor.

The two analog ports are as described in Figs 12.9 and 12.13. Figure 13.1 does not show any necessary filtering and buffering.

Quad 3-state buffer U9/10 provides input port facilities for four switches. This 74HCT125 is directly enabled from the address decoder. A 74HCT377 connected

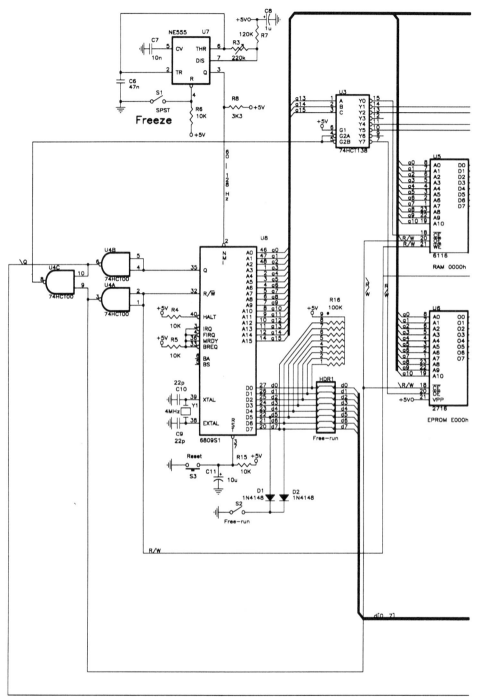

Figure 13.1: The 6809-based embedded microprocessor implementation (*continued next page*).

6809 – Target Hardware 353

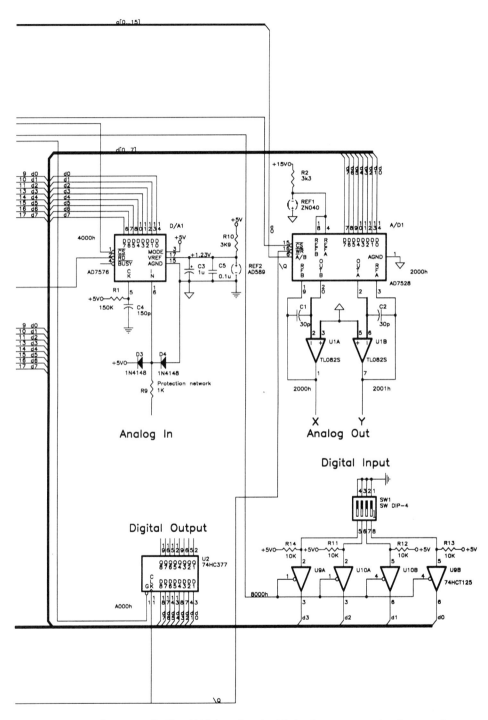

Figure 13.1: (*continued*) The 6809-based embedded microprocessor implementation.

354 The Target Microcomputer

as described in Fig. 1.7 gives a byte-sized digital output port. One of the lines can be used to blank out the CRO during flyback, and the others are free. Some CROs require large negative voltages, typically $-40\,V$, to perform this function. In such cases a suitable transistor buffer and power supply will be required.

A free-run facility, HDR1, R16, D1, D2 and SW2 is shown in Fig. 13.1. This allows the user to exercise the processor before software is available for the EPROM and without using an in-circuit emulator. Its action is described on page 405.

The complete circuit requires $+5\,V$ at typically $250\,mA$ and ± 12 to $\pm 15\,V$ at $25\,mA$. The analog $\pm 15\,V$ is conveniently supplied from a dual d.c./d.c. converter, such as the Citec BC5151S $+5V$ to $\pm 15\,V$ device. Care should be taken, as most converters are not short-circuit proof. Any analog grounds should be returned to this power supply $0\,V$ together with the $+5\,V$'s ground return. The supplies should be decoupled using a mixture of $1\,\mu F$ tantalum and $0.1\,\mu F$ ceramic capacitors at around one capacitor each two devices.

Any suitable wiring technique may be used for the prototype. We use wire-wrap with considerable success. This avoids close parallel paths for the clock and bus signals and reduces crosstalk. It is especially important to keep the analog signals as far away from such digital lines as possible. Whatever technique is used,

```
Title    6809 TCM PAL Address decoder
Chip     PAL16L8
;pin list
A15 A14 A13 R_W Q NC NC NC NC VCC
NC /OE RAM_RW /ROM_EN /DIG_OUT /DIG_IN /AN_IN /AN_OUT /RAM_EN VCC

Equations
/RAM_EN  = /A15*/A14*/A13          ; RAM enabled at 0000
/AN_OUT  = /A15*/A14* A13*/R_W*Q   ; Analog output on 2000h and Write and Q
/AN_IN   = /A15* A14*/A13*R_W      ; Analog input on 4000h and Read
/DIG_IN  =  A15*/A14*/A13*R_W      ; Digital in on 8000h and Read
/DIG_OUT=   A15*/A14* A13*/R_W*Q   ; Digital out on A000h and Write and Q
/ROM_EN  =  A15* A14* A13*R_W      ; EPROM at E000h and Read
RAM_EN   = R_W+/R_W*Q              ; Modified R/W
/OE      = /R_W                    ; For memory OE

/RAM_EN.TRST  = VCC
/AN_OUT.TRST  = VCC
/AN_IN.TRST   = VCC
/DIG_IN.TRST  = VCC
/DIG_OUT.TRST= VCC
/ROM_EN.TRST  = VCC
RAM_RW.TRST   = VCC
/OE.TRST      = VCC

; Tie 74377's /CE to /DIG_OUT
; Tie AD7528,s /CS and /WR to AN_OUT
```

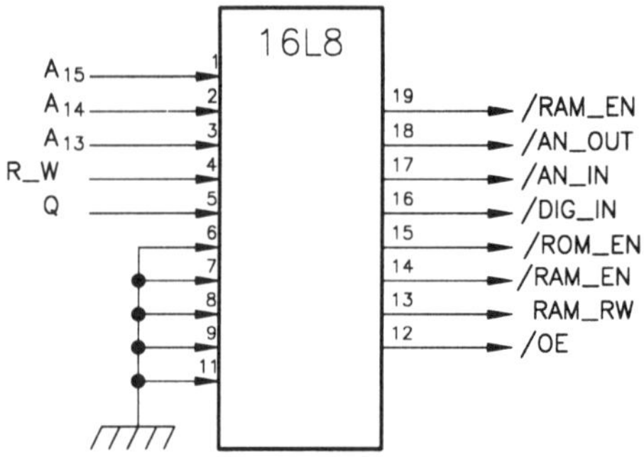

Figure 13.2: A PAL-based 6809 address decoder implementation.

it is important to color-code any wiring to aid in the debug phase. Several strands should be used for the +5 V and its return paths.

One final point refers to the address decoder. Most current circuitry uses a PAL (Programmable Array Logic) implementation to reduce the chip count. The circuit of Fig. 13.1 is so simple that it is unlikely to be commercially viable to do so. However, the design is straightforward and if you have access to a PAL programmer and CAD software, then a PAL16L8 provides for ten inputs and eight active-low outputs in a single 20-pin package. Connection details and the requisite equations in the PALASM2 language [2] are given in Fig. 13.2. Other languages use a similar, but not identical, notation. Note that the analog and digital output ports are qualified internally by Q. The absence of an explicit $\overline{Q}$ (indicated as /Q in PALASM language), means slightly rewiring those devices using this qualifier, as indicated on the equation comments. A complete design of a PAL-based 6809 system is given in Example 6.2 of reference [3].

13.2 68008 – Target Hardware

The implementation of Fig. 13.3 is based on a 68008 MPU, running at 8 MHz. Recall from Fig. 3.3 that the 68008 device is a full 68000 processor, but with an 8-bit data bus, single Data Strobe ($\overline{DS}$) and a commoned $\overline{IPL0}/\overline{IPL2}$ interrupt request line.

The 68008 is externally reset when both $\overline{Halt}$ and $\overline{Reset}$ lines are asserted. Although an active period of only ten clock cycles (1.25 μs for an 8 MHz clock) is required for a successful initialization, at least 100 ms is required on power up. This provides for stabilization of the system clock and on-chip circuitry. The situation is further complicated by the fact that both $\overline{Halt}$ and $\overline{Reset}$ can be used by the MPU as an output. $\overline{Halt}$ is asserted if the processor detects a double-bus fault, for example where the the initial **PC** or **SSP** addresses in the vector table are odd. Remember that odd addresses are illegal. The privileged instruction RESET forces the $\overline{Reset}$ pin low, which is used to initialize external peripheral devices.

These requirements are met in Fig. 13.3 using a 555 timer connected as a monostable, through 3-state buffers U4A and B to $\overline{Reset}$ and $\overline{Halt}$. The active period of the monostable is set by R3/C1, according to the relationship [4]:

$$t_p = 1.1(\text{R3 C1}) \approx 500\,\text{ms}$$

The monostable is triggered when $\overline{TR}$ (pin 2) rises above $\frac{2}{3}V_{CC}$, and is delayed on power-up by R4/C2.

Sampling is regulated by using an astable to drive interrupt lines $\overline{IPL0/2}\,\overline{IPL1}$ in parallel. This gives a level 7 non-maskable interrupt, at a rate varying between nominally 60 and 250 Hz. Design details are given on page 351. No external interrupt flag is required due to the edge-triggered nature of the interrupt. Gate U2A decodes Function Code 111, the interrupt acknowledge condition, and by driving $\overline{VPA}$ ensures that autovector 31 is used as the pointer to the interrupt service routine (see Fig. 6.8).

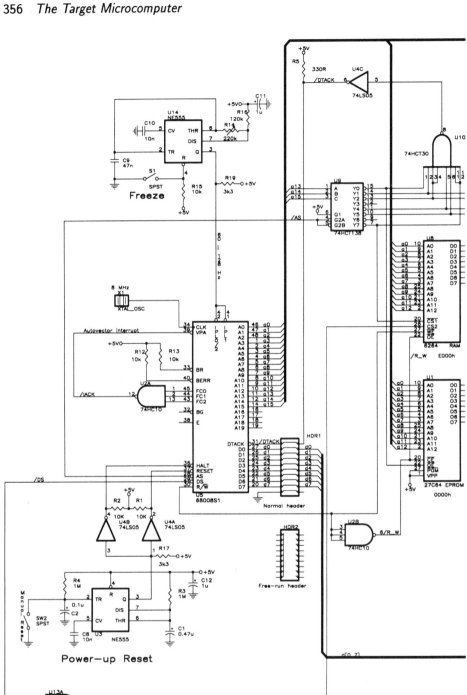

Figure 13.3: The 68008-based embedded microprocessor implementation (*continued next page*).

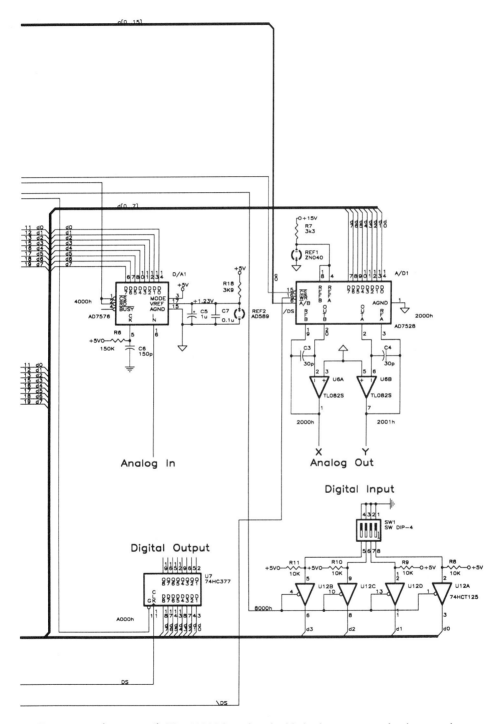

Figure 13.3: (*continued*) The 68008-based embedded microprocessor implementation.

358 *The Target Microcomputer*

The memory map for the system is:

00000–01FFFh	27C64 EPROM (250 ns or better)
02000h	Analog output Channel X
02001h	Analog output Channel Y
04000h	Analog input channel
08000h	Digital input port
0A000h	Digital output port
0E000–0FFFFh	6264 RAM (or 6116)

The address decoder comprises U9, U10 and U4C. The 74HCT138 splits memory up into eight 8 kbyte pages. Address lines $a_{19}-a_{16}$ are ignored by this scheme, and this of course gives 15 images of each page. Gates U10/U4C detect wherever a memory access is made, and activate $\overline{\text{DTACK}}$. All peripheral devices are fast enough

```
Title   68008 TCM Address decoder
Chip    PAL20L10
;pin list
A13 A14 A15 /AS /DS R_W FC0 FC1 FC2 NC NC GND
NC /DTACK /IACK DS /R_W /RAM_EN /DIG_OUT /DIG_IN /AN_IN /AN_OUT /ROM_EN VCC


Equations
/DTACK   = /RAM_EN+/DIG_OUT+/DIG_IN+/AN_IN+/AN_OUT+/ROM_EN    ; Any legitimate device selected
/IACK    = FC0*FC1*FC2                                        ; FC = 111 for an interrupt
DS       = DS                                                 ; Complement of /DS
/R_W     = /R_W                                               ; Complement of R/W
/ROM_EN  = /A15*/A14*/A13*/AS*R_W                             ; EPROM from 0000-1FFFFh and Read
/AN_OUT  = /A15*/A14* A13*/R_W                                ; Analog output on 2000h and Write
/AN_IN   = /A15* A14*/A13*/AS*R_W                             ; Analog input on 4000h and Read
/DIG_IN  =  A15*/A14*/A13*/AS*R_W                             ; Digital in on 8000h and Read
/DIG_OUT =  A15*/A14* A13*/AS*/R_W                            ; Digital out on A000h and Write
/RAM_EN  =  A15* A14* A13*/AS                                 ; RAM enabled at E000-FFFFh


/DTACK.TRST    = VCC
DS.TRST        = VCC
/R_W.TRST      = VCC
/ROM_EN.TRST   = VCC
/AN_OUT.TRST   = VCC
/AN_IN.TRST    = VCC
/DIG_IN.TRST   = VCC
/DIG_OUT.TRST  = VCC
/RAM_EN.TRST   = VCC
```

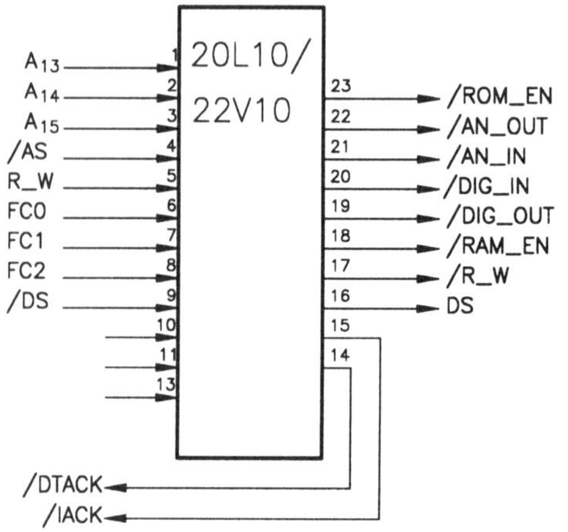

Figure 13.4: A PAL-based 68008 address decoder implementation.

to support direct feedback in this manner, without the necessity of introducing a delay as shown in Fig. 3.9. Care should be taken that the EPROM has an access time of 250 ns or better, and the RAM has a 120 ns maximum access time (see Section 3.3). Alternatively, a lower-frequency clock oscillator (minimum 2 MHz) can be used with slower devices. The digital output port is clocked by the falling edge of $\overline{DS}$, at which time data on the bus has stabilized (point 5 in Fig. 3.7). Both RAM and analog output ports are enabled when $\overline{DS}$ is active. The EPROM is only enabled when R/$\overline{W}$ is high, to prevent an accidental Write-to operation. The RAM's output buffers are similarly disabled when R/$\overline{W}$ is high. Interface details for the AD7528 and AD7576 are given in Figs 12.9 and 12.13 respectively.

A free-run facility, HDR1 and HDR2 is shown between the data bus/$\overline{DTACK}$ and the MPU. By substituting the two headers, the user can check out the system before software is available for the EPROM and without using an in-circuit emulator. It is also a useful diagnostic aid when the system is in service. Its action is described on page 405.

The complete circuit requires +5 V at typically 300 mA and ±12/ ± 15 V for the analog circuitry, at 25 mA. Normal power supply and decoupling practice, as described in the last section, should be followed. However, the data sheet indicates that the 68008 MPU can take current peaks of 1.5 A [5]. Thus a direct connection using heavier or multiple wiring between the 68008's power pins and the power supply is recommended, as is local decoupling.

If you have access to a PAL programmer, a PAL20L10 or 22V10 can be used to implement the address decoder and other glue logic. Chips U9, U10 and U4C/R2 are replaced by the one 24-pin device. Connection details and the requisite equations are given in Fig. 13.4.

References

[1] Berlin, H.M.; *The 555 Timer Applications Sourcebook with Experiments*, H.W. Sams, 1976, Chapter 3.

[2] Alford, R.C.; *Programmable Logic Designer's Guide*, H.W. Sams, 1989, Chapter 5.

[3] Cahill, S.J.; *Digital and Microprocessor Engineering*, Ellis Horwood/Simon and Schuster, 2nd. ed., 1993, Section 6.1.

[4] Berlin, H.M.; *The 555 Timer Applications Sourcebook with Experiments*, H.W. Sams, 1976, Chapter 2.

[5] Wilcox, A.D.; *68000 Microprocessor Systems: Designing and Troubleshooting*, Prentice-Hall, 1987, Section 9.1.1.

CHAPTER 14

Software in C

From Section 11.1, two main tasks can be identified.

Task 1:
BEGIN:
    Forever do:
        Scan and send out to the Y-plates the 256 stored array values from oldest to newest, while incrementing and sending out the X count to the X-plates (left to right).
    Flyback:
        End each scan with a flyback procedure.
END:

Task 2:
BEGIN:
    Forever do:
        At regular intervals interrogate input and place sampled value into array, overwriting oldest value.
END:

In the remainder of this chapter we will develop the necessary data structures and interaction between these tasks. From this, a general program in C is developed; followed by topics specific to the two chosen targets.

14.1 Data Structure and Program

Central to the software implementation of our time-compressed memory is the data organization. This comprises an array of 256 bytes, each holding a sample of the analog input. This array is to be scanned at high speed from the oldest to newest sample. At the same time, at around 128 times per second, a new value is to overwrite the oldest element, and the pointer-to oldest moved on one place.

Treating the array as a circular structure, as shown in Fig. 14.1, emphasizes the repetitive nature of the scan and update. Of course this closed data organization is conceptual only; the array is stored in RAM in the normal linear manner.

Data Structure and Program 361

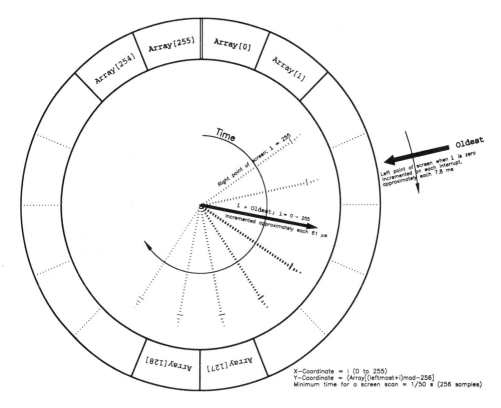

Figure 14.1: Data stored as a circular array.

Two tasks have been identified. The scanning task is sequenced by the variable i, which counts from 0 to 255. By adding this index to the pointer Oldest, elements are accessed from the oldest member (i = 0) to the newest member (i = 255). At the same time, i is converted to its analog equivalent and hence drives the X spot from left (oldest) to right (newest).

The job of the updating task is to fetch a sample into the array, to where Oldest points, and move that index on one. Thus the element just before Oldest is the most juvenile sample. When a whole scan of 256 samples has been completed, flyback occurs and the process begins again, but this time beginning from the current most ancient element. The circular manner of this scan is simulated by wrapping around the sum of Oldest plus i modulo-256, that is from 255 back to 0 ($11111111b + 1 = 00000000b$).

The software implementation of our time-compressed memory is given in Table 14.1. The two tasks are assigned to different functions. main() implements the initialization, repetitive scan and flyback. New samples are acquired and the array and Oldest pointer updated by the function update(). This is designed to be entered via an interrupt, and so no data is sent or returned from it. Communication between tasks is via the global data array Array[] and global index Oldest. Both

362 *Software in C*

Table 14.1: The fundamental C coding.

```
/* Version 16/11/89                                                         */
#include <hard.h>
unsigned char Array [256];          /* Global array holding display data    */
unsigned char Oldest;  /* Index to the Oldest inserted data byte (left point on screen)*/

main()
{
register short int i;                  /* Scan counter                              */
register unsigned char leftmost;       /* The initial array index when x is 0       */
unsigned char * const x = ANALOG_X;    /* x points to a byte @ (address) ANALOG_X   */
unsigned char * const y = ANALOG_Y;    /* y points to a byte @ (address) ANALOG_Y   */
unsigned char * const z = Z_BLANK;     /* The z-mod port (digital port)             */
Oldest = 0;                            /* Start New index at beginning of the array */
for(i=0; i<256; i++)                   /* Clear array                               */
    {Array[i] = 0;}
while(1)                               /* Do forever display contents of array      */
    {
    leftmost = Oldest;          /* Make leftmost point on the screen the oldest sample */
    for (i=0; i<256; i++)
        {
        *x = (unsigned char)i;          /* Send x co-ordinate to X plates       */
        *y = Array[(leftmost+i)&0xOff]; /* and the display byte to the Y D/A    */
        }
    *z = BLANK_ON;                      /* Blank out for flyback                */
    *x = 0;                             /* Move to right of screen              */
    *y = Array[Oldest];                 /* Y value at left of screen            */
    for(i=0; i<5; i++) {;}              /* Delay                                */
    *z = BLANK_OFF;                     /* Blank off                            */
    }                                   /* Do another scan                      */
}

/******************************************************************************
 * This is the NMI interrupt service routine which puts the analog sample in the array *
 * and updates the New index                                                   *
 * ENTRY : Via NMI and startup                                                 *
 * ENTRY : Array[] and Oldest are global                                       *
 * EXIT  : Value held at a_d in Array[Oldest], Oldest incremented with wraparound *
 *         at 256 (modulo-256)                                                 *
 ******************************************************************************/

void update(void)
{
volatile unsigned char * const a_d = ANINPUT; /* This is the Analog input port  */
Array[Oldest++] = *a_d;      /* Overwrite oldest sample in Array[] & inc Oldest index */
}
```

are defined before main(), and therefore are known to both functions. The former is defined as having 256 unsigned char (byte) elements, whilst the latter is a single unsigned char. Each element therefore can vary from 0 to 255. Details of the entry to update() and the header file hard.h, included at the beginning of the file, are discussed in Sections 14.2 and 14.3. The header file contains hardware-related detail, such as the locations of the various peripheral devices.

main() begins by defining five local variables. Both i, the integer scan counter, and leftmost, the char element indicating the most ancient array entry, are defined as being of type register. Both are used inside the scan loop, and will benefit from being stored internally. Processors with insufficient registers will ignore this request. The variables x, y and z are defined as being fixed pointers to unsigned chars (bytes), and are assigned as ANALOG_X, ANALOG_Y and Z_BLANK, which are given values (addresses) in the header file. As they are qualified as const, any subsequent attempt to change them will be reported by the compiler as an error.

The program proper commences by zeroing the global variables Oldest and Array[]. Strictly this run-time initialization is not necessary, as ANSII C specifies that global variables are to be considered zero if not explicitly initialized in their definition. To simulate this situation, the relevant RAM locations could be zeroed in the startup routine. In this case we have chosen to do this in the C coding. Actually the system will operate perfectly satisfactorily if not cleared, but there would be a 2-second transient display while the array was being filled with the first 256 samples.

After initialization, an endless loop is entered inside the body of while(1). At the commencement of this loop the local variable leftmost is equated to Oldest. This prevents changes in Oldest during the scan (i.e. via update()), altering the display.

The scan itself uses a for loop construction, with i acting as the loop counter. i has been defined as an int, so that the condition i < 256 False can be used as a loop terminator. If i was a char, it would wrap around at 255. In this situation a break on i == 255 at the closing brace should be used as the out condition (see Table 14.8).

The for body simply assigns the contents of x (the ANALOG_X output) to i (0 to 255), and the contents of y (the ANALOG_Y output) to the array element. The index of the array is the sum of the X co-ordinate (i.e. i) plus the leftmost value, truncated to 8-bits (modulo-256) by ANDing with 0000111111111b (0xFF). This stratagem achieves a wrap around at 255. For example if leftmost were 180 and i were 159, then Array[83] is the value sent to the Y-plates (180 + 159 is 83 when added modulo-256). A similar result could be obtained if the sum was given an independent int-sized existence and then cast to char. I have used such a cast in equating the char-sized contents of x to the integer i, x = (unsigned char)i;. In practice the compiler will truncate the r_value in assigning to a small l_value (see page 227).

Flyback is generated by sending the correct patterns to the Z port (BLANK_ON is defined in the header), zero to the X-plates and the initial array value to the Y-plates. A short null for-loop gives a delay, before the BLANK_OFF pattern is sent out to Z. After this, the scan begins again.

Function update() is very short. The local pointer variable a_d is defined as being the const address ANINPUT, whose absolute value is given in the header. This pointer is to an unsigned char (byte) which is volatile (changes spontaneously) and is const (read-only). The value read from this port is then put into the array at the oldest index, and the global variable Oldest automatically incremented. We are relying here on the char nature of Oldest wrapping around at 255. An explicit wraparound would be necessary for other array lengths.

Function update() assumes that the analog to digital converter can be treated as a simple read-only input port. In that respect the program is not portable. Normally, a separate function is used for more complex parts, frequently called getchar(). Such a function would be part of an input/output library, which was hardware specific, or would appear in the header file. Similar assumptions have also been made for output in main().

Portability has been further compromised by the assumption that char objects are 8-bit wide. In practice this is true for the vast majority of microprocessor targeted compilers. However, 9-bit character systems do exist, and the use of complex character sets, such as Japanese, requires 16-bit characters. ANSII C makes no guarantees regarding the 8-bit nature of char objects.

In the next two sections we look at machine-specific details regarding our two target circuits of Figs 13.1 and 13.3.

14.2 6809 – Target Code

The first of our targets is the 6809 hardware implementation of Fig. 13.1. The header file hard_09.h of Table 14.2 gives a memory map of the various peripherals and defines the two digital patterns BLANK_ON and BLANK_OFF for the Z output. ROM and RAM details are given for use by the diagnostic software of Section 15.2. Including this header customizes the C program of Table 14.1 to the 6809 target board, and no further modification is required.

A complete listing of 6809 assembly-level code intermingled with the original C source-code is shown in Table 14.3. This was produced using the Intermetrics/COSMIC 6809-C cross-compiler V3.3. I have tidied up the original compiler output, for example removing remarks inserted by the optimizer, and added comments, indicated by the prefix ;##.

The added comments are self-explanatory and will not be discussed in the text. There are, however, several points to note. Firstly the **register** qualifier

Table 14.2: The hard_09.h header file.

```
#define ANALOG_X    (unsigned char *) 0x2000   /* Analog  output to X amplifier    */
#define ANALOG_Y    (unsigned char *) 0x2001   /* Analog  output to Y amplifier    */
#define ANINPUT     (unsigned char *) 0x6000   /* Analog  input port at 6000h      */
#define SWITCH      (unsigned char *) 0x8000   /* Digital input port at 8000h      */
#define Z_BLANK     (unsigned char *) 0xA000   /* Digital output port at A000h     */
#define RAM_START   (unsigned char *) 0x0000   /* 6116 chip starts at location 0000h */
#define RAM_LENGTH                    0x800    /* 6116 byte capacity is 2K or 800h */
#define ROM_START   (unsigned short *)0xE000   /* 2716 chip starts at location E000h */
#define ROM_LENGTH                    0x800    /* 2716 byte capacity is 2K or 800h */
#define BLANK_ON                      0xFF     /* Bit pattern to blank out beam    */
#define BLANK_OFF                     0        /* Bit pattern to enable beam       */
```

for the variables i and leftmost have been ignored by the compiler. This is common for 8-bit MPU targets, as most of these devices are characterized by a paucity of registers. This is rather a pity, as the Y register is not used and would conveniently hold the loop variable i. This would remove the necessity for the double-precision incrementation for i++ (e.g. lines 76–78 could be replaced by LEAY 1,Y) and the addition of lines 68–73 could be replaced by LDB -3,U: CLRA: LEAY D,Y: LDB _Array,Y. As both processes are done in each loop pass, the savings are obvious. Table 14.9 shows code where the register qualifier is obeyed.

ANSI C specifies that chars and shorts are promoted to ints during processing (see Fig. 8.4). Objects larger than bytes (chars) are handled with difficulty in most 8-bit processors. The Intermetrics/COSMIC 6809 C cross-compiler permits processing of chars in their byte form. Thus the assignment leftmost = Oldest; is simply implemented in lines 54 and 55 using Accumulator_B only. However, the usefulness of this option is not fully realized in this particular instance, as i is a 16-bit object, and most arithmetic involves this variable.

Communication between the background function main() and interrupt function update() is handled via the two global objects Oldest and Array[]. By defining these outside any function (lines C3 and C4) the compiler has placed their base labels _Oldest and _Array (lines 134–138) in absolute memory. One byte has been reserved for the former and 256 for the latter. Both labels have been declared public, and thus are known to all, including files compiled/assembled externally. Both objects are in the _bss program sector, which is used by this compiler for static and extern data with no initial values. C specifies that these should be pre-initialized by default to zero, and as they lie in RAM, this should be done by the startup routine. However, in this instance I have chosen explicitly to clear them at the C level in main() at lines C13–C15.

The linker has been configured to commence the _bss section at 0001h (as location 0000h, the null pointer, should never be used), which locates _Oldest at 0001h and _Array at 0002h. Similarly _text begins at E000h. The program shown in Table 14.3, however, commences at E00Dh. The missing 12 bytes are taken up by the startup routine, which is an assembly-level routine linked in before the compiled file.

The startup routine, shown in Table 14.4(a) has three functions. The first is to set the Stack Pointer to the top of the System stack. Hence in line 7 I have put this at the top of the 6116 RAM. If the library routines malloc()[1] (Memory ALLOCate) and other related functions are being used, then this can be lowered somewhat and memory above used as a general storage pool (called the heap).

The second purpose of this startup routine is to go to the main C function. This is implemented as a simple JSR _main in line 8. In this case, startup.s does not pass any parameters to main(). main() is an endless loop and so no return should occur, but if it does, a skip back to the beginning is actioned. The re-entry point is labelled _exit, and can be reached from the C level by calling the library routine exit(). exit() is supposed to return True or False to indicate an error condition, but no use is made of this in our situation.

366 Software in C

The final function deals with NMI interrupt handling. Function update() is terminated with a Return From Subroutine operation (implemented in line 131 of Table 14.3 with a PULS PC) and therefore cannot be directly entered from an interrupt. Instead, the startup routine has a stub in lines 12 and 13, which is labelled NMI. This stub simply jumps to update() (JSR _update) and terminates on return with RTI. If the address NMI is placed in the NMI vector, then on receiving

Table 14.3: 6809 code resulting from Tables 14.1 and 14.2 (*continued next page*).

```
 1  ; Compilateur C pour MC6809 (COSMIC-France)
 2                  .list     +
 3                  .psect _text
 4  ;  1  /* Version 16/11/89                                                          */
 5  ;  2  #include <hard_09.h>

 6  ;  1  #define ANALOG_X   (unsigned char *)0x2000 /* Analog  output to X amplifier  */
 7  ;  2  #define ANALOG_Y   (unsigned char *)0x2001 /* Analog  output to Y amplifier  */
 8  ;  3  #define ANINPUT    (unsigned char *)0x6000 /* Analog  input port  at 6000h   */
 9  ;  4  #define SWITCH     (unsigned char *)0x8000 /* Digital input port  at 8000h   */
10  ;  5  #define Z_BLANK    (unsigned char *)0xA000 /* Digital output port at A000h   */
11  ;  6  #define RAM_START  (unsigned char *)0x0000 /* 6116 chip starts at 0000h      */
12  ;  7  #define RAM_LENGTH                  0x800  /* 6116 byte capacity is 2K or 800h*/
13  ;  8  #define ROM_START  (unsigned short *)0xE000/* 2716 chip starts at E000h      */
14  ;  9  #define ROM_LENGTH                  0x800  /* 2716 byte capacity is 2K or 800h*/
15  ; 10  #define BLANK_ON                    0xFF   /* Bit pattern to blank out beam  */
16  ; 11  #define BLANK_OFF                   0      /* Bit pattern to enable beam     */

17  ;  3  unsigned char Array [256];       /* Global array holding display data */

18  ;  4  unsigned char Oldest;  /* Index to the Oldest inserted data byte (left point on)
19     5
20  ;  6  main()
21  ;  7  {
22  E00D 3440  _main:  pshs  u          ;## Open a frame
23  E00F 33E4          leau  ,s         ;## U is the Top Of Frame (TOF)
24  E011 3277          leas  -9,s       ;## Nine bytes deep
25  ;  8  register short int i;            /* Scan counter                             */
26  ;  9  register unsigned char leftmost; /* The initial array index when x is 0      */
27  ; 10  unsigned char * const x = ANALOG_X;/* x points to a byte @ (address) ANALOG_X*/
28  E013 CC2000        ldd   #2000h     ;## Put constant 2000h in frame at FP-5/-4
29  E016 ED5B          std   -5,u
30  ; 11  unsigned char * const y = ANALOG_Y;/* y points to a byte @ (address) ANALOG_Y*/
31  E018 CC2001        ldd   #2001h     ;## Put constant 2001h in frame at FP-7/-6
32  E01B ED59          std   -7,u
33  ; 12  unsigned char * const z = Z_BLANK; /* The z-mod port (digital port)          */
34  E01D CCA000        ldd   #0A000h    ;## Put constant A000h in frame at FP-9/-8
35  E020 ED57          std   -9,u
36  ; 13  Oldest = 0;              /* Start New index at beginning of the array */
37  E022 7F0001        clr   _Oldest
38  ; 14  for(i=0; i<256;i++)              /* Clear array                             */
39  E025 4F            clra
40  E026 5F            clrb
41  E027 ED5E          std   -2,u       ;## i lives in FP-2/-1; is cleared, i=0
42  E029 AE5E   L1:    ldx   -2,u       ;## Get i into X
43  E02B 8C0100        cmpx  #256       ;## i<256?
44  E02E 2C0C          jbge  L14        ;## IF not THEN jump out of for loop
```

6809 – Target Code   367

Table 14.3: 6809 code resulting from Tables 14.1 and 14.2 (*continued next page*).

```
45  ; 15       {Array[i]=0;}
46  E030 6F890002      clr    _Array,x   ;## EA is Array[0]+i, clear it
47  E034 6C5F          inc    -1,u       ;## Double-precision increment of 16-bit int i, i++
48  E036 2602          jbne   L4
49  E038 6C5E          inc    -2,u
50  E03A 20ED          jbr    L1
51  ; 16    while(1)                     /* Do forever display contents of array    */
52  ; 17    {
53  ; 18        leftmost = Oldest;/* Make leftmost point on the screen the oldest sample*/
54  E03C F60001 L4:    ldb    _Oldest    ;## Put Oldest array index
55  E03F E75D          stb    -3,u       ;## in FP-3/-2, where leftmost lives in the frame
56  ; 19        for (i=0; i<256; i++)
57  E041 4F            clra
58  E042 5F            clrb
59  E043 ED5E          std    -2,u       ;## Again i=0
60  E045 AE5E   L16:   ldx    -2,u       ;## Get i into X
61  E047 8C0100        cmpx   #256       ;## i<256?
62  E04A 2C1C          jbge   L17        ;## IF not THEN jump out of for loop
63  ; 20        {
64  ; 21            *x = (unsigned char)i;    /* Send x co-ordinate to X plates    */
65  E04C EC5E          ldd    -2,u       ;## Get i into D
66  E04E E7D8FB        stb    [-5,u]     ;## Put lower byte (char) indirectly into X D/A
67  ; 22            *y = Array[(leftmost+i)&0xOff];/* and the display byte to the Y D/A*/
68  E051 E65D          ldb    -3,u       ;## Get leftmost out of the frame into B
69  E053 4F            clra              ;## extended to 16 bits (int)
70  E054 E35E          addd   -2,u       ;## Add to int i; leftmost+i
71  E056 4F            clra              ;## Neat way of ANDing with 0000 0000 1111 1111b!
72  E057 1F01          tfr    d,x        ;## X holds (leftmost+i)&0xff
73  E059 E6890002      ldb    _Array, x  ;## EA is Array[0]+(leftmost+i)&0xff; get element
74  E05D E7D8F9        stb    [-7,u]     ;## Put Array[(leftmost+i)&0xff] indirectly into Y
75  ; 23        }
76  E060 6C5F          inc    -1, u      ;## Double-precision increment of 16-bit int i, i++
77  E062 2602          jbne   L6
78  E064 6C5E          inc    -2,u
79  E066 20DD   L6:    jbr    L16
80  ; 24        *z = BLANK_ON;                /* Blank out for flyback           */
81  E068 C6FF   L17:   ldb    #255       ;## Send out indirectly 1111 1111b to Z
82  E06A E7D8F7        stb    [-9,u]
83  ; 25        *x = 0;                       /* Move to right of screen         */    */
84  E06D 6FD8FB        clr    [-5,u]     ;## Send out indirectly 00h to X D/A; i.e. flyback
85  ; 26        *y = Array[Oldest];           /* Y value at left of screen       */
86  E070 8E0002        ldx    #_Array    ;## While this is happening get Array[Oldest]
87  E073 F60001        ldb    _Oldest
88  E076 4F            clra
89  E077 E68B          ldb    d,x
90  E079 E7D8F9        stb    [-7,u]     ;## and put it indirectly into the Y D/A converter
91  ; 27        for(i=0;i<5;i++) {;}          /* Delay                           */
92  E07C 5F            clrb
93  E07D ED5E          std    -2,u       ;## i=0
94  E07F AE5E  L121:   ldx    -2,u       ;## Get i into X
95  E081 8C0005        cmpx   #5         ;## i<5?
96  E084 2C08          jbge   L131       ;## IF not THEN jump out of for delay loop
97  E086 6C5F  L141:   inc    -1,u       ;## Double-precision increment of 16-bit int i, i++
98  E088 2602          jbne   L01
99  E08A 6C5E          inc    -2,u
100 E08C 20F1  L01:    jbr    L121
101 ; 28        *z = BLANK_OFF;               /* Blank off                       */    */
102 E08E 6FD8F7 L131:  clr    [-9,u]     ;## Send out indirectly 0000 0000b to Z
103 ; 29    }
104 E091 20A9          jbr    L14        ;## Do another scan; forever
105 ; 30  }
```

368 Software in C

Table 14.3: (*continued*) 6809 code resulting from Tables 14.1 and 14.2.

```
106 ; /************************************************************************
107 ; 32 * This is the NMI interrupt service routine which puts the analog sample in the
108 ; 33 * ENTRY : Via NMI and startup
109 ; 34 * ENTRY : Array[] and Oldest are global
110 ; 35 * EXIT  : Value held at a_d in Array[Oldest], Oldest incremented with wraparound
111 ; 36 ************************************************************************
112 ; 37
113 ; 38  void update(void)
114 ; 39  {
115 E093 3440   _update:pshs  u              ;## Open a frame
116 E095 33E4           leau  ,s             ;## With U as TOF
117 E097 327E           leas  -2,s           ;## of two bytes
118 ; 40  volatile unsigned char * const a_d = ANINPUT;/* This is the Analog input port*/
119 E099 CC6000         ldd   #6000h         ;## to locate the constant 6000
120 E09C ED5E           std   -2,u
121 ; 41  Array[Oldest++] = *a_d;             /* Overwrite oldest sample in Array[] & inc
122 E09E 8E0002         ldx   #_Array        ;## Point x to Array[0]
123 E0A1 F60001         ldb   _Oldest        ;## Get Oldest
124 E0A4 7C0001         inc   _Oldest        ;## Oldest++
125 E0A7 4F             clra                 ;## 16-bit Oldest++
126 E0A8 308B           leax  d,x            ;## Point X to Array[0]+Oldest++; ie Array[Oldest++]
127 E0AA E6D8FE         ldb   [-2,u]         ;## Get indirectly the contents of A/D; ie of 6000h
128 E0AD E784   L61:    stb   ,x             ;## Put it away as the latest entry into the array
129 ; 42  }
130 E0AF 32C4           leas  ,u             ;## Close the frame
131 E0B1 35C0           puls  u,pc
132                     .public _update
133                     .public _main

134                     .psect _bss          ;## Space on RAM for
135     0001   _Oldest: .byte  [1]           ;## the 1-byte (char) object Oldest
136                     .public _Oldest
137     0002   _Array:  .byte  [256]         ;## and 256-byte array
138                     .public _Array       ;## both of which are external, that is public
139                     .end
```

such an interrupt, the processor will jump to NMI (E009h here) and again jump to update(). The way back is a similar RTS–RTI double hop. As all registers are saved on entry, no other action need be taken.

The vector routine of Table 14.4(b) is linked in after the C code and begins at E7F6h, which is the FIRQ vector in the 2716 EPROM. All vectors are specified to point to the beginning of the startup routine (E000h), except the NMI vector. The addresses start and NMI have been broadcast by the startup routine as public and declared external by the vector routine.

The end production of the compilation/assembly and linkage of these three files is the Intel-format machine-code file of Table 14.5. This is used as the input to the EPROM programmer or in-circuit emulator. In total there are 178 bytes of EPROM text plus the ten Vector bytes.

The double-hop interrupt handling technique will work with any compiler. However, most compilers specifically designed to produce ROMable code support exten-

Table 14.4: The 6809 Time Compressed Memory Startup.

```
1                           .processor m6809
2  ;
3  ; C STARTUP FOR 6809 Time Compressed Memory
4  ; With primitive interrupt handling and no initialization of statics & globals
5                           .external _main, _update
6                           .public   _exit, NMI, start
7     E000  10CE0400 start: lds     #0400h    ; Set Stack Ptr to top of 6116 RAM
8     E004  BDE00D          jsr     _main     ; Execute main()
9     E007  20F7     _exit: bra     start     ; IF return THEN repeat
10 ; Now follows the NMI stub leading to update()
11 ; It is reached from the vector table
12    E009  BDE093    NMI:  jsr     _update   ; Go to update()
13    E00C  3B              rti               ; On return terminate service
14                          .end
```

(a) Startup code.

```
1  ; Table of vectors, all point into the startup module
2                           .processor m6809
3                           .list     +.text
4                           .psect    _text    ; Link in at E7F6h for a 2716 EPROM
5                           .external NMI, start
6  ; The NMI service routine stub is in the startup routine, as is start
7     E7F6  E000            .word     start, start, start, NMI, start
8     E7F8  E000
9     E7FA  E000
10    E7FC  E009
11    E7FE  E000
12 ; Five vectors; namely FIRQ, IRQ, SWI, NMI, Reset.  All go to the start
13 ; except NMI, which goes to the NMI stub
14                          .end
```

(b) Vector table linked in after the C code.

sions to the ANSII standard, enabling the user to declare a function as an interrupt handler (See Section 10.2). The function name, in our case update, is then entered into the Vector table directly in the normal way. This direct entry should decrease the response time to an interrupt and at the same time reduce the code emitted by the compiler.

This particular compiler uses the directive @port to designate a function in this way, the function header then becoming @port update(). It is instructive to look at the code produced, which is shown in Table 14.6(a). Here we can see the RTI in line 120, but also the RTS in line 133. What has happened, is that the original code has been cocooned by the RTI at the end and a library subroutine c_cstk at the beginning. As you will remember, the 6809 has three interrupt inputs: $\overline{\text{NMI}}$,

370 *Software in C*

Table 14.5: The machine-code file for the 6809-based time-compressed memory.

```
e093   _update
e00d   _main
e009   NMI
e007   _exit
e000   start
0001   _Oldest
0002   _Array
e0b8   __prog_top
0001   __data_top
0102   __stack_bottom
0400   __stack_top
e7f6   a:vecttcm9.o
$
:20E0000010CE0400BDE00D20F7BDE0933B344033E43277CC2000ED5BCC2001ED59CCA000EB
:20E02000ED577F00014F5FED5EAE5E8C01002C0C6F8900026C5F26026C5E20EDF60001E7B0
:20E040005D4F5FED5EAE5E8C01002C1CEC5EE7D8FBE65D4FE35E4F1F01E6890002E7D8F91A
:20E060006C5F26026C5E20DDC6FFE7D8F76FD8FB8E0002F600014FE68BE7D8F95FED5EAED2
:20E080005E8C00052C086C5F26026C5E20F16FD8F720A9344033E4327ECC6000ED5E8E0048
:18E0A00002F600017C00014F308BE6D8FEE78432C435C03B32C435C0B0
:0AE7F600E000E000E000E009E000B0
:03E0B800E0BB00CA
:00E000011F
```

$\overline{\text{IRQ}}$ and $\overline{\text{FIRQ}}$. The two former save all internal registers on the System stack and retrieve them, whilst the latter saves only the **CCR** and **PC**. c_cstk, shown disassembled (see page 392) in Table 14.6(b), first checks the E flag. If **E** is clear then a NMI or IRQ interrupt service is in progress and nothing further needs doing. If not, the E flag is cleared and all the registers are put into the System stack to pretend that the $\overline{\text{FIRQ}}$ is really an $\overline{\text{IRQ}}$/$\overline{\text{NMI}}$ type interrupt.

Table 14.6 shows us that although @port is deceptively simple at the C level, it neither improves the speed nor reduces the size of the resulting code. Knowing, as we do, that update() is entered via an $\overline{\text{NMI}}$, we could simply alter line 131 of Table 14.3 in its source form to PULS U : RTI, before letting it through to the assembler. This is messy, and as an alternative the function:

_asm ("LEAS, U\n PULS U\n RTI\n")

is used to insert three assembly-level instructions. The first two close the frame, whilst an RTI terminates the interrupt routine. This is shown in Table 14.7, line C42. The principle could be extended to $\overline{\text{FIRQ}}$ by saving registers at the beginning, and pushing them out at the end. Incidentally, _asm() could also be used to implement the startup routine as a front to the C code.

All three approaches are non-portable and error prone, so in the majority of cases a stub approach is best, if rather slow. The @port solution gives 192 bytes whilst using _asm() yields 179 bytes. Creatively editing the source file is the most efficient of all, giving a total of 175 bytes. These figures take account of the removal

Table 14.6: The @port directive.

```
106 ;         /******************************************************************
107 ;   32    * This is the NMI interrupt service routine which puts the analog sample in
108 ;   33    * ENTRY : Via NMI and startup1                                       *
109 ;   34    * ENTRY : Array[] and Oldest are global                              *
110 ;   35    * EXIT  : Value held at a_d in Array[Oldest], Oldest incremented with wrap
111 ;   36    ******************************************************************/
112 ;   37
113 ;   38    @port update()
114 ;   39    {
115   E093 BDE0B2 _update:     jsr     c_cstk  ;## Save registers if FIRQ
116   E096 33E4                leau    ,s      ;## Open a frame
117   E098 327E                leas    -2,s    ;## With U as TOF
118   E09A 8D03                jbsr    L02     ;## Do the core code
119   E09C 3262                leas    2,s     ;## Close frame
120   E09E 3B                  rti             ;## Return from interrupt
121 ;   40    volatile unsigned char * const a_d = ANINPUT;/* This is the Analog input port*/
122   E09F CC6000  L02:        ldd     #6000h  ;## The constant 6000h put into frame
123   E0A2 ED5E                std     -2,u
124 ;   41    Array[Oldest++] = *a_d; /* Overwrite oldest sample in Array[] & inc Oldest idx
125   E0A4 8E0002              ldx     #_Array ;## Point x to Array[0]
126   E0A7 F60001              ldb     _Oldest ;## Get Oldest
127   E0AA 7C0001              inc     _Oldest ;## Oldest++
128   E0AB 4F                  clra            ;## 16-bit Oldest++
129   E0AC 308B                leax    d,x     ;## Point X to Array[0] + Oldest++
130   E0AE E6D8FE              ldb     [-2,u]  ;## Get contents of A/D, that is of 6000h
131   E0B0 E784    L61:        stb     ,x      ;## Put it away as the latest entry
132 ;   42    }
133   E0B1 39                  rts             ;## Return to stub above
134                            .public _update
135                            .public _main
136                            .psect  _bss    ;## Space on RAM for
137   0001           _Oldest:  .byte   [1]     ;## the 1-byte (char) object Oldest
138                            .public _Oldest
139   0002           _Array:   .byte   [256]   ;## and 256-byte array
140                            .public _Array  ;## both of which are external
141                            .external c_cstk
142                            .end
```

(a) Resulting code.

```
0xe0a9 6d62    c_cstk:    tst    2,s         ;## Check E flag
0xe0ab 2b13               bmi    0xe0c0      ;## IF 0 THEN forget about the rest
0xe0ad 327f               leas   -1,s        ;## ELSE make a copy on Stack
0xe0af 341f               pshs   cc,d,dp,x   ;## of the registers used by compiler
0xe0b1 a669               lda    9,s         ;## in the correct order
0xe0b3 8a80               ora    #0x80       ;## setting E = 1
0xe0b5 a7e4               sta    0,s         ;## to mimic a IRQ/NMI type Stack
0xe0b7 ae67               ldx    7,s
0xe0b9 10af66             sty    6,s
0xe0bc ef68               stu    8,s
0xe0be 6e84               jmp    0,x         ;## Return without altering SP
0xe0c0 39                 rts                ;## Exit point for IRQ/NMI type interrupt
```

(b) A disassembly of the library routine c_cstk.

372 Software in C

Table 14.7: Using _asm() to terminate a NMI/IRQ type interrupt service function.

```
; 31/******************************************************************************
; 32 * This is the NMI service routine which puts the analog sample in the array and update
; 33 * ENTRY : Via NMI and startup
; 34 * ENTRY : Array[] and Oldest are global
; 35 * EXIT  : Value held at a_d in Array[Oldest], Oldest incremented with wraparound @ 256
; 36 ******************************************************************************
; 37
; 38 void update(void)
; 39 {
_update:        pshs    u
                leau    ,s
                leas    -2,s
; 40 volatile unsigned char * const a_d = ANINPUT; /* This is the Analog input port    */
                ldd     #6000h
                std     -2,u
; 41 Array[Oldest++] = *a_d;/* Overwrite oldest sample in Array[] and inc Oldest index */
                ldx     #_Array
                ldb     _Oldest
                inc     _Oldest
                clra
                leax    d,x
                ldb     [-2,u]
L61:            stb     ,x
; 42 _asm("LEAS ,U\nPULS U,PC\nRTI            ; Wrap up frame and return to main \n");
                LEAS    ,U              ;## Three inserted assembler-level instructions
                PULS    U,PC            ;## to wrap up frame
                RTI                     ;## and return from interrupt
; 43 }
                leas    ,u              ;## These 2 instructions are now dead code; ie never entered
                puls    u,pc
                .public _update
                .public _main
                .psect  _bss
_Oldest:        .byte   [1]
                .public _Oldest
_Array:         .byte   [256]
                .public _Array
                .end
```

of the stub from startup.s, but not the vector table. Creative editing, whilst being efficient, is the most dangerous, as it does not show up in the source of any of the constituent files, and, unless extremely well documented, will cause havoc if any but the original designer tries to make subsequent changes.

We will compare the resulting machine code to a hand-assembled version in Section 16.1, but the question must be asked here: can the resulting machine code be reduced in size, knowing the way the compiler produces such code.

Two possibilities spring to mind. As we have said the 6809 does not handle 16-bit quantities with any finesse. If we could use a char-sized i, instead of short, a considerable economy should be achieved. This can be done, if rather inelegantly, by defining i as unsigned char and replacing the statement:

```
for (i=0; i<256, i++)
    {body;}
```

by:

```
i = 0;
do
    {
    body;
    i++;
    } while (i!=0);
```

Here i will be 1 after the first pass, and the while argument will be True. When i reaches 255, then i++ will wrap around to 0 and the while argument will return False, causing the do...while loop to exit.

This structure is of course only relevant to loops of 256 iterations on an 8-bit machine, and presupposes an 8-bit char.

A further reduction can be obtained if the compiler's treatment of pointer constants, such as a_d in lines 26–33 of Table 14.3 is studied. There are four such constants in our program, and each is put into the frame on entry to the function, for example:

```
119    LDD   #6000h         ; the constant a_d
120    STD   -2,U           ; in the frame at TOF-2 and TOF-1
```

Once in the frame they can be used as a pointer via Indirect addressing, for instance $[-2,U] = 6000h$. With main() this stack initialization is done only once on entry, and execution proceeds to the core endless loop. The same setup occurs on each entry to update(); however, this will happen around 128 times per second!

It is not necessary to store constants in the essentially dynamic frame; it is better to use absolute locations. This can be done by defining such pointers as static; for example:

```
static   volatile unsigned char  *  const a_d = ANINPUT;
```

which reads a_d is a const pointer/ to a volatile unsigned char/ is stored statically/ and has an initial value of ANINPUT (i.e. $6000h$). The combination of the qualifiers static and const tells the compiler to put constants in ROM, that is the _text section. The compile-time nature of these constants is clearly seen in lines 6–8 of Table 14.8, where they are placed in EPROM at locations E00D–E012h. This saves 4×3 bytes and results in quicker execution (it also makes the code easier to read). Defining const pointers externally is an alternative to a static declaration, see Table 15.5. Notice that update() no longer requires a frame.

Defining Oldest to lie in zero page (with the @dir prefix in line C4) saves another few bytes, giving a total size of 132 bytes, plus vectors. Table 14.8 uses the startup stub entry for the interrupt entry to update(). A further few bytes may be saved at the expense of portability by using _asm().

374   *Software in C*

Table 14.8: Optimized 6809 code (*continued next page*).

```
; Compilateur C pour MC6809 (COSMIC-France)
        .list    +
        .psect   _text
;  1  /* Version 07/12/89                                                   */
;  2  #include <hard_09.h>

L5_x:   .word 2000h    ;## 1st word in the txt sect (ROM) holds the pointer constant 2000h
L51_y:  .word 2001h    ;## Next in absolute memory is the constant 2001h (Y amplifier port)
L52_z:  .word 0A000h   ;## and A000h the Z-blank port
;  3  unsigned char Array [256]; /* Global array holding display data        */
;  4  @dir unsigned char Oldest; /* Index to the Oldest inserted data byte (left point on
;  5
;  6  main()
;  7  {
_main:  pshs    u
        leau    ,s
        leas    -2,s
;  8  unsigned char i;                  /* Scan counter                      */
;  9  unsigned char leftmost;           /* The initial array index when x is 0 */
; 10  static unsigned char * const x = ANALOG_X; /* x points to a byte @ (address) ANALOG_X
; 11  static unsigned char * const y = ANALOG_Y; /* y points to a byte @ (address) ANALOG_Y
; 12  static unsigned char * const z = Z_BLANK;  /* The z-mod port (digital port)     */
; 13  Oldest = 0;                       /* Start New index at beginning of the array */
        clr     _Oldest
; 14  i=0;
        clr     -1,u
; 15  do                                /* Clear array                       */
; 16      {Array[i]=0; i++;} while(i!=0);
L1:     ldx     #_Array     ;## First do the body statements
        ldb     -1,u        ;## i is now a char; and is at U-1
        clra
        clr     d,x
        inc     -1,u        ;## i++
        lda     -1,u        ;## Then do the test i != 0
        jbne    L1
; 17  while(1)                          /* Do forever display contents of array */
; 18      {
; 19      leftmost = Oldest; /* Make the leftmost point on the screen the oldest sample
L13:    ldb     _Oldest
        stb     -2,u
; 20      i=0;
        clr     -1,u
; 21      do
; 22          {
; 23          *x = (unsigned char)i;   /* Send x co-ordinate to X plates      */
L15:    ldb     -1,u
        stb     [L5_x]
; 24          *y = Array[(leftmost+i)&0x0ff]; /* and the display byte to the y D/A */
        clra
        addb    -2,u
        rola
        clra
        tfr     d,x
        ldb     _Array,x
        stb     [L51_y]
; 25          } while(++i!=0);
        inc     -1,u        ;## i++
```

Table 14.8: Optimized 6809 code (*continued next page*).

```
            ldb    -1,u          ;## Once again note the test after the body is executed
            jbne   L15
; 26          *z = BLANK_ON;         /* Blank out for flyback                  */
            ldb    #255
            stb    [L52_z]
; 27          *x = 0;                /* Move to right of screen                */
            clr    [L5_x]
; 28          *y = Array[Oldest];    /* Y value at left of screen              */
            ldx    #_Array
            ldb    _Oldest
            ldb    d,x
            stb    [L51_y]
; 29          for(i=0; i<5; i++) {;} /* Delay                                  */
            clr    -1,u
L101:       ldb    -1,u
            cmpb   #5
            jbhs   L111
L121:       inc    -1,u
            jbr    L101
; 30          *z = BLANK_OFF;        /* Blank off                              */
L111:       clr    [L52_z]
; 31        }
            jbr    L13
L53_a_d:  .word    6000h         ;## The static pointer constant 6000h is also stored in EPROM
; 32     }
; 33 /********************************************************************************
; 34  * This is the NMI interrupt service routine which puts the analog sample in the array
; 35  * ENTRY : Via NMI and startup
; 36  * ENTRY : Array[] and Oldest are global
; 37  * EXIT  : Value held at a_d in Array[Oldest], Oldest incremented with wraparound @ 256
; 38  ********************************************************************************/
; 39
; 40 void update(void)
; 41 {
; 42 static volatile unsigned char * const a_d = ANINPUT;/* This is the Analog input port*/
; 43 Array[Oldest++] = *a_d;/* Overwrite oldest sample in Array[] & inc Oldest index modulo
_update:    ldx    #_Array
            ldb    _Oldest
            inc    _Oldest
            clra
            leax   d,x
            ldb    [L53_a_d]
L01:        stb    ,x
; 44 }
            rts
            .public _update
            .public _main
_Oldest:  .psect  zpage         ;## Oldest is stored in page zero (direct page) at 0001h
          .byte   [1]
          .public _Oldest
          .psect  _bss
_Array:   .byte   [256]         ;## Array is stored in the normal _bss area, starting @ 0100h
          .public _Array
          .end
```

(a) Resulting assembly code.

Table 14.8: (*continued*) Optimized 6809 code.

```
e073    _update
e013    _main
e009    NMI
e007    _exit
e000    start
0100    _Array
e084    __prog_top
0100    __data_top
0200    __stack_bottom
0400    __stack_top
e7f6    a:vecttcm9.o
0001    _Oldest
$
:20E0000010CE0400BDE01320F7BDE0733B20002001A000344033E4327E0F016F5F8E010083
:20E02000E65F4F6F8B6C5FA65F26F2D601E75E6F5FE65FE79FE00D4FEB5E494F1F01E68909
:20E040000100E79FE00F6C5F26E7C6FFE79FE0116F9FE00D8E0100D601E68BE79FE00F6F80
:20E060005FE65FC10524046C5F20F66F9FE01120BA60008E0100D6010C014F308BE69FE012
:04E0800071E7843987
:0AE7F600E000E000E000E009E000B0
:01E08C000093
:08E08400E08C7A0001E08D0040
:00E000011F
```

(b) Executable code.

Another possibility, not implemented in Table 14.8, is to replace the array representations in the three loops by equivalent pointer constructions. As these loops walk through the array, this procedure should be more effective (see Section 9.2). However, the saving is illusionary in this rather efficient compiler [2]. See Table 15.5 for an example of this technique.

14.3 68008 – Target Code

Although the target of Fig. 13.3 is based on a 68008 processor, the hardware and address map were chosen to resemble that of the 6809 equivalent of Fig. 13.1. This is reflected in the header file hard_68k.h included in Table 14.9, which is similar to hard_09.h. If there were changes in the memory map, then the header file would be suitably altered, whilst the remainder of the C code would remain unchanged (see also Section 15.2). Major changes in the input/output circuitry could be handled by including the I/O functions appropriate to the hardware. In such cases the main body of C code still remains portable (see Section 10.4).

A complete listing of 68000 assembly-level code intermingled with the original C source, as produced by the Intermetrics/COSMIC 68000 C cross-compiler V3.2, is given in Table 14.9.

I have added self explanatory comments, as indicated by the prefix *##, and thus this code will not be discussed in any detail in the text. There are, nevertheless, some points which should be noted. First the register variables i and leftmost have indeed been placed as requested in registers **D5[15:0]** and **D4[7:0]** respectively. As i is a short variable, 16 bits have been reserved, whereas 8 bits is sufficient for char leftmost.

ANSI C specifies that chars and shorts are promoted to ints during processing (see Fig. 8.4). This can clearly be seen in lines 55–59, where the unsigned 8-bit char leftmost is added to the (signed) 16-bit short i. The former is first unsigned promoted to 32 bits (i.e. a 32-bit int) as follows:

```
57      MOVEQ.L     #0,D7       * A 32-bit clear
58      MOVE.B      D4,D7       * Lower 8 bits to D7; D7 = 000000|leftmost
```

Then the 16-bit i is sign extended:

```
59      MOVE.W      D5,D6       * 16-bit i to D6[15:0]
60      EXT.L       D6          * Sign extended to 32 bits; that is D6[31:0]
```

After all this 32-bit 'fiddling around', the sum of these two is truncated by ANDing with 11111111b (0xFF):

Table 14.9: 68000 code resulting from Tables 14.1 and 14.2 (*continued next page*).

```
~~1WSL 3.0 as68k    Sat Dec 02 14:13:38 1989
 1 *  1 /* Version 02/12/89                                                     */
 2 *  2 #include <hard_68k.h>
 3 *  1 #define ANALOG_X   (unsigned char *)0x2000 /* Analog output to X amplifier  */
 4 *  2 #define ANALOG_Y   (unsigned char *)0x2001 /* Analog output to Y amplifier  */
 5 *  3 #define ANINPUT    (unsigned char *)0x6000 /* Analog   input port at 6000h  */
 6 *  4 #define SWITCH     (unsigned char *)0x8000 /* Digital  input port at 8000h  */
 7 *  5 #define Z_BLANK    (unsigned char *)0xA000 /* Digital output port at A000h  */
 8 *  6 #define RAM_START  (unsigned char *)0xE000 /* 6264 chip starts at location E000h*/
 9 *  7 #define RAM_LENGTH             0x2000 /* 6264 byte capacity is 8K or 2000h */
10 *  8 #define ROM_START  (unsigned short *)0x0000 /* 2764 chip starts at location 0000h*/
11 *  9 #define ROM_LENGTH             0x2000/* 2764 byte capacity is 8K or 2000h  */
12 *10 #define BLANK_ON                 0xFF  /* Bit pattern to blank out beam    */
13 *11 #define BLANK_OFF                0     /* Bit pattern to enable beam       */
14 *  3 unsigned char Array [256];            /* Global array holding display     */
15 *  4 unsigned char Oldest;                 /* Index to the Oldest inserted     */
16 *  5
17 *  6 main()
18 *  7 {
19                          .text
20                          .even
21 00418 4e56 fff4   _main:  link    a6,#-12     *## Frame of 3 words with A6 as FP (TOF)
22 0041C 48e7 0c00           movem.l d5/d4,-(sp) *## D4/D5 not to be changed by any ftn
23 *  8 register short int i;                /* Scan counter                      */
24 *  9 register unsigned char leftmost;     /* The initial array index when x is 0 */
25 *10 unsigned char * const x = ANALOG_X;   /* x points to a byte @ (address) ANALOG_X*/
26 00420 2d7c 00002000fffc  move.l #0x2000,-4(a6) *## Pointer constant 2000h @ TOF -4/-1
27 *11 unsigned char * const y = ANALOG_Y;   /* y points to a byte @ (address) ANALOG_Y*/
28 00428 2d7c 00002001fff8  move.l #0x2001,-8(a6) *## Likewise constant 2001h @ TOF-8/-5
29 *12 unsigned char * const z = Z_BLANK;    /* The z-mod port (digital port)     */
30 00430 2d7c 0000a000fff4  move.l #0xa000,-12(a6)*## Likewise constant A000h @ TOF-12/-9
31 *13 Oldest = 0;                           /* Start New index at beginning of array */
32 00438 4239 0000e000      clr.b   _Oldest     *## _Oldest lives in absolute memory @ E000h
```

378 *Software in C*

Table 14.9: 68000 code resulting from Tables 14.1 and 14.2 (*continued next page*).

```
33 *14 for(i=0; i<256; i++)                    /* Clear array                                         */
34 0043E 4245                   clr.w   d5           *## D5(15:0) holds register short i
35 00440 0c45 0100     L1:      cmpi.w  #256,d5      *## Is i beyond 255?
36 00444 6c   0e                bge.s   L14          *## IF yes THEN exit clear for loop
37 *15           {Array[i]=0;}
38 00446 227c 0000e002           move.l  #_Array,a1   *## ELSE point A1 to Array[0] each time thru
39 0044C 4231 5000              clr.b   (a1,d5.w)    *## Clear Array[i]
40 00450 5245                   addq.w  #1,d5        *## i++
41 00452 60   ec                bra.s   L1           *## and repeat
42 *16 while(1)                                /* Do forever display contents of array         */
43 *17           {
44 *18           leftmost = Oldest; /* Make the leftmost point on screen the oldest sample */
45 00454 1839 0000e000 L14: move.b _Oldest,d4  *## Now make leftmost (in reg D4) = _Oldest
46 *19           for (i=0; i<256; i++)
47 0045A 4245                   clr.w   d5           *## i=0
48 0045C 0c45 0100     L16:     cmpi.w  #256,d5      *## i>255 yet?
49 00460 6c   2a                bge.s   L17          *## IF yes THEN end scan for loop
50 *20           {
51 *21             *x = (unsigned char)i;     /* Send x co-ordinate to X plates              */
52 00462 226e fffc               move.l  -4(a6),a1   *## Get pointer constant 2000h (ie x) to A1
53 00466 1e05                   move.b  d5,d7        *## Move lower 8 bits of i into D7[7:0]
54 00468 1287                   move.b  d7,(a1)      *## and then send it to x
55 *22             *y = Array[(leftmost+i)&0x0ff]; /* and the display byte to the Y D/A */
56 0046A 226e fff8               move.l  -8(a6),a1   *## Get pointer constant 2001h (y) to A1
57 0046E 7e00                   moveq.l #0,d7        *## Move 8-bit leftmost extended to 32-bit
58 00470 1e04                   move.b  d4,d7        *## int to D7
59 00472 3c05                   move.w  d5,d6        *## Get i to D6[15:0]
60 00474 48c6                   ext.l   d6           *## and extend to 32-bit int
61 00476 de86                   add.l   d6,d7        *## Add them in int form = leftmost+i
62 00478 0287 000000ff          andi.l  #255,d7      *## Reduce to 8-bit (leftmost+i)&0xff
63 0047E 2447                   move.l  d7,a2        *## Put this array index in A2
64 00480 d5fc 0000e002          add.l   #_Array,a2   *## + to Array gives address of Array[index]
65 00486 1292                   move.b  (a2),(a1)    *## Move to y
66 *23           }
67 00488 5245                   addq.w  #1,d5        *## i++
68 0048A 60   d0                bra.s   L16          *## and repeat scan
69 *24             *z = BLANK_ON;               /* Blank out for flyback                              */
70 0048C 226e fff4     L17: move.l -12(a6),a1   *## Get constant pointer to z into A1
71 00490 12bc 00ff              move.b  #-1,(a1)     *## Send 1111 1111 to z
72 *25             *x = 0;                      /* Move to right of screen                           */
73 00494 226e fffc               move.l  -4(a6),a1   *## Get constant pointer to x into A1
74 00498 4211                   clr.b   (a1)         *## x=0
75 *26             *y = Array[Oldest];          /* Y value at left of screen                         */
76 0049A 226e fff8               move.l  -8(a6),a1   *## A1 now points to y
77 0049E 247c 0000e002          move.l  #_Array,a2   *## A2 now points to Array[0]
78 004A4 7e00                   moveq.l #0,d7        *## Extend _Oldest to 32-bit int size
79 004A6 1e39 0000e000          move.b  _Oldest,d7
80 004AC 12b2 7800              move.b  (a2,d7.l),(a1)*## Send array[Oldest] to y
81 *27           for(i=0; i<5; i++) {;}         /* Delay                                              */
82 004B0 4245                   clr.w   d5           *## i=0
83 004B2 0c45 0005     L121:    cmpi.w  #5,d5        *## i<5?
84 004B6 6c   04                bge.s   L131         *## IF yes THEN exit from delay for loop
85 004B8 5245          L141:    addq.w  #1,d5        *## ELSE i++
86 004BA 60   f6                bra.s   L121         *## and repeat
87 *28             *z = BLANK_OFF;              /* Blank off                                          */
88 004BC 226e fff4     L131: move.l -12(a6),a1  *## A1 now points to z
89 004C0 4211                   clr.b   (a1)         *## Send 0000 0000 to z
90 *29           }
91 004C2 60   90                bra.s   L14          *## Repeat the complete scan
92                              *fnsize=86
93 *30 }
```

Table 14.9: (*continued*) 68000 code resulting from Tables 14.1 and 14.2.

```
 .94 *31 /***********************************************************************
 95 *32  * This is the NMI interrupt service routine which puts the analog sample in the
 96 *33  * ENTRY : Via NMI and startup
 97 *34  * ENTRY : Array[] and Oldest are global
 98 *35  * EXIT  : Value held at a_d in Array[Oldest], Oldest incremented with wraparound
 99 *36  ************************************************************************
100 *37
101 *38 void update(void)
102 *39 {
103                        .even
104 004C4 4e56 fffc _update: link     a6,#-4       *## Make frame of 1 word for pointer const
105 *40 volatile unsigned char * const a_d = ANINPUT; /* This is the Analog input port */
106 004C8 2d7c 00006000fffc move.l  #0x6000,-4(a6)*## Constant pointer ANINPUT in TOF-4/-1
107 *41 Array[Oldest++] = *a_d;               /* Overwrite oldest sample in Array[] & inc */
108 004D0 1e39 0000e000     move.b   _Oldest,d7   *## _Oldest to D7[7:0]
109 004D6 5239 0000e000     addq.b   #1,_Oldest   *## _Oldest++
110 004DC 0287 000000ff     and.l    #255,d7      *## Expand to 32-bit int
111 004E2 2247             move.l    d7,a1        *## A1 now holds array index
112 004E4 d3fc 0000e002     add.l    #_Array,a1   *## A1 now points to Array[Oldest]
113 004EA 246e fffc         move.l   -4(a6),a2    *## A2 now points to ANINPUT
114 004EE 1292             move.b    (a2),(a1)    *## Put [ANINPUT] into Array[Oldest]
115 *42 }
116 004F0 4e5e             unlk      a6           *## Close up frame
117 004F2 4e75             rts                    *## and return
118                        *fnsize=110
119                        .globl   _update
120                        .globl   _main
121                        .bss
122                        .even
123 0E000       _Oldest:   .=.+1                  *## Reserve one byte for char Oldest
124                        .globl   _Oldest
125                        .even
126 0E002       _Array:    .=.+256                *## Reserve 256 bytes for Array[256]
127                        .globl   _Array

            no assembler errors
            code segment size = 220
            data segment size = 0


61       ADD.L    D6,D7       * 32-bit leftmost+i
62       ANDI.L   #0FFh,D7    * Truncated to 8-bits
63       MOVEA.L  D7,A2       * and moved as a 32-bit offset to A2.L
```

It is clear from this discussion that nothing has been gained in making these two register variables char and short. Unlike its 6809 counterpart, no provision to buck the ANSII promotion requirement is provided by this compiler. We will return to this point later.

Communication between the background main() (strictly void main(void)) and the interrupt function update() is handled via the two global objects, Oldest and Array[]. By defining these outside any function (lines 14 and 15 in Table 14.9), the compiler has placed their base labels _Oldest and _Array in absolute memory (lines 123–127). One byte has been reserved for the former and 256 for the latter. However, as Array[] is not a byte object, a hole of one byte is left after _Oldest, to ensure that it starts at an even address (i.e. .EVEN). Both labels have been declared .GLOBL, and thus are known to all, through the linker. The two labels have been placed in the _bss program section (directive .BSS), which is used by this compiler for static and extern data with no initial values. C specifies that these should be load-time initialized by default to zero, and that, by inference, this should be done in the startup routine. However, in this instance I have chosen to do this at run time in the main() function at the C level, in lines 33–41.

The linker has been configured to commence the _bss sector at E000h, which locates _Oldest at E000h and _Array at E002h. Program section _text actually begins at 0000h, but the startup routine vector table of Table 14.10 brings _main up above the vector table top (03FFh).

The startup routine has three functions. The first is to place the initial System Stack Pointer address in locations 00000–00003h and Reset address (i.e. initial Program Counter value) in 00004–7h. In addition, the level-7 interrupt auto vector, which points to the startup NMI stub, is placed in 0007C–0007Fh. Other vectors could of course be filled in the same manner (see Table 10.8). Space is then reserved up to 003FFh.

The startup program proper begins at 00400h. This has two purposes. The first is to go to the main C routine, which is implemented as a simple JSR _main in line 12. No flags require changing in the Status register before this move, as we are remaining within the Supervisor state, and the initial Interrupt mask setting of 111b still permits edge triggered non-maskable interrupts. In our situation, no parameters are passed (i.e. through the stack) to main(), and, as this is an endless loop, there should be no return. If there is, a move back to the beginning is actioned. This re-entry point is labelled _exit, and can be reached from the C level by calling the ANSII library routine exit() [3]. exit() is supposed to return True or False, to indicate an error condition, but no use is made of this in our implementation.

The final function deals with the level-7 interrupt handler. The update() function is terminated with RTS, in line 117 of Table 14.9, and so cannot be directly entered via an interrupt. Instead, the startup has a stub, in lines 14–17, which is labelled NMI. This address was placed in the vector table earlier in line 8. When a level-7 interrupt occurs, the processor goes to this stub. All that happens here is that the registers **D7**, **A1** and **A2** are pushed onto the System stack, and a subroutine Jump (JSR _update) is made to update(). The way back is a similar double-hop, with update()'s RTS returning the processor to the stub, the registers then being pulled off the stack followed by a terminating RTE. This compiler's house rule always preserves **D3**, **D4**, **D5**, **A3**, **A4**, **A5** (all of which are used for register variables) and **A6**, **A7** (the Frame and Stack Pointers) on return from a function. Thus a general interrupt stub need only save **D0**, **D1**, **D2**, **D6**, **D7**, **A0**, **A1**, **A2**. However, specifically update() only uses **D7**, **A1** and **A2**.

Table 14.10: The 68000 Time Compressed Memory Startup.

```
~~1WSL 3.0 as68k    Fri Dec 08 15:09:06 1989
1                            * Startup code for 68008-based Time-Compressed Memory
2                            * S.J.Cahill Version 07/02/89
3                                    .text
4                                    .even
5  00000   00010000   STARTUP: .long   0x10000  * Initial System Stack Pointer.
6  00004   00000400            .long   START    * The startup code below
7  00008   00                  .       =.+116   * 116 bytes down to IRQ-7 vector
8  0007c   00000408            .long   NMI      * IRQ7 service routine below
9                            * The startup routine on reset follows
10 00080   00                  .       =.+896   * 896 bytes on at 400h
11                                   *
12 00400   4eb9 00000418 START: jsr   _main     * Go to main().
13 00406   60   f8       _exit: bra.s START     * IF returns THEN restart
14 00408   48e7 0160     NMI:   movem.l d7/a1/a2,-(sp) * Save used regs
15 0040c   4eb9 000004C4        jsr   _update   * Go to function update()
16 00412   4cdf 0680            movem.l (sp)+,d7/a1/a2 * Retrieve regs
17 00416   4e73                 rte             * return to caller
18                                   *
19                                   *
20                                   .public _main, _update, _exit
21 * Make main(), update() and _exit known to the linker (i.e. global)
```

The linker places the startup code before the output from the C compiler, giving the Intel-coded machine-code file of Table 14.11. This is used as the input to an EPROM programmer or in-circuit emulator. In total there are 244 bytes of EPROM text (excluding the fixed vector table). It is interesting to compare this with the 178 bytes produced by the 6809 equivalent in the last section. Although there are less lines, 68000 instructions tend to be longer than their 6809 counterparts.

The double-hop interrupt handling technique will work with any compiler. However, most compilers with aspirations to produce ROMable code, support extensions to the ANSII standard, enabling the user to declare a function as an interrupt handler (see Section 10.2). The function name, in our case _update, should then be placed directly in the vector table, rather than the stub label. This direct entry should decrease the response period to an interrupt, and at the same time possibly reduce the code emitted by the compiler.

This compiler uses the directive @port to designate a function in this way, the function heading becoming @port update(). The code produced by this stratagem is shown in Table 14.12. Here, four instructions have been inserted into the function code, lines 104–107. These instructions are virtually identical to the stub of Table 14.10, but of course are directly entered at _update. In reality no time is saved, as the main body of update() is unchanged, and is simply treated as a subroutine. Thus a double hop still occurs on entry and exit.

382 *Software in C*

Table 14.11: Machine-code file from Tables 14.9 and 14.10.

```
000004c4  _update
00000418  _main
00000000  STARTUP
00000408  NMI
00000406  _exit
00000400  START
00000440  L1
0000048c  L17
0000045c  L16
00000454  L14
000004b8  L141
000004bc  L131
000004b2  L121
0000e002  _Array
0000e000  _Oldest
000004f4  __prog_top
0000e000  __data_top
0000e102  __stack_bottom
00010000  __stack_top
$
:200000000001000000000400000000000000000000000000000000000000000000DE  Stack/Reset
:20002000000000000000000000000000000000000000000000000000000000000000C0  vectors
:200040000000000000000000000000000000000000000000000000000000000000A0
:2000600000000000000000000000000000000000000000000000000000000004087 4  Level-7
:20008000000000000000000000000000000000000000000000000000000000000060  autovector
:2000A0000000000000000000000000000000000000000000000000000000000000040
:2000C0000000000000000000000000000000000000000000000000000000000000020
:2000E0000000000000000000000000000000000000000000000000000000000000000
:200100000000000000000000000000000000000000000000000000000000000000DF
:20012000000000000000000000000000000000000000000000000000000000000BF
:200140000000000000000000000000000000000000000000000000000000000009F
:200160000000000000000000000000000000000000000000000000000000000007F
:200180000000000000000000000000000000000000000000000000000000000005F
:2001A000000000000000000000000000000000000000000000000000000000003F
:2001C000000000000000000000000000000000000000000000000000000000001F
:2001E00000000000000000000000000000000000000000000000000000000000FF
:200200000000000000000000000000000000000000000000000000000000000DE
:2002200000000000000000000000000000000000000000000000000000000000BE
:200240000000000000000000000000000000000000000000000000000000000009E
:200260000000000000000000000000000000000000000000000000000000000007E
:200280000000000000000000000000000000000000000000000000000000000005E
:2002A0000000000000000000000000000000000000000000000000000000000003E
:2002C00000000000000000000000000000000000000000000000000000000000001E
:2002E0000000000000000000000000000000000000000000000000000000000000FE
:20030000000000000000000000000000000000000000000000000000000000000DD
:20032000000000000000000000000000000000000000000000000000000000000BD
:2003400000000000000000000000000000000000000000000000000000000000009D
:2003600000000000000000000000000000000000000000000000000000000000007D
:2003800000000000000000000000000000000000000000000000000000000000005D
:2003A00000000000000000000000000000000000000000000000000000000000003D
:2003C00000000000000000000000000000000000000000000000000000000000001D
:2003E00000000000000000000000000000000000000000000000000000000000FD

:200400004EB90000041860F848E701604EB90000004C44CDF06804E734E56FFF448E70C00BE  Startup
:200420002D7C00002000FFFC2D7C00002001FFF82D7C0000A000FFF442390000E000424519  and main()
:200440000C4501006C0E227C0000E00242315000524560EC18390000E00042450C450100A0
:200460006C2A226EFFFC1E051287226EFFF87E001E043C0548C6DE860287000000FF2447D2
:20048000D5FC0000E0021292524560D0226EFFF412BC00FF226EFFFC4211226EFFF8247CE9
:2004A000000000E0027E001E390000E00012B2780042450C4500056C04524560F6226EFFF4AC
:2004C000421160904E56FFFC2D7C00006000FFFC1E390000E00052390000E000028700000B  update()
:1404E00000FF2247D3FC0000E002246EFFFC12924E5E4E754F
:00000001FF
```

Bonding with the update() function code can be improved by eschewing the use of @port and using the (once again non-standard) _asm() function to insert the relevant assembly-level code, as shown in Table 14.13. Thus:

_asm("movem.l d7/a0/a1,-(sp)   * Save used regs on Stack\n");

after the opening brace and

_asm("movem.l (sp)+,d7/a0/a1   * Pull registers\n");
_asm("unlk a6\n rte   * Close frame and exit \n");

at the close, tightly couples the additional interrupt code to the compiler-emitted code. Care must be taken to mirror any registers pushed out on to the stack by the

Table 14.12: The @port directive.

```
~~1WSL 3.0 as68k    Sat Dec 02 14:20:42 1989

 94 * 31 /*******************************************************************************
 95 * 32  * This is the NMI interrupt service routine which puts the analog sample in the
 96 * 33  * ENTRY : Via NMI and startup
 97 * 34  * ENTRY : Array[] and Oldest are global
 98 * 35  * EXIT  : Value held at a_d in Array[Oldest], Oldest incremented with wraparound
 99 * 36  *******************************************************************************
100 * 37
101 * 38     @port update()
102 * 39     {
103                                  .even
104 004B4 48e7 e3e0        _update: movem.l d0-d2/d6/d7/a0-a2,-(sp)  *## Save all registers
105 004B8 4eb9 000000bc             jsr     L6                       *## Go to update() proper
106 004BE 4cdf 07c7                 movem.l (sp)+,d0-d2/d6/d7/a0-a2  *## Restore regs before
107 004C2 4e73                      rte

108 004C4 4e56 fffc        L6: link    a6,#-4                        *## As Table 14.9
109 * 40     volatile unsigned char * const a_d = ANINPUT; /* This is the Analog input port*/
110 004C8 2d7c 00006000fffc         move.l  #0x6000,-4(a6)
111 * 41     Array[Oldest++] = *a_d; /* Overwrite oldest sample in Array[] and inc index   */
112 004D0 1e39 0000e000            move.b  _Oldest,d7
113 004D6 5239 0000e000            addq.b  #1,_Oldest
114 004DC 0287 000000ff            and.l   #255,d7
115 004E2 2247                     move.l  d7,a1
116 004E4 d3fc 0000e002            add.l   #_Array,a1
117 004EE 246e fffc                move.l  -4(a6),a2
118 004F0 1292                     move.b  (a2),(a1)
119 004F2 4e5e                     unlk    a6
120 004F4 4e75                     rts
121                                *fnsize=110
122                                .globl _update
123                                .globl _main
124                                .bss
125                                .even
126 0E000            _Oldest: .     =.+1
127                                .globl _Oldest
128                                .even
129 0E002            _Array:  .     =.+256
130                                .globl _Array
```

compiler. _asm() could also, in principle, be used to implement the startup code as a front end to the C code.

These latter two approaches are non-portable, and can be error prone. Thus, in the majority of cases a startup stub approach is best, if rather slow. If speed and/or space is extremely tight, then the compiler generated assembly-level file can be creatively edited. Thus any MOVEM instruction emitted by the compiler can be augmented with the registers not left untouched by the routine, and RTS replaced by RTE. But this approach is dangerous, as it does not show up in the source of any constituent file, and, unless extremely well documented, will cause havoc if any but the original designer tries to make subsequent changes. In any case, tinkering with intermediate files is not what compiling is all about.

Table 14.13: Using _asm() to terminate an interrupt service function.

```
* 31   /*******************************************************************************
* 32   * This is the NMI interrupt service routine which puts the analog sample in the array
* 33   * ENTRY : Via NMI and startup
* 34   * ENTRY : Array[] and Oldest are global
* 35   * EXIT  : Value held at a_d in Array[Oldest], Oldest incremented with wraparound @ 256
* 36   *******************************************************************************
* 37
* 38   void update(void)
* 39   {
              .even
_update:      link    a6,#-4
* 40   volatile unsigned char * const a_d = ANINPUT; /* This is the Analog input port    */
              move.l  #0x6000,-4(a6)
* 41   _asm("movem.l d7/a0/a1,-(sp)     * Save used regs on Stack \n");
              movem.l d7/a0/a1,-(sp)    * Save used regs on Stack
* 42   Array[Oldest++] = *a_d;   /* Overwrite oldest sample in Array[] & increment index */
              move.b  _Oldest,d7
              addq.b  #1,_Oldest
              and.l   #255,d7
              move.l  d7,a1
              add.l   #_Array,a1
              move.l  -4(a6),a2
              move.b  (a2),(a1)
* 43   _asm("movem.l (sp)+,d7/a0/a1     * Pull registers \n");
              movem.l (sp)+,d7/a0/a1    * Pull registers
* 44   _asm("unlk a6 \n rte             * Close frame and exit \n");
              unlk    a6
              rte                       * Close frame and exit
* 45   }
              unlk    a6      *## These two lines are now dead code
              rts
*fnsize=125
              .globl _update
              .globl _main
              .bss
              .even
_Oldest:      .       =.+1
              .globl _Oldest
              .even
_Array:       .       =.+256
              .globl _Array
```

In the last section we were able to fine tune our C source file, knowing the characteristics of the target processor. The increase in speed and size is of course at the expense of portability. Can we do this for the 68000-target version? For example, we have previously observed that the use of **register short** and **char** objects is counterproductive, as such objects are extended to **int** during most arithmetic processes. Neither **short i** nor **char leftmost** rely on modulo-256 wraparound, so they can profitably be redefined as **int**s to overcome this additional processing.

Most 68000-targeted compilers can be persuaded to define **int** as either a 16 or 32-bit word. All previous examples have been based on 32-bit **int**s. Using 16-bit **int**s will speed up memory access and ALU processes. However, any address arithmetic, such as the calculation of the position of an array element, will require conversion to the 32-bit pointer size.

Where constants are being stored, for example pointers to fixed hardware ports, it is not necessary to locate these values dynamically in the frame. We can see this run-time setup in line 110 of Table 14.12, where the constant $6000h$ (the address of the A/D) is put into the frame on each entry to update(). Constants are best stored in absolute locations, preferably in ROM along with the program text. In the case of constant pointers, this can be done by defining such objects as **static**, for example:

```
static    volatile unsigned char   *  const a_d = ANINPUT;
```

which reads a_d is a constant pointer/ to a **volatile unsigned** char/ is stored statically/ and has an initial value of ANINPUT (i.e. $6000h$ from the header). The combination of **static** and **const** tells the compiler to put constants in ROM, that is the _text section. The compile-time nature of these constants is clearly seen in lines 3–9 of Table 14.14, where they are placed in the EPROM at locations 00418–0041Fh. This saves four bytes for each of the four pointer constants. Furthermore, neither main() nor update() require a frame, as no **auto** variables are used.

Table 14.14 shows the tuned version of our software. It differs from Table 14.9 in the following respects:

1. The compiler has been configured for a 16-bit **int**. This obviates conversion to 32-bit for arithmetic processes, and suits the 16-bit ALU used by the 68000/8 processors. However, it is a double-edged sword, in that pointers are still 32-bit, and the use of an **int** to generate an address (e.g. as an array index) will require a promotion (see code between C22 and C23).

2. The **register** variables i and leftmost have been redefined as **int** types. This avoids conversion extensions in arithmetic processes.

3. Pointer constants have been defined as **static**, which places them in ROM. The alternative of defining them externally (see Table 15.5) does the same thing. The compiler then uses absolute addressing to get these values (see code between lines C21 and C22). Some compilers (not this) have a small

386 *Software in C*

Table 14.14: Optimized 68000-based code (*continued next page*).

```
* 1     /* Version 08/03/90                                                           */
* 2     #include <hard_68k.h>
            .text
            .even
L5_x:       .long   0x2000  *##The 1st long word in text section (ROM) holds pntr constant 2000h
            .even
L51_y:      .long   0x2001  *## Next in absolute memory is the constant 2001h (Y amplifier port)
            .even
L52_z:      .long   0xa000  *## and A000h, the Z-blank port
* 3     unsigned char Array [256];              /* Global array holding display data       */
* 4     unsigned char Oldest;/* Index to the Oldest inserted data byte (left point on scrn)*/
* 5
* 6     main()
* 7     {
            .even
_main:      movem.l d5/d4/a5,-(sp)
* 8     register unsigned char * array_prt;     /* Pointer into array                      */
* 9     register int i;                         /* Scan counter                            */
*10     register int leftmost;                  /* The initial array index when x is 0     */
*11     static unsigned char * const x = ANALOG_X; /* x points to a byte @ (addr) ANALOG_X */
*12     static unsigned char * const y = ANALOG_Y; /* y points to a byte @ (addr) ANALOG_Y */
*13     static unsigned char * const z = Z_BLANK;  /* The z-mod port (digital port)        */
*14     Oldest = 0;                             /* Start New index at beginning of the array */
            clr.b   _Oldest
*15     for(array_ptr=Array; array_ptr<Array+256; *array_ptr++ = 0) {;} /* Clear array    */
            move.l  #_Array,a5      *## Make array_ptr (in A5.L) point to bottom of Array
L1:         cmp.l   #_Array+256,a5
            bge.s   L14
            move.l  #_Array,a1      *## array_ptr < Array+256?
            bcc.s   L14             *## IF not then exit for loop
            clr.b   (a5)+           *## ELSE clear array element & inc array pntr all at once
            bra.s   L1              *## and again
*16     while(1)                                /* Do forever display contents of array    */
*17     {
*18         leftmost = Oldest; /* Make the leftmost point on the screen the oldest sample */
            clr.w   d4
            move.b  _Oldest,d4
*19         for (array_ptr=Array, i=0; array_ptr<Array+256;)
            move.l  #_Array,a5      *## Again make array_ptr in A5.L point to bottom of array
            clr.w   d5              *## i (in D5.W) = 0
L16:        cmp.l   #_Array+256,a5  *## array_ptr < Array+256?
            bcc.s   L17             *## IF not THEN exit for loop
*20             {
*21             *x = (unsigned char)i;  /* Send x co-ordinate to X plates                 */
            move.l  L5_x,a1         *## L5_x in abs memory, thus abs mode used to get pntr to X
            move.b  d5,d7
            move.b  d7,(a1)
*22             *y = *(array_ptr++ +leftmost)&0xff; /* and the display byte to the y D/A */
            move.l  L51_y,a1        *## Same for pntr to Y. Note pntr constants aren't in Stack
            move.w  a5,a2           *## Incrementing array_ptr for next time
            addq.l  #1,a5
            move.b  (a2,d4.w),d7    *## array_ptr + leftmost into D7.B
            andi.b  #0xff,d7        *## Reduce to modulo-256 (8-bit)
            move.b  d7,(a1)         *## Put it out to Y port (address of which is in A1)
*23             }
```

Table 14.14: Optimized 68000-based code (*continued next page*).

```
                bra.s     L16             *## and again
*24         *z = BLANK_ON;                  /* Blank out for flyback            */
L17:            move.l    L52_z,a1        *## Use absolute addressing mode to get pointer to Z
                move.b    #0xff,(a1)      *## Make Z port all 1s
*25         *x = 0;                         /* Move to right of screen          */
                move.l    L5_x,a1         *## Move X back to start
                clr.b     (a1)
*26         *y = Array[Oldest];             /* Y value at left of screen        */
                move.l    L51_y,a1
                move.l    #_Array,a2
                moveq.l   #0,d7
                move.b    _Oldest,d7
                move.b    (a2,d7.l),(a1)
*27         for(i=0; i<5; i++) {;}          /* Delay                            */
                clr.w     d5
L101:           cmpi.w    #5,d5
                bge.s     L131
L121:           addq.w    #1,d5
                bra.s     L121
*28         *z = BLANK_OFF;                 /* Blank off                        */
L111:           move.l    L52_z,a1
                clr.b     (a1)
*29         }
                bra.s     L14
*fnsize=78

                .even
L53_a_d:.long   0x6000          *## Pointer constant to A_D here in ROM
*30      }


*31  /*****************************************************************************
*32   * This is the NMI interrupt service routine which puts the analog sample in the array
*33   * ENTRY : Via NMI and startup
*34   * ENTRY : Array[] and Oldest are global
*35   * EXIT  : Value held at a_d in Array[Oldest], Oldest incremented with wraparound @ 256
*36   *****************************************************************************
*37
*38  void update(void)
*39  {
                .even
*40      static volatile unsigned char * const a_d = ANINPUT;/* This is the Analog i/p port*/
*41      Array[Oldest++] = *a_d; /* Overwrite oldest sample in Array[] & inc Oldest index   */
_update: move.b  _Oldest,d7     *## Notice, no frame is made for this function as no autos
                addq.b    #1,_Oldest
                and.l     #255,d7
                move.l    d7,a1
                add.l     #_Array,a1
                move.l    L53_a_d,a2
                move.b    (a2),(a1)
*43      }
                rts
*fnsize=99
                .globl   _update
                .globl   _main
                .bss
                .even
_Oldest:  .              =.+1
                .globl   _Oldest
                .even
_Array:   .              =.+256
                .globl   _Array
```

(a) Resulting assembly code.

Table 14.14: (*continued*) Optimized 68000-based code.

```
000004bc    _update
00000424    _main
00000000    STARTUP
00000408    NMI
00000406    _exit
00000400    START
00000434    L1
000004b8    L53_a_d
00000478    L17
00000450    L16
00000440    L14
0000043c    L12
000004aa    L121
000004ae    L111
000004a4    L101
0000e002    _Array
00000418    L5_x
00000420    L52_z
0000e000    _Oldest
0000041c    L51_y
000004e0    __prog_top
0000e000    __data_top
0000e102    __stack_bottom
00010000    __stack_top
$
:20000000000100000000040000000000000000000000000000000000000000000000DB
:200020000000000000000000000000000000000000000000000000000000000000C0
:200040000000000000000000000000000000000000000000000000000000000000A0
:2000600000000000000000000000000000000000000000000000000000000040874
:200080000000000000000000000000000000000000000000000000000000000000060
:2000A000000000000000000000000000000000000000000000000000000000000040
:2000C000000000000000000000000000000000000000000000000000000000000020
:2000E000000000000000000000000000000000000000000000000000000000000000
:200100000000000000000000000000000000000000000000000000000000000000DF
:200120000000000000000000000000000000000000000000000000000000000000BF
:2001400000000000000000000000000000000000000000000000000000000000009F
:2001600000000000000000000000000000000000000000000000000000000000007F
:2001800000000000000000000000000000000000000000000000000000000000005F
:2001A00000000000000000000000000000000000000000000000000000000000003F
:2001C00000000000000000000000000000000000000000000000000000000000001F
:2001E0000000000000000000000000000000000000000000000000000000000000FF
:200200000000000000000000000000000000000000000000000000000000000000DE
:200220000000000000000000000000000000000000000000000000000000000000BE
:2002400000000000000000000000000000000000000000000000000000000000009E
:2002600000000000000000000000000000000000000000000000000000000000007E
:2002800000000000000000000000000000000000000000000000000000000000005E
:2002A00000000000000000000000000000000000000000000000000000000000003E
:2002C00000000000000000000000000000000000000000000000000000000000001E
:2002E0000000000000000000000000000000000000000000000000000000000000FE
:200300000000000000000000000000000000000000000000000000000000000000DD
:200320000000000000000000000000000000000000000000000000000000000000BD
:2003400000000000000000000000000000000000000000000000000000000000009D
:2003600000000000000000000000000000000000000000000000000000000000007D
:2003800000000000000000000000000000000000000000000000000000000000005D
:2003A00000000000000000000000000000000000000000000000000000000000003D
:2003C00000000000000000000000000000000000000000000000000000000000001D
:2003E0000000000000000000000000000000000000000000000000000000000000FD
:200400004EB90000042460F848E701604EB9000004BC4CDF06804E7300002000000020014B
:200420000000A00048E70C0442390000E0002A7C0000E002BBFC0000E1026404421D60F445
:2004400042441839000E0002A7C0000E0024245BBFC0000E10264202279000004181E05DE
:20046000128722790000041C244D528D1E324000020700FF128760D822790000042012BCE2
:20048000000FF22790000418421122790000041C247C0000E0027E001E390000E00012B29D
:2004A000780042450C4500056C04524560F62279000004204211608800006000E1390000D9
:2004C000E00052390000E0000287000000FF2247D3FC0000E0022479000004B812924E756F
:00000001FF
```

(b) Executable code.

model mode where the Short Absolute address mode is used. Where this is available, two bytes are saved for each absolute access.

With these alterations, the total size is now down to 224 bytes plus data storage and the fixed-size vector table. I have used the startup stub entry to `update()`, which is the most portable technique. A few bytes may be saved at the expense of this portability by direct editing of the assembly-level code. For comparison, a hand-assembled 68000-based version is given in Table 16.2.

Another possibility, not implemented here, is to use pointers to implement the three array handling loops. As these loops walk through the array, the process may be more efficient (see Section 9.2). However, with this compiler savings are minimal, as its array-handling code is quite efficient [2]. See Table 15.5 for an example of this technique.

References

[1] Banahan, M.; *The C Book*, Addison-Wesley, 1988, Chapter 5.

[2] Sutherland, D.; Compiled Thoughts, Letters to the Editor, *Embedded Systems*, **6**, no. 5, May 1993, pp. 11.

[3] Banahan, M.; *The C Book*, Addison-Wesley, 1988, Section 9.15.4.

CHAPTER 15

Looking For Trouble

The process from program text to binary bits in ROM has already been charted in Fig. 7.5. But what then? It would be naive to presume that the production of a machine-code file and programmed ROM is the end of the story. Just inserting this ROM into the target hardware, switching on and hoping for the best is unlikely to be productive of anything except frustration. Invariably testing and debugging this software will take far longer than the writing of the original code [1].

Testing involves executing the software in a controlled environment, to exercise the various responses to typical input stimuli. It is impossible to test every possible pathway in all but the simplest of routines, but a range of typical and boundary values should help and ensure that the program behaves properly. Decomposing the program into functional modules helps to facilitate this process.

Malfunctions are said to be caused by bugs (after an alleged incident where a moth was caught in a relay of an early electro-mechanical computer [2]). Bugs are normally found by applying a series of tests which focuses down onto the area of software (or hardware) which is exhibiting the erroneous behavior. Hardware testing and debugging is aided by a range of tools, varying in complexity from meters through to the logic analyser. Similarly, software debug tools are available in various levels of sophistication, to enable the tester to 'see into the works' while the program executes.

The easiest scenario arises when a general-purpose computer is used to generate (i.e. compile and assemble) code which it will itself subsequently run. Its many resources, such as operating system, VDU and keyboard, can be utilized with resident debug software to test the operation of the application software. Virtually total ignorance of the underlying hardware is possible.

The situation is very different when the target system is a dedicated ROM-based stand-alone system, usually with a different processor to the code-generating computer. In this cross environment (see Fig. 7.3(b)), gone is any resident debug software or superfluous peripheral devices. Interaction with the hardware at the machine code level looms large. Problems are compounded where a high-level language is used as the source, as the correlation between the executing code and source code is tortuous. At the time of writing, high-level simulators and emulators are the fastest growing area of cross software support.

In this chapter we will look at some of the debug tools available for cross-target support. Our time-compressed memory will be used to illustrate the characteristics of these aids.

15.1 Simulation

Given that a program has been written in a high-level language for another target, how is it to be tested? We have already observed that a naked target will carry no debug overhead to permit meaningful monitoring. Furthermore, it will execute (obviously) at machine-code level.

A first approach to the problem is to use a native compiler running on the host machine. Thus, if an IBM PC is utilized as the development system, then use a compiler which produces native executable files. Obviously the environment of the host is very different to that of the naked target. However, gross algorithmic problems can often be eliminated using this technique. Function input/output parameters can be simulated by using operating system input/output functions. Table 10.14 shows a simple example, where the sum_of_n function is emulated using keyboard input and VDU output.

Monitoring high-level objects can be accomplished by using output functions to display or print their values. In Table 10.14, a printf() statement inside the loop would be suitable. Many native compilers, especially (but not exclusively) targeted to MSDOS 80x86 family hosts, can be run in conjunction with a debug package [3]. Such packages allow the operator to watch a selection of variables as the program advances in a single-step or trace mode. Alternatively a breakpoint may be inserted (e.g. stop at line 26, or/and when Array[6] = 0), which permits high-speed operation to a predetermined point, at which time execution ceases and variables can be examined.

The usefulness of this native technique is enhanced if the native and cross compilers belong to the same family of products. Several compiler vendors provide suitable products, such as Aztec and the Intermetrics/COSMIC Whitesmiths group. In these circumstances, native and cross products usually share common characteristics, such as libraries.

Where there is a great deal of interaction between software and hardware, native debugging is of limited use. This is particularly the case where the target processor is different to the host. For example, a 68030 MPU-based Hewlett Packard workstation hosting a Z80-based target. Monitoring machine-level code will often be necessary to reveal the more subtle problems, especially where hardware interaction is involved.

One way of tackling this problem is to use the host to **simulate** the target MPU [4]. Such a cross-simulator, sometimes known as a low-level symbolic debugger, is particularly of use in testing cross-assembler code. However, languages whose compilers produce assembly-level code, can also be tested in this manner. One major advantage of the use of a simulator is that no target hardware is involved. Thus the hardware and software design stages can stay apart longer. This takes the load off expensive equipment, such as an in-circuit emulator, which

392   *Looking For Trouble*

can then be used for the really obscure problems and final testing. By their nature, simulators cannot run in real time and they still leave a lot to be desired when interaction with hardware is problematical.

Most simulators take their output from the linker in terms of the machine code, location data and symbol tables. Part of the host's memory space is used to hold this machine code, and the target's data memory space is likewise mapped. The major facilities offered by a simulator are:

Disassembly
Displays the contents of simulated target memory as instruction mnemonics — a sort of reverse assembler.

Register and memory examine/change
To be able to examine any internal register(s) or memory location(s) and make necessary changes.

Step execution
To execute the target program one or more instructions at a time, usually displaying registers after each one.

Trace execution
Similar to the previous item, but as fast as can be displayed.

Breakpoints
Insertion of conditions, such as reaching a certain address, which causes execution to pause or stop.

Execute
Similar to Trace, but as fast as the simulator can operate with no screen output. Normally stops when a breakpoint is encountered.

The operation of a simulator is very much product specific. The COSMIC/Intermetrics MIMIC range of simulators have been used for the following three examples.

Our first simulation is our old friend the sum of n integers, Table 4.10. Table 15.1 is a log of a simulation session, with comments added later for clarity. After loading in the file, the process was:

1. Disassemble program mnemonics from the beginning (e SUM_OF_N or e 0x400).

2. Change **D0** to 0xFF0003 to simulate **D0.W** = 0003h ($D0 = 0xFF0003).

3. Single step until S_EXIT is reached (s or s1). Note that Step goes from the current value of **PC**, here initialized to 400h when the object file was loaded. Thus to start again, do $pc = SUM_OF_N or $pc = 0x400.

The second example is more elaborate. Here the target is the 6809 equivalent of Tables 2.9 and 2.10. We wish to trace the execution down to where the simulated processor attempts to fetch the final instruction (RTS) at S_EXIT. Thus we have to set up a breakpoint at this address.

This time the log shown in Table 15.2 was generated as follows:

Table 15.1: Simulating the program of Table 4.10. User input shown in quotes, comments bracketed.

```
"e SUM_OF_N #5"
(Disassemble from SUM_OF_N for five instructions)

    0x000400              SUM_OF_N:
--> 0x000400 02800000ffff             andi.l    #0xffff,d0
    0x000406 4281                     clr.l     d1
    0x000408              SLOOP:
    0x000408 d280                     add.l     d0,d1
    0x00040a 51c8fffc                 dbf       d0,SLOOP
    0x00040e              S_EXIT:
    0x00040e 4e75                     rts

"$d0 = 0xff0003"
(Set D0.L to 00FF0003h)

"s"
(Single step for as long as desired)
                                                            TS I XNZVC
d0:00ff0003 d1:00000000 d2:00000000 d3:00000000 ssp:00000000 sr:../0/..... |
d4:00000000 d5:00000000 d6:00000000 d7:00000000 usp:00000000 pc:00000400 | (Before
a0:00000000 a1:00000000 a2:00000000 a3:00000000                          | ANDI.L #FFFFh)
a4:00000000 a5:00000000 a6:00000000 a7:00000000 andi.l    #0xffff,d0
"s"
d0:00000003 d1:00000000 d2:00000000 d3:00000000 ssp:00000000 sr:../0/..... |
d4:00000000 d5:00000000 d6:00000000 d7:00000000 usp:00000000 pc:00000406 | (After execution.
a0:00000000 a1:00000000 a2:00000000 a3:00000000                          | Next instruction
a4:00000000 a5:00000000 a6:00000000 a7:00000000 clr.l     d1             | is CLR.L D1)
"s"
d0:00000003 d1:00000000 d2:00000000 d3:00000000 ssp:00000000 sr:../0/..Z.. |
d4:00000000 d5:00000000 d6:00000000 d7:00000000 usp:00000000 pc:00000408 | (Status register
a0:00000000 a1:00000000 a2:00000000 a3:00000000                          | showing the
a4:00000000 a5:00000000 a6:00000000 a7:00000000 add.l     d0,d1          | Z flag setting)
"s"
d0:00000003 d1:00000003 d2:00000000 d3:00000000 ssp:00000000 sr:../0/.....
d4:00000000 d5:00000000 d6:00000000 d7:00000000 usp:00000000 pc:0000040a
a0:00000000 a1:00000000 a2:00000000 a3:00000000
a4:00000000 a5:00000000 a6:00000000 a7:00000000 dbf       d0,SLOOP
"s"
d0:00000002 d1:00000003 d2:00000000 d3:00000000 ssp:00000000 sr:../0/..... | (PC goes back to
d4:00000000 d5:00000000 d6:00000000 d7:00000000 usp:00000000 pc:00000408 | 408h, i.e. SLOOP
a0:00000000 a1:00000000 a2:00000000 a3:00000000                          | as D0.W isn't
a4:00000000 a5:00000000 a6:00000000 a7:00000000 add.l     d0,d1          | -1)
"s"
d0:00000002 d1:00000005 d2:00000000 d3:00000000 ssp:00000000 sr:../0/.....
d4:00000000 d5:00000000 d6:00000000 d7:00000000 usp:00000000 pc:0000040a
a0:00000000 a1:00000000 a2:00000000 a3:00000000
a4:00000000 a5:00000000 a6:00000000 a7:00000000 dbf       d0,SLOOP
"s"
d0:00000001 d1:00000005 d2:00000000 d3:00000000 ssp:00000000 sr:../0/.....
d4:00000000 d5:00000000 d6:00000000 d7:00000000 usp:00000000 pc:00000408
a0:00000000 a1:00000000 a2:00000000 a3:00000000
a4:00000000 a5:00000000 a6:00000000 a7:00000000 add.l     d0,d1
"s"
d0:00000001 d1:00000006 d2:00000000 d3:00000000 ssp:00000000 sr:../0/.....
d4:00000000 d5:00000000 d6:00000000 d7:00000000 usp:00000000 pc:0000040a
a0:00000000 a1:00000000 a2:00000000 a3:00000000
a4:00000000 a5:00000000 a6:00000000 a7:00000000 dbf       d0,SLOOP
"s"
d0:00000000 d1:00000006 d2:00000000 d3:00000000 ssp:00000000 sr:../0/.....
d4:00000000 d5:00000000 d6:00000000 d7:00000000 usp:00000000 pc:00000408
a0:00000000 a1:00000000 a2:00000000 a3:00000000
a4:00000000 a5:00000000 a6:00000000 a7:00000000 add.l     d0,d1
"s"
d0:00000000 d1:00000006 d2:00000000 d3:00000000 ssp:00000000 sr:../0/..... | (D0.W has been
d4:00000000 d5:00000000 d6:00000000 d7:00000000 usp:00000000 pc:0000040a | decremented to
a0:00000000 a1:00000000 a2:00000000 a3:00000000                          | -1, so end of
a4:00000000 a5:00000000 a6:00000000 a7:00000000 dbf       d0,SLOOP       | DBF loop)
"s"
d0:0000ffff d1:00000006 d2:00000000 d3:00000000 ssp:00000000 sr:../0/..... | (Ans is in D1.L
d4:00000000 d5:00000000 d6:00000000 d7:00000000 usp:00000000 pc:0000040e | at end of
a0:00000000 a1:00000000 a2:00000000 a3:00000000                          | subroutine. RTS
a4:00000000 a5:00000000 a6:00000000 a7:00000000 (rts         )           | isn't executed)
```

1. Set a breakpoint at S_EXIT (b S_EXIT or b 0xE00C). Note, br S_EXIT sets a break when reading from S_EXIT, which is an alternative in this situation. Breaks on a Write and over a range of addresses are possible. For example bw 0xE000, 0xFFFF breaks when a Write is attempted in memory between E000h and FFFFh, which is one way of simulating ROM. An unlimited number of breakpoints can be set.

2. Change Accumulator_B to 03h to simulate the passing of n ($b = 3).

3. Trace to breakpoint from SUM_OF_N (t SUM_OF_N).

Breakpoints can be a great deal more sophisticated than shown here. For example, every time data is stored at, say, A000h (simulating an output port) the

Table 15.2: Tracing the program of Table 2.9.

```
"b S_EXIT"
(Set a breakpoint at S_EXIT)

"$b = 03"
(Make Accumulator B 03)

"t SUM_OF_N"
(Trace from sum_of_N down to breakpoint)

        EFHINZVC       Register states before execution of      this instruction
CCR     --------       ---------------------------------------  ---------------
cc:........ dp:00 a:00 b:03 x:0000 y:0000 u:0000 s:0000 pc:e000 ldx    #0x0000
cc:.....Z.. dp:00 a:00 b:03 x:0000 y:0000 u:0000 s:0000 pc:e003 tstb
cc:........ dp:00 a:00 b:03 x:0000 y:0000 u:0000 s:0000 pc:e004 beq    SEND
cc:........ dp:00 a:00 b:03 x:0000 y:0000 u:0000 s:0000 pc:e006 abx
cc:........ dp:00 a:00 b:03 x:0003 y:0000 u:0000 s:0000 pc:e007 decb
cc:........ dp:00 a:00 b:02 x:0003 y:0000 u:0000 s:0000 pc:e008 bra    SLOOP
cc:........ dp:00 a:00 b:02 x:0003 y:0000 u:0000 s:0000 pc:e003 tstb
cc:........ dp:00 a:00 b:02 x:0003 y:0000 u:0000 s:0000 pc:e004 beq    SEND
cc:........ dp:00 a:00 b:02 x:0003 y:0000 u:0000 s:0000 pc:e006 abx
cc:........ dp:00 a:00 b:02 x:0005 y:0000 u:0000 s:0000 pc:e007 decb
cc:........ dp:00 a:00 b:01 x:0005 y:0000 u:0000 s:0000 pc:e008 bra    SLOOP
cc:........ dp:00 a:00 b:01 x:0005 y:0000 u:0000 s:0000 pc:e003 tstb
cc:........ dp:00 a:00 b:01 x:0005 y:0000 u:0000 s:0000 pc:e004 beq    SEND
cc:........ dp:00 a:00 b:01 x:0005 y:0000 u:0000 s:0000 pc:e006 abx
cc:........ dp:00 a:00 b:01 x:0006 y:0000 u:0000 s:0000 pc:e007 decb
cc:.....Z.. dp:00 a:00 b:00 x:0006 y:0000 u:0000 s:0000 pc:e008 bra    SLOOP
cc:.....Z.. dp:00 a:00 b:00 x:0006 y:0000 u:0000 s:0000 pc:e003 tstb
cc:.....Z.. dp:00 a:00 b:00 x:0006 y:0000 u:0000 s:0000 pc:e004 beq    SEND
cc:.....Z.. dp:00 a:00 b:00 x:0006 y:0000 u:0000 s:0000 pc:e00a tfr    x,d
cc:.....Z.. dp:00 a:00 b:06 x:0006 y:0000 u:0000 s:0000 pc:e00c (rts   )

breakpoint (1)  0xe00c
```

time since the last store and a register dump may be output to the display and the program continued. The syntax here would be:

bw 0xA000 ? {"time is %d\n" .time - last_time; last_time = .time; g}

which reads: break at a write to A000h/ then print out "time is zz", where zz is the value of the reserved label .time less the value of last_time/ then make last_time equal to the present time/ go on. Adding m 0xA000 would also print out the data sent to the port! The label .time is predefined by the simulator to give the number of cycles taken since last set up.

Assembly-level simulators can also be used to debug C programs. As our example, we will simulate Table 7.14(d), as shown in Table 15.3. This time the parameter n is in absolute memory (for demonstration purposes, it is not passed to the function), but otherwise the process is similar:

1. Set a breakpoint at __prog_top-1 (or 0xE021) and cause it to print out the value of .time:

   br __prog_top-1 ? "Execution time = %d\n" .time

2. Initialize n to 03h, that is set memory byte at L3_n to 03h (mb L3_n 03).

3. Trace to breakpoint from _sum_of_n (t _sum_of_n).

Cross-simulators are little better than native debug packages in their relationship with target hardware. Memory-mapped port registers/buffers can be simulated as memory locations. On a break, their value can be changed from the keyboard and execution recommenced. Interrupts can similarly be simulated from the keyboard. In MIMIC the predefined symbol .irq has a bit correspondence to the various interrupt lines. Thus for the 68000 MPU, making .irq = 01000000b (e.g. typing .irq = 0x40) during a break pause makes the simulator respond to a level-7 interrupt when processing recommences. As an example:

b ? .time == 20000 {.time=0; .irq=0x40; g}

generates a level-7 interrupt each 20,000 cycles (100 interrupts per simulated second with an 8 MHz clock) and continues on.

Simulating high-level sourced programs of any size at this level is tedious at the very least. Table 7.14 was deliberately chosen to have only static variables, so that each variable has a meaningful label attached. Most variables in C are dynamic (i.e. auto), and have no fixed abode. Of course in a simulator, their position relative to the Frame Pointer can be found and therefore accessed. However, determining by hand where many variables are in the frame is time consuming. Some compilers produce a report on the size and location of each variable. An example of such a report is given in Table 15.4. Not only is this useful for assembly-level simulation, but it is the first step towards high-level cross simulation.

Table 15.3: Tracing a C function.

```
"br __prog_top-1 ? {"Execution time = %d\n" .time}"
(Set breakpoint and printout time)

".time = 0"
(Set time symbol to zero)

"mw L3_n"
(Set word in memory = int n to 0003)

0x0001: 0x0000 = "03"
0x0003: 0x0000 = "."

"l"
(List all labels (not predefined))

0x0001: L3_n
0x0003: L31_sum
0x0005: __stack_bottom
0x0400: __stack_top
0xe000: _sum_of_n
0xe022: __prog_top

"t _sum_of_n"

cc:.......C dp:00 a:00 b:06 x:0000 y:0000 u:0000 s:2000 pc:e000 clra
cc:....Z.. dp:00 a:00 b:06 x:0000 y:0000 u:0000 s:2000 pc:e001 clrb
cc:....Z.. dp:00 a:00 b:00 x:0000 y:0000 u:0000 s:2000 pc:e002 std    L31_sum
cc:....Z.. dp:00 a:00 b:00 x:0000 y:0000 u:0000 s:2000 pc:e005 ldx    L3_n
cc:........ dp:00 a:00 b:00 x:0003 y:0000 u:0000 s:2000 pc:e008 beq    0xe01e
cc:........ dp:00 a:00 b:00 x:0003 y:0000 u:0000 s:2000 pc:e00a ldd    L3_n
cc:........ dp:00 a:00 b:03 x:0003 y:0000 u:0000 s:2000 pc:e00d addd   L31_sum
cc:........ dp:00 a:00 b:03 x:0003 y:0000 u:0000 s:2000 pc:e010 std    L31_sum
cc:........ dp:00 a:00 b:03 x:0003 y:0000 u:0000 s:2000 pc:e013 ldd    #0xffff
cc:....N... dp:00 a:ff b:ff x:0003 y:0000 u:0000 s:2000 pc:e016 addd   L3_n
cc:........ dp:00 a:00 b:02 x:0003 y:0000 u:0000 s:2000 pc:e019 std    L3_n
cc:.......C dp:00 a:00 b:02 x:0003 y:0000 u:0000 s:2000 pc:e01c bra    0xe005
cc:.......C dp:00 a:00 b:02 x:0003 y:0000 u:0000 s:2000 pc:e005 ldx    L3_n
cc:.......C dp:00 a:00 b:02 x:0002 y:0000 u:0000 s:2000 pc:e008 beq    0xe01e
cc:.......C dp:00 a:00 b:02 x:0002 y:0000 u:0000 s:2000 pc:e00a ldd    L3_n
cc:.......C dp:00 a:00 b:02 x:0002 y:0000 u:0000 s:2000 pc:e00d addd   L31_sum
cc:........ dp:00 a:00 b:05 x:0002 y:0000 u:0000 s:2000 pc:e010 std    L31_sum
cc:........ dp:00 a:00 b:05 x:0002 y:0000 u:0000 s:2000 pc:e013 ldd    #0xffff
cc:....N... dp:00 a:ff b:ff x:0002 y:0000 u:0000 s:2000 pc:e016 addd   L3_n
cc:........ dp:00 a:00 b:01 x:0002 y:0000 u:0000 s:2000 pc:e019 std    L3_n
cc:.......C dp:00 a:00 b:01 x:0002 y:0000 u:0000 s:2000 pc:e01c bra    0xe005
cc:.......C dp:00 a:00 b:01 x:0002 y:0000 u:0000 s:2000 pc:e005 ldx    L3_n
cc:.......C dp:00 a:00 b:01 x:0001 y:0000 u:0000 s:2000 pc:e008 beq    0xe01e
cc:.......C dp:00 a:00 b:01 x:0001 y:0000 u:0000 s:2000 pc:e00a ldd    L3_n
cc:.......C dp:00 a:00 b:01 x:0001 y:0000 u:0000 s:2000 pc:e00d addd   L31_sum
cc:........ dp:00 a:00 b:06 x:0001 y:0000 u:0000 s:2000 pc:e010 std    L31_sum
cc:........ dp:00 a:00 b:06 x:0001 y:0000 u:0000 s:2000 pc:e013 ldd    #0xffff
cc:....N... dp:00 a:ff b:ff x:0001 y:0000 u:0000 s:2000 pc:e016 addd   L3_n
cc:....Z.C dp:00 a:00 b:00 x:0001 y:0000 u:0000 s:2000 pc:e019 std    L3_n
cc:....Z.C dp:00 a:00 b:00 x:0001 y:0000 u:0000 s:2000 pc:e01c bra    0xe005
cc:....Z.C dp:00 a:00 b:00 x:0001 y:0000 u:0000 s:2000 pc:e005 ldx    L3_n
cc:....Z.C dp:00 a:00 b:00 x:0000 y:0000 u:0000 s:2000 pc:e008 beq    0xe01e
cc:....Z.C dp:00 a:00 b:00 x:0000 y:0000 u:0000 s:2000 pc:e01e ldd    L31_sum
cc:.......C dp:00 a:00 b:06 x:0000 y:0000 u:0000 s:2000 pc:e021 (rts   )

Execution time = 617

*****breakpoint(1) 0xe021
```

Table 15.4: A report on the variables used in the 68008 TCM system of Table 15.5.

```
Information extracted from a:diag_682.xeq

SOURCE FILE : a:diag_68k.c                                  <- Added comments

FILE VARIABLES :                                            <- These are the globals
    extern unsigned char Array[256]    at 0xe002            <- known throughout file
    extern unsigned char Oldest        at 0xe000
    extern unsigned char *x            at 0x418
    extern unsigned char *y            at 0x41c
    extern unsigned char *z            at 0x420
    extern unsigned char *a_d          at 0x424
    extern unsigned char *diag_port    at 0x428

FUNCTION : extern int main() lines 11 to 40 at 0x42c-0x4d4  <- Function main()
    VARIABLES:                                              <- All its local variables
        register unsigned char *array_ptr at reg. a5
        register unsigned char i at reg. d5
        register unsigned char leftmost at reg. d4

FUNCTION : extern void update() lines 50 to 53 at 0x4d4-0x4f8  <- Function update() has
                                                               <- no local variables

FUNCTION : extern void diagnostic() lines 61 to 74 at 0x4f8-0x548  <- Nor has diagnostic()

FUNCTION : extern void output_test() lines 77 to 86 at 0x548-0x572 <- Ftn output_test has
    VARIABLES:                                                     <- only 1 reg. variable
        register unsigned char count at reg. d5

FUNCTION : extern void input_test() lines 88 to 92 at 0x572-0x590  <- Ftn input_test() has
                                                                   <- no local variables

FUNCTION : extern void RAM_test() lines 94 to 111 at 0x590-0x5d0   <- Function RAM_test()
    VARIABLES:                                                     <- All its local variables
        register unsigned int i at reg. d5
        register unsigned char temp at reg. d4
        register unsigned char *memory at reg. a5

FUNCTION : extern void ROM_test() lines 113 to 120 at 0x5d0-0x606  <- Function ROM_test()
    VARIABLES:                                                     <- All its local variables
        register unsigned char *address at reg. a5
        register unsigned short sum at reg. d5

000004dc    _update           00000554   L133         0000053c   L103
0000042c    _main             00000514   L142         00000598   _RAM_test
00000000    STARTUP           000004ce   L151         0000049c   L101
00000408    NMI               000005d6   L114         0000e002   _Array
00000406    _exit             000004d2   L141
00000400    START             000005a6   L104
00000444    L1                00000500   L122
00000468    L17               000004c8   L131
00000460    L15               00000420   _z
00000550    _output_test      00000428   _diag_port
0000057a    _input_test       0000041c   _y
0000044c    L11               00000418   _x
000005dc    _ROM_test         0000e000   _Oldest
0000060c    L165              00000500   _diagnostic
000005fe    L135              00000424   _a_d
000005d0    L144              00000612   __prog_top
00000528    L162              0000e000   __data_top
000005ec    L125              0000e102   __stack_bottom
                              00010000   __stack_top
```

Simulators used to debug realistic high-level sourced software must provide the facility to monitor directly high-level objects and instructions as well as addresses, registers and assembly-level instructions. The next few examples are based on the Intermetrics/COSMIC CXDB (C cross DeBugger) products. These are high-level front ends to the MIMIC simulator (renamed MICSIM for MICroprocessor SIMulator) we have just used. The user can move down to MICSIM at any time to perform any machine-level task, for example to set up a breakpoint on an attempt to write to simulated ROM, and move back up again. MICSIM's instructions are the same as those illustrated in Tables 15.1–15.3.

At the high level, the following core features are available:

Listing
Displays the C source with or without the resulting assembly code, around the current execution point in the source window.

Monitor
Displays state of C objects or values of C expressions continuously in the monitor window.

Update
Allows a C data object to be altered at any time.

Step execution
Steps through the C program, each step executing one or more C source lines. During this time, variables may be continually monitored, and the function window shows which function the program is in and what values were passed to it. Also the state of the frame and any variables in that function can be examined.

Breakpoints
Insertion of high-level conditions, such as executing a C line or entering a function, which causes execution to pause. Actions may be taken automatically on pause, such as changing the value of a variable.

Execute
Runs simulation at full speed from halt point, normally to the next breakpoint.

For our first example, consider the screen dump of Fig. 15.1(a). This shows the sum-of-integers program of Table 7.14 in the central code window. The cursor is at line 6, which has yet to be executed. The variables n and sum appear in the top left monitor window. To get to this point, the following commands were entered:

1. Step from line 1 (entry point) to line 5 (s5 or five single steps, s).

2. Command that the variables n and sum be monitored (m n,sum).

3. Set (Update) the value of n to 20 to simulate a passed parameter (u n 20).

Simulation

```
┌──────────── CXDB Copyright 1989 (c) COSMIC (France) ────────────┐
│(1) n = 19                          │                            │
│(2) sum = 20                        │                            │
│                                    │                            │
│                                    │sum_of_n()                  │
├────────────────────────────────────┴────────────────────────────┤
│1  unsigned int sum_of_n()                                       │
│2  {                                                             │
│3  static unsigned int n;                                        │
│4  static unsigned int sum;                                      │
│5  sum=0;                                                        │
│6  ▓while(n>0)▓                                                  │
│7      {                                                         │
│8      sum=sum+n;                                                │
│9      n=n-1;                                                    │
│10     }                                                         │
│11 return(sum);                                                  │
│12 }                                                             │
│                                                                 │
│                                          sum_of_n()  list7_13.c │
│CXDB>s                                                           │
│CXDB>s                                                           │
│CXDB>s                                                           │
│CXDB>                                                            │
└─────────────────────────────────────────────────────────────────┘
```

(a) Checking the loop operation.

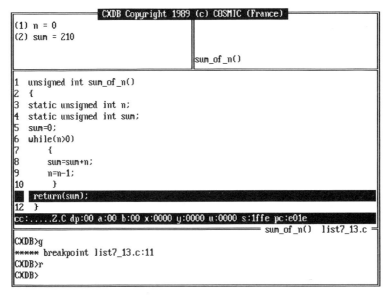

(b) Going to the termination.

Figure 15.1: Tracing function sum_of_n().

4. Single step around the loop once, that is four steps (s4). After doing this, n has been added to sum, thus sum is 20. Also n has been decremented, thus n is 19.

With the loop operation checked, the algorithm can be verified by either single stepping until line 12, or more conveniently, setting a breakpoint and executing at full speed. The screen dump of Fig. 15.1(b) is the result, with the following additional commands:

1. Set a breakpoint at line 11 (b :11)

2. Go from current cursor (g :11 or just g)

3. Look at register state at breakpoint (r)

Now n has been decremented to zero and sum is correctly read as 210. Also shown in the code window is the state of the registers. If we could have gone back to the calling function, Accumulator_B would have been D2h (i.e. sum returned). The register display can be toggled on and off by entering r.

Figure 15.1(b) showed that the underlying target was 6809 code. If the register display had not been toggled, this fact would not have been known (actually Fig. 15.1(a) was generated using the 68000 simulator, just to illustrate this point!). Machine independence is a feature of high-level symbolic simulators. It is possible to step showing the mixture of high and low-level codes, but stepping and monitoring are still at source level.

The top right window shows both the current function and the function path taken to arrive at the current point. This is not very informative in the simple situation simulated in Fig. 15.1, but is useful in realistic situations. Figure 15.2 depicts the exponentiation program of Table 9.1. Three functions are coded, namely main(), power() and abs(). I have stepped through this program until exp is 1. The function window then shows abs(15625) at the bottom, which says you are now in function abs() which has had 15,625 passed to it (i.e. result). Above this is power(25,3), which says the entry point was from power(), to which had been passed the values 25 and 3. Finally above this is main(), the caller of power(). The 68000 simulator was used for this diagram.

Finally let us examine our time-compressed memory software of Table 14.1. In Fig. 15.3(a) I have stepped around the loop so that i is ten. The monitor window shows the contents of x (the X D/A converter), which is just i, the contents of y (the Y D/A converter), the variable Oldest (changed by update(), here = 0) and the array value currently being sent out to y (here Array[10]). This shows that expressions and indirection can be monitored, as well as simple variables. If pointers are monitored, they display in hexadecimal. Ordinary objects default to decimal, but using the mx command (examine Memory in heXadecimal) forces hexadecimal.

In Fig. 15.3(b) I have converted the code window to a view box of the variables in function main(). The top five (i, leftmost, *x, *y and *z) are all auto variables

```
┌──────────────────CXDB Copyright 1989 (c) COSMIC (France)──────────────────┐
│ (1) power():exp = 1                                                        │
│ (2) power():old_result = 625         main()                                │
│ (3) power():result = 15625           power(25,3)                           │
│                                      abs(15625)                            │
├────────────────────────────────────────────────────────────────────────────┤
│ 16      for(result=1; exp>0; exp--)                                        │
│ 17         {                                                               │
│ 18         old_result = result;                                            │
│ 19         result*=y;       /* Repetitive multiplication by y    */        │
│ 20         if(abs(result)<=abs(old_result)) {return 0;}  /* Overflow error*│
│ 21         }                                                               │
│ 22      return result;                                                     │
│ 23      }                                                                  │
│ 24                                                                         │
│ 25 /* Here follows the definition of abs()                      */         │
│ 26                                                                         │
│ 27 int abs(int z)                                                          │
│ 28      {return (z>=0 ? z:-z);}                                            │
│                                              ══════════ abs()   list9_1.c ═│
│ CXDB>s                                                                     │
│ CXDB>s                                                                     │
│ CXDB>s                                                                     │
│ CXDB>                                                                      │
└────────────────────────────────────────────────────────────────────────────┘
```

Figure 15.2: Illustrating the function path in reaching line 27.

and their address is given in the System stack (set to 400h by the startup code in the 6809 version). As *x, *y and *z are pointers, their value is given in hexadecimal.

Array[] is an external variable and is listed under file variables. All 256 values are given. If the window is scrolled down, the file variable Oldest is given as:

at 0x1 extern unsigned char Oldest = 0

The screen dumps shown in Fig. 15.3(a) and (b) were taken on different runs and thus *y and Array[10] vary between the two situations.

Both Array[] and Oldest objects are updated by an interrupt service routine. How is such an interrupt simulated with CXDB? To 'generate' an interrupt at any time (typically in-between steps or at a break point) requires a move down to the underlying assembly-level simulator (MICSIM with this product). The sequence of commands to generate the screen dump shown in Fig. 15.4 was:

1	/	Move to MICSIM (assembly level)
2	.irq = 1	Making this variable = 1 simulates NMI
3	/	Return to the high level
4	s	Step
5	mx Oldest, Array[Oldest-1]	Monitor these variables
6	s	Step
7	u &a_d 0x55	Simulate the A/D converter's output
8	s	Step

402 Looking For Trouble

```
┌──────────────── CXDB Copyright 1989 (c) COSMIC (France) ────────────────┐
│(1) Oldest = 0                                                           │
│(2) *x = 10                                                              │
│(3) *y = 115                                                             │
│(4) Array[(leftmost+i)&0xff] = 115    │main()                            │
├──────────────────────────────────────┴──────────────────────────────────┤
│15          {Array[i]=0;}                                                │
│16        while(1)                    /* Do forever display contents of array │
│17        {                                                              │
│18          leftmost = Oldest;        /* Make the leftmost point on the scre │
│19          for (i=0; i<256; i++)                                        │
│20          {                                                            │
│21            *x = (unsigned char)i;   /* Send x coordinate to X plates  │
│22            *y = Array[(leftmost+i)&0x0ff]; /* and the display byte to the │
│23          }                                                            │
│24          *z = BLANK_ON;            /* Blank out for flyback           │
│25          *x = 0;                   /* Move to right of screen         │
│26          *y = Array[Oldest];       /* Y value at left of screen       │
│27          for(i=0;i<5;i++) {;}      /* Delay                           │
│                                              ═══════ main()  tcmuc092.c ═│
│CXDB>                                                                    │
│CXDB>                                                                    │
│CXDB>                                                                    │
│CXDB>                                                                    │
└─────────────────────────────────────────────────────────────────────────┘
```

(a) Stepping around the loop until i is 10.

```
┌──────────────── CXDB Copyright 1989 (c) COSMIC (France) ────────────────┐
│(1) Oldest = 0                                                           │
│(2) *x = 10                                                              │
│(3) *y = 127                                                             │
│(4) Array[(leftmost+i)&0xff] = 127    │main()                            │
├──────────────────────────────────────┴──────────────────────────────────┤
│vars in main():                                                          │
│  at 0x3fa   auto short          i = 10                                  │
│  at 0x3f9   auto unsigned char  leftmost = 0                            │
│  at 0x3f7   auto unsigned char  *x = 0x2000                             │
│  at 0x3f5   auto unsigned char  *y = 0x2001                             │
│  at 0x3f3   auto unsigned char  *z = 0xa000                             │
│vars in tcmuc092.c:                                                      │
│  at 0x2     extern unsigned char  Array[256] = 56 61 69 75 81 88 95 103 110 1│
│19 127 120 116 108 0 0 0 0 0 0 0 0 0 0 0 0 0 0 0 0 0 0 0 0 0 0 0 0 0 0 0 0│
│ 0 0 0 0 0 0 0 0 0 0 0 0 0 0 0 0 0 0 0 0 0 0 0 0 0 0 0 0 0 0 0 0 0 0 0 0 0│
│ 0 0 0 0 0 0 0 0 0 0 0 0 0 0 0 0 0 0 0 0 0 0 0 0 0 0 0 0 0 0 0 0 0 0 0 0 0│
│ 0 0 0 0 0 0 0 0 0 0 0 0 0 0 0 0 0 0 0 0 0 0 0 0 0 0 0 0 0 0 0 0 0 0 0 0 0│
│ 0 0 0 0 0 0 0 0 0 0 0 0 0 0 0 0 0 0 0 0 0 0 0 0 0 0 0 0 0 0 0 0 0 0 0 0 0│
│                                              ═══════ main()  tcmuc092.c ═│
│CXDB>s                                                                   │
│CXDB>s                                                                   │
│CXDB>v&p                                                                 │
└─────────────────────────────────────────────────────────────────────────┘
```

(b) Viewing the variables.

Figure 15.3: Simulating the time-compressed memory software.

```
┌──────────────────── CXDB Copyright 1989 (c) COSMIC (France) ─────────────────┐
│(1) Oldest = 5                                                                │
│(5) Array[Oldest-1] = 0x55                                                    │
│                                          main()                              │
│                                          update()                            │
├──────────────────────────────────────────────────────────────────────────────┤
│30     }                                                                      │
│31  /*****************************************************************       │
│32   * This is the NMI interrupt service routine which puts the analog sample │
│33   * ENTRY  : Via NMI and startup1                                        * │
│34   * ENTRY  : Array[] and Oldest are global                                 │
│35   * EXIT   : Value held at a_d in Array[Oldest], Oldest incremented with ura│
│36   *****************************************************************       │
│37                                                                            │
│38  void update(void)                                                         │
│39  {                                                                         │
│40      volatile unsigned char * const a_d = ANINPUT;  /* This is the Analog  │
│41      Array[Oldest++] = *a_d;                /* Overwrite oldest sample in Arra│
│42  ▓▓▓▓▓▓}▓▓▓▓▓▓▓▓▓▓▓▓▓▓▓▓▓▓▓▓▓▓▓▓▓▓▓▓▓▓▓▓▓▓▓▓▓▓▓▓▓▓▓▓▓▓▓▓▓▓▓▓▓▓▓▓▓▓▓▓▓▓▓▓▓▓│
│                                                     update()  tcmuc092.c    │
├──────────────────────────────────────────────────────────────────────────────┤
│CXDB>u *a_d 0x55                                                              │
│    static unsigned char *a_d = 0   => 85                                     │
│CXDB>s                                                                        │
│CXDB>                                                                         │
└──────────────────────────────────────────────────────────────────────────────┘
```

Figure 15.4: Simulating an interrupt entry into update().

The monitor screen shows the array element has taken on the simulated value $55h$ and Oldest has been incremented to 5. Another step and the simulator returns to main().

In moving between levels, care must be taken. For example the compiler linker places the NMI vector at E7FC:Dh, assuming a 2716 EPROM (see Table 14.4). However, the simulator has memory all the way up to FFFFh, and thus goes to FFFC:Dh on a simulated NMI interrupt. Hence a manual setting up of the vectors is needed (mW 0xFFFC NMI). Also some high-level simulators do not execute assembly-level startup routines, and these may require execution at the low-level simulation before beginning the high-level process.

Our closing example shows a slightly more sophisticated high-level simulator, based around the 68000-family Microtec compilers. This XRAY68K simulator is a true window-based package in that an on-screen cursor can be moved to any window and used to scroll up or down [5]. Also windows may be altered in size and even removed if desired. Both high- and low-level simulation is provided without having to move between packages. Figure 15.5 shows a simulation of the array-clearing routine from the time-compressed memory main routine. The simulator is set to its low-level mode. Here simulation is done at assembly-level, each step being one machine instruction. Nevertheless, the Data window shows the C-level variables being monitored. The state of the System stack is shown in window 14, which can be entered and scrolled up or down as desired. The startup routine

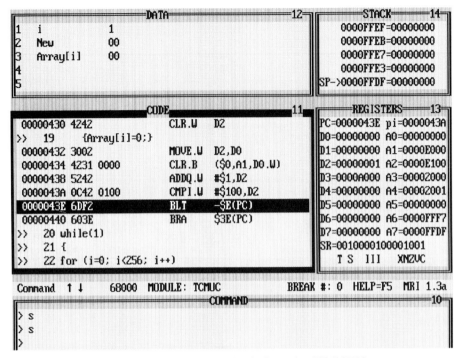

Figure 15.5: Mixed-mode simulation using XRAY68K.

sets **A7** (i.e. the SSP) to $10000h$ before going to main(). All registers and flags are shown in window 13. The pseudo register pi gives the Previous Instruction address. This particular software is set up to operate at a level-2 interrupt, which explains the setting of the three interrupt mask bits to $001b$. The simulator can be changed between high- and low-levels by using the mode command or toggling the F3 function key (MSDOS version).

One interesting feature of XRAY68K, is the ability to simulate input and output ports. In the former case the command inport <address> stops when <address> is read and accepts data from the keyboard before continuing; for example inport 6000 would simulate the A/D converter. The command outport sends data, for example to a printer, when a specified memory location is written to.

15.2 Resident Diagnostics

Unless a commercial system is going to be used as the target, a major difficulty lies in the verification of the hardware integrity. Although the application software may have been checked using simulation, its testing in its intended environment cannot easily be carried out if the latter's operation is uncertified. Even with bought-in hardware, malfunction may occur in service. In such a situation, is it a hardware or software fault?

In practice little more than d.c. and continuity tests can be carried out on the hardware as it stands. In order to isolate the testing of the hardware and the application software, it is necessary to introduce a package of programs specifically designed to exercise the various components. Such diagnostic software could be transported to the hardware environment by using an in-circuit emulator, as discussed in the next section. Alternatively, the EPROM(s) may be removed and replaced by a diagnostic EPROM-based package. Where sufficient capacity exists, this package may co-reside with the application package, and this will be particularly convenient for field servicing, especially if accessible by the customer.

Following along the same path, hardware should be built at the outset for testability. In a microprocessor-based system, this usually means designing in a **free-run** facility. A free-running microprocessor has a null instruction jammed on to its data bus, and it spends all its time running this phantom software. During this endless execution, the address bus is incrementing, fetching down the next null instruction. Thus all address logic, Chip Enable connections and some control signals, such as Reset and Clocks, can be monitored dynamically. Simple test equipment, such as a logic probe or oscilloscope are adequate for this purpose. Free running is especially useful for signature analysis [6].

A null instruction has certain common characteristics, irrespective of the target MPU. Firstly it should not execute any Write cycles, as the data bus has been hijacked. This means that it will either do something on an internal register or even nothing at all. Secondly, its op-code should be the size of the data bus, or if larger, a repetitive multiple.

Figure 15.6(a) and (b) shows the free-run facility applied to the 6809 MPU. Normally the switch is open, and the two back-to-back diodes do not conduct. With the switch closed and the data bus isolated from the outside world, the pattern 01011111b (5Fh) is jammed onto the bus. Thus on Reset, the 6809 fetches down the two bytes at FFFE:Fh, 5F5Fh in this case, and commences execution at this address. Its first instruction is 5Fh, or CLRB. Once this has been done (all Read cycles), the instruction at 5F60h is fetched, again CLRB... ad infinitum. CLRB rather than NOP was chosen, as the latter's op-code of 01h would require seven diodes.

As long as the 6809 free runs, its address bus acts as a 16-bit counter, cycling from 0000h to FFFFh. Assuming a 1 MHz clock (4 MHz crystal), a_{15} will cycle in $2^{16} \times 2 = 0.131$ s, a_{14} in 0.0655 s, down to 2 μs for a_0. During this time R/$\overline{\text{W}}$ and E and Q can be monitored using an oscilloscope.

The address decoder outputs will last $\frac{1}{8}$ of this cycle time, and will appear in the correct sequence. Some typical examples are shown in Fig. 15.7. These can in turn be traced to the appropriate Chip Enables. Although the data bus is disconnected from the MPU during this time, it will still be activated by any enabled input device. Thus using a 2-beam oscilloscope, monitoring the **Switch_Port_Enable** (i.e. 8000h) and d_0, will enable the state of Switch 0 to be seen, gated through to the data line. Similarly, activity at the time of the EPROM and RAM Chip Enables can be viewed.

The requirements for a 68000 null instruction are more stringent, as its op-code must be even. This is because of the requirements that both **PC** and **SP** must be

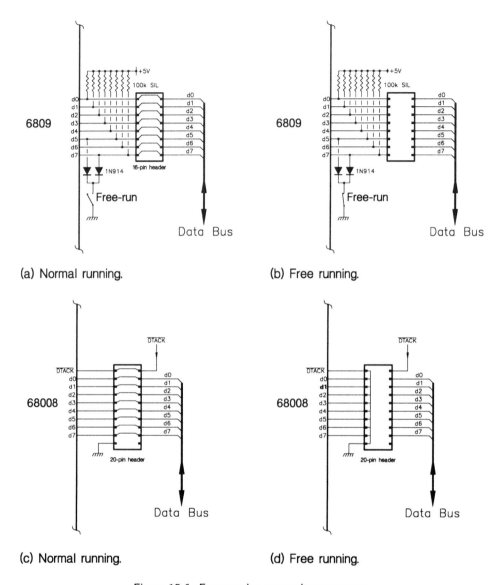

(a) Normal running. (b) Free running.

(c) Normal running. (d) Free running.

Figure 15.6: Free-running your microprocessor.

even, and these will be equal to the null op-code after Reset. Should an odd word be fetched for these, then a fatal Double-Bus fault will occur [7]. The word size of a 68000 op-code puts further restrictions on the choice of a null instruction for the 68008 processor, as this will fetch the op-code down in two identical bytes. Both considerations rule out the NOP instruction, with its op-code of 4E-71h. Fortunately the op-code for ORI.B #0,D0 is 00-00h, and this fulfils all these requirements.

The free-run circuitry shown in Figs 15.6(c) and 15.6(d) comprises two head-

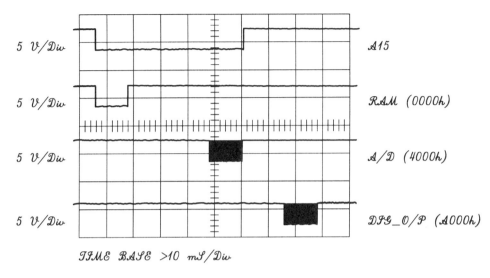

Figure 15.7: One free-run cycle, showing RAM, A/D and DIG_O/P Enables.

ers. The normal header simply connects the eight data lines and $\overline{\text{DTACK}}$ directly through. Free running is accomplished by replacing this by a header shorting these lines to ground.

As in the 6809 case, the **PC** and hence the address bus repetitively cycle through the entire address space. As the null instruction takes 16 cycles, then at 8 MHz a 2 μs instruction time is obtained. Address line a_{19} takes $2^{20} \times 2 = 2.1$ s to complete a sweep. During this time both $\overline{\text{AS}}$ and $\overline{\text{DS}}$ operate in the normal way, and $\text{R}/\overline{\text{W}}$ is high (Read).

The address decoder outputs cycle with a repetition rate of $2^{16} \times 2 = 0.131$ s, as a_{15} is the highest decoded line. Waveforms are similar to those for the 6809, shown in Fig. 15.7, but decoder outputs are qualified by $\overline{\text{AS}}$, giving striated Chip Enables. As previously described, these can be used with a 2-beam oscilloscope to monitor activity on the data bus. The $\overline{\text{DTACK}}$ generator can also be monitored, although this is trivial in simple circuits such as shown in Fig. 13.3. Logic analyzer traces showing the 68000 in this free-run mode are shown in reference [8].

The free-run facility is useful, as it requires a minimum of built-in test hardware. It is possible to take the process a stage further, and incorporate a hardware single-step facility [8]. However, this isn't often done, as the extra testability rarely justifies the expense.

With a reasonable assurance that the target hardware is functioning, the diagnostic software can be loaded in. This can be done by using a romulator (ROM emulator) or programming an EPROM and inserting into its socket. The former uses a block of dual-port RAM to take the place of ROM memory. One port is connected to the ROM socket in the target system via a ribbon cable and DIL plug. The other port is controlled from the terminal, typically the workstation on which the compiler/assembler runs. A driver software package downloads hex files to this

RAM, usually through a serial link. With the loading completed, the ROMulator can be switched to emulate mode, and will appear as a programmed ROM [9]. The use of an in-circuit emulator for this purpose is the subject of the following section.

The circuits of Figs 13.1 and 13.3 have a 4-bit switch port available to choose between normal and diagnostic modes. With this port at zero at power-up, the normal application program is run. In the diagnostic mode, one of four tests are made, as follows:

Switch 0:   Check analog and digital output ports
Switch 1:   Check analog input port
Switch 2:   Check RAM chip
Switch 3:   Check ROM chip

Once one of the two modes have been entered, change-over can only be implemented through a reset. After Reset the switch port can be used as a normal run-time port.

The application software shown on the first page of Table 15.5 is basically a modified version of Table 14.14. There are two significant changes. Firstly, all ports (now including the switch port, named `diag_port`) are defined externally, that is before `main()`. This is because they are needed by the various diagnostic routines, besides `main()` and `update()`. Although it is considered poor programming practice to use public objects unnecessarily, hardware ports are by nature global. As can be seen from Table 15.6, such objects are stored as constants in ROM in the same manner as the `static const` equivalents of Table 14.14. They could still be qualified as `static`, in which case they would not be declared globally known to the linker.

The second change checks the state of `diag_port` (ANDed with 00001111b to zero the undefined upper four bits). If non-zero (True), then execution is transferred to function `diagnostic()`. Otherwise the time-compressed memory endless loop is entered. The diagnostic software thus adds nothing to the execution time of the applications software.

Function `diagnostic()` comprises a main body having an endless loop selecting one of four subfunctions, depending on which switch is set. Notice from Table 15.6(b), lines C69–C72, that the BTST instruction is used to check the state of the target switch, rather than use the less efficient AND or BIT instruction.

The `output_test()` function simply counts up from 0 to 255 and sends each value to the Z digital and X analog output ports. The complement is sent to the Y analog port. Using an oscilloscope, the X and Y ports give ramps, up and down respectively, as shown in Fig. 15.8. The Z port acts as an 8-bit counter.

The `input_test()` function 'connects' the analog input port to the two analog outputs. Thus using a sinewave generator as an input should give two quantized copies at the output. The switch input port is of course implicitly tested by getting to this routine in the first place.

Testing the RAM chip is in essence a matter of sending out a test pattern (10101010b in this case) to each cell in turn and checking that it gets there [10]. This is of course a destructive test, so the original value must be fetched and

Table 15.5: Complete 68008 package, including resident diagnostics (*continued next page*).

```c
/* Version 01/02/90                                                             */
#include <hard_68k.h>
unsigned char Array [256]; /* Global array holding display data                 */
unsigned char Oldest;      /* Index to the Oldest inserted data byte (left point on scr*/
unsigned char * const x = ANALOG_X; /* x points to a byte @ (address) ANALOG_X  */
unsigned char * const y = ANALOG_Y; /* y points to a byte @ (address) ANALOG_Y  */
unsigned char * const z = Z_BLANK;  /* The z-mod port (digital port)            */
volatile unsigned char * const a_d = ANINPUT;     /* This is the Analog input port */
volatile unsigned char * const diag_port = SWITCH; /* The z-mod port (digital port) */

main()
{
register unsigned char * array_ptr; /* Pointer into array                       */
register unsigned char i;           /* Scan counter                             */
register unsigned char leftmost;    /* The initial array index when x is 0      */
void diagnostic (void);             /* Define the diagnostic function           */

if(*diag_port&0x0f)    /* Call the diagnostic function if switch port set to non-zero */
    {diagnostic();}

Oldest = 0;                         /* Start New index at beginning of the array */

for(array_ptr=Array; array_ptr<Array+256; *array_ptr++=0)    {;}    /* Clear array */

while(1)                            /* Do forever display contents of array      */
    {
    leftmost = Oldest;              /* Make leftmost point on screen the oldest sample */
    for (array_ptr=Array, i=0; array_ptr<Array+256;)
        {
        *x = i;                     /* Send x co-ordinate to X plates            */
        *y = *(array_ptr++ +leftmost)&0x0ff;   /* and the display byte to the y D/A */
        }

    *z = BLANK_ON;                  /* Blank out for flyback                     */
    *x = 0;                         /* Move to right of screen                   */
    *y = Array[Oldest];             /* Y value at left of screen                 */
     for(i=0; i<5; i++) {;}         /* Delay                                     */
    *z = BLANK_OFF;                 /* Blank off                                 */
     }                              /* Do another scan                           */
}


/********************************************************************************
 * This is the NMI ISR which puts the analog sample in the array & updates the New index*
 * ENTRY : Via NMI and startup                                                  *
 * ENTRY : Array[] and Oldest are global                                        *
 * EXIT  : Value held at a_d in Array[Oldest], Oldest incremented with wraparound at 256*
 ********************************************************************************/

void update(void)
{
Array[Oldest++] = *a_d;/* Overwrite oldest sample in Array[] & inc Oldest index mod-256 */
}
```

410 Looking For Trouble

Table 15.5: (*continued*) Complete 68008 package, including resident diagnostics.

```c
/*******************************************************************************
* The diagnostic routine calling up 1 of 4 tests depending on the state of the switches*
* ENTRY : When switches are non-zero                                           *
* EXIT  : Endless loop, use Reset to exit, first setting all switches to zero  *
*******************************************************************************/
void diagnostic(void)
{
void output_test(void);          /* Declare each diagnostic sub-function       */
void input_test(void);
void RAM_test(void);
void ROM_test(void);
while(1)                         /* Do forever the diagnostic tests            */
    {
    if(*diag_port&0x01)      {output_test();}
    else if(*diag_port&0x02) {input_test();}
    else if(*diag_port&0x04) {RAM_test();}
    else if(*diag_port&0x08) {ROM_test();}
    }
}

void output_test(void)
{
register unsigned char count = 0;
do
    {
    *x = count;                  /* Send count out to X D/A converter, i.e. ramp up */
    *z = count;                  /* and to the Z digital port                       */
    *y =~count;                  /* ramp Y output down                              */
    } while(++count != 0);
}
void input_test(void)
{
*x = *a_d;                       /* Get input from a_d and send to X d/a       */
*y = *a_d;                       /* and to Y d/a                               */
}

void RAM_test(void)
{
register unsigned int i;
register unsigned char temp;
register unsigned char * address;/* Address of the memory byte being tested    */
*z = 0;                          /* Set digital port to all zeros (pass)       */
for(address=RAM_START; address<RAM_START+RAM_LENGTH;)
    {
    temp = *address;             /* Get ith memory byte                        */
    *address = 0xAA;             /* Send out 10101010b to it                   */
    if(*address != 0xAA)
        {                        /* IF not this value THEN signal failure by sending out 11111111b*/
        *z = 0xFF;
        break;
        }
    *address++ = temp;           /* Restore original value                     */
    }
}

void ROM_test(void)
{
register unsigned short * address; /* Address points to 16_bit word in EPROM   */
register unsigned short sum=0;
*z = 0;                          /* Set digital output to all zeros to signal pass */
for(address=ROM_START; address<ROM_START+ROM_LENGTH; sum+=*address++)   {;}
if(sum)   {*z = 0xFF;}/*IF a non-zero sum THEN signal error by digital output = 10101010*/
}
```

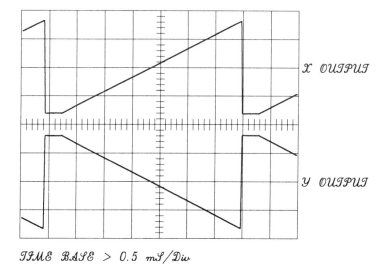

Figure 15.8: The output_test() traces.

saved, before each cell is checked (line C102) and returned afterwards (line C109). The pointer variable address is used to move up through memory. The values RAM_START (a pointer) and RAM_LENGTH are defined for the circuit in the header file (Tables 14.2 and 14.9). The digital Z port is used to indicate pass or fail by being set respectively to all ones or all zeros. Of course exercising with such a simplistic test pattern is not a fully comprehensive verification; for example, it will not detect a stuck at bit 0 error. However, the principle is the same for a more sophisticated set of test patterns.

Checking the ROM is something that is likely to be carried out as part of a field test. This is normally done by replacing an unused word by a 16-bit error checking code, which is such that the sum of all memory contents is zero. Most EPROM programmers will give the 16-bit sum of all word locations (i.e. 16 bits at a time). Unprogrammed locations are usually FFh, and so one must be added to this sum to compensate for the overwritten word (FFFFh is -1). Thus we have for this checksum (CS):

$$CS + (sum + 1) = 0$$

or

$$CS = -sum - 1 = (sum' + 1) - 1 = sum'$$

where sum' is the 1's complement and sum'+1 is the 2's complement, that is −sum. Hence all that is needed is to invert the modulo-65536 (16-bit) summation of all EPROM words and overwrite a convenient unprogrammed word. In cases where unprogrammed locations are zero, then one should be subtracted from the inverted summation.

Function ROM_test() walks through the contents of the EPROM using a pointer-to short moving from ROM_START to ROM_START + ROM_LENGTH. Each word is added to the 16-bit variable sum, which eventually will give the modulo-65536 check digit, which is hopefully zero. Care must be taken as the checksum is not always calculated in this way. For example the modulo-65536 sum of all *bytes* does not give the same answer.

Code generated by the diagnostics() source is given in Table 15.6. The time compressed code is virtually the same as in Table 14.14 and is not reproduced here. The assembly-level code is commented and is straightforward. One interesting point concerns lines C90 and C91. The reader might suppose the textbook equivalent:

```
*x = *y = *a_d;
```

to be the same. Not necessarily so. This compiler implemented this as follows:

1. Get value from a_d and send out to y.

2. Get value from y and send out to x.

While this may be logically correct, y is a write-only port; it cannot be read! There is no way in ANSII C to designate an object write-only. A read-only object is of course designated as const. Designating such an object volatile may help, in that it should signal to the compiler that it cannot depend on what it reads, but this is a grey area of compiler design.

Table 15.5 declared that this was the software for the 68008 implementation. In fact it is applicable to the 6809 target except for RAM_test(). Testing RAM in C is dangerous, as the stack and frame area is of course in this part of memory. Even though I have made RAM_test() non-destructive, problems can arise. By making temp a register variable, the original value of any RAM location can temporarily be stored (in **D4.B**) out of harm's way. However, the 6809 compiler ignores any register qualifications (as do implementations for most 8-bit targets) and puts temp as an auto variable in a frame, that is in RAM. When that particular address is tested, temp will be overwritten by the test pattern. Similarly the pointer address itself is in RAM.

In practice this non-register implementation may not cause problems; much depends on the code a compiler produces. Rather than run a risk, Table 15.7 shows an alternative RAM_test() with an embedded assembly-level routine. This time the pointer equivalent to address is held in Index register_Y, and temp resides in Accumulator_A.

A diagnostic package is intimately tangled up in the hardware, and, of course, should be above reproach. Because of this, special care needs to be taken if diagnostic software is written in **C**. Including some assembly-level code for the purpose of diagnosis may well be desirable, as the uncertainty of compiler allocation and action is eliminated.

Table 15.6: Code for the 68008 implementation (*continued next page*).

```
*  5  unsigned char * const x = ANALOG_X;   /* x points to a byte @ (address) ANALOG_X   */
            .text
            .even
_x:         .long     0x2000
*  6  unsigned char * const y = ANALOG_Y;   /* y points to a byte @ (address) ANALOG_Y   */
            .even
_y:         .long     0x2001
*  7  unsigned char * const z = Z_BLANK;    /* The z-mod port (digital port)             */
            .even
_z:         .long     0xa000
*  8  volatile unsigned char * const a_d = ANINPUT; /* This is the Analog input port    */
            .even
_a_d:       .long     0x4000
*  9  volatile unsigned char * const diag_port = SWITCH; /* The z-mod port (digital port) */
            .even
_diag_port: .long     0x8000
* 10
* 11  main()
```

(a) Defining the ports as external constants.

```
* 61  void diagnostic (void)
* 62  {
            .even
* 63      void output_test(void);         /* Declare each diagnostic sub-function  */
* 64      void input_test(void);
* 65      void RAM_test(void);
* 66      void ROM_test(void);
* 67      while(1)                        /* Do forever the diagnostic tests       */
* 68      {
* 69         if(*diag_port&0x01)          {output_test();}
_diagnostic:
L151:       move.l    _diag_port,a1   *## Point A1 to switch port
            btst      #0,(a1)         *## Test switch 0
            beq.s     L171            *## IF zero THEN try next switch
            jsr       _output_test    *## ELSE do output test
* 70         else if(*diag_port&0x02)  {input_test();}
            bra.s     L151            *## and redo the switch scan on return
L171:       move.l    _diag_port,a1   *## Repeat the above for switch 1
            btst      #1,(a1)
            beq.s     L112
            jsr       _input_test     *## which if set commands the input analog port test
* 71         else if(*diag_port&0x04)  {RAM_test();}
            bra.s     L151
L112:       move.l    _diag_port,a1
            btst      #2,(a1)         *## IF switch 2 is set THEN
            beq.s     L132
            jsr       _RAM_test       *## go do the RAM test
* 72         else if(*diag_port&0x08)  {ROM_test();}
            bra.s     L151
L152:       move.l    _diag_port,a1
            btst      #3,(a1)         *## IF switch 3 is set THEN
            beq.s     L151
            jsr       _ROM_test       *## go do the ROM test
            bra.s     L151
* 73      }
* 74  }
* 75
* 76
```

Table 15.6: Code for the 68008 implementation (*continued next page*).

```
*  77   void output_test(void)
*  78   {
               .even
_output_test:  move.l     d5,-(sp)
*  79     register unsigned char count = 0;
               clr.b      d5              *## count lives in D5.B and is zeroed
*  80     do
*  81     {
*  82        *x = count;                   /* Send count out to X D/A converter; ie ramp up */
L162:          move.l     _x,a1           *## A1 points to X port
               move.b     d5,(a1)         *## send count out to this port
*  83        *z = count;                   /* and to the Z digital port                     */
               move.b     _z,a1           *## A1 points to Z port
               move.b     d5,(a1)         *## send count out to this port
*  84        *y = ~count;                  /* ramp Y output down                            */
               move.l     _y,a1           *## Point to Y analog port
               clr.w      d7
               move.b     d5,d7           *## Get count again!
               not.w      d7              *## Invert it (ie ~count)
               move.b     d7,(a1)         *## and send it out
*  85     } while(++count != 0);
               addq.b     #1,d5           *## First increment count
               bne.s      L162            *## IF not folded over to zero (ie 256) THEN repeat
               move.l     (sp)+,d5
               rts
*  86   }
*  87


*  88   void input_test(void)
*  89   {
               .even
*  90        *x = *a_d;                    /* Get input from a_d and send to X d/a          */
_input_test:   move.l     _x,a1           *## Point A1 to X d/a output port
               move.l     _a_d,a2         *## Point A2 to a/d input port
               move.b     (a2),(a1)       *## Send input data to output X port
*  91        *y = *a_d;                    /* and to Y d/a                                  */
               move.l     _y,a1           *## Point A1 to Y d/a output port
               move.l     _a_d,a2         *## Point A2 to a/d input port
               move.b     (a2),(a1)       *## Send input data to output Y
               rts
*  92   }
*  93
*  94   void RAM_test(void)
*  95   {
_RAM_test:     movem.l    d5/d4/a5,-(sp)
*  96     register unsigned int i;
*  97     register unsigned char temp;
*  98     register unsigned char * address     /* Address of the memory word being tested */
*  99     *z = 0;                              /* Set digital port to all zeros (pass)    */
               move.l     _z,a1           *## A1 points to Z port
               clr.b      (a1)            *## Send out 00000000b
```

Table 15.6: (*continued*) Code for the 68008 implementation.

```
* 100  for(address=RAM_START; address<RAM_START+RAM_LENGTH;)
             move.l    #0xe000,a5    *## A5 holds the constant E000h (RAM_START from hdr)
L113:        move.l    a5,d7         *## Also put into D7
             cmpi.l    #0x10000,d7   *## Passed the end of ROM (i.e. > FFFFh?)
             bcc.s     L123          *## IF yes THEN finish
* 101    {
* 102        temp = *address;                /* Get ith memory byte                      */
             move.b    (a5),d4       *## Get byte pointed to by address in safe keeping
* 103        *address = 0xAA;                /* Send out 10101010b to it                 */
             move.b    #0xaa,(a5)    *## by adding the constant E000h (RAM_START) to i
* 104        if(*address != 0xAA)
             move.b    (a5),d7       *## Get new contents of RAM
             cmp.b     #0xaa,d7      *## Is it 10101010b?
             beq.s     L133          *## IF not THEN there's something wrong, ELSE moveon
* 105        {   /* IF not this value THEN signal failure by sending out 11111111b */
* 106            *z = 0xFF;
             move.l    _z,a1         *## Point A1 to Z port
             move.b    #0xff,(a1)    *## Make it 11111111b to signal an error
* 107            break;
             bra.s     L123          *## and exit the loop
* 108        }
* 109        *address++ = temp;              /* Restore original value                   */
L133:        move.b    d4,(a5)+      *## IF ok, restore old RAM byte & move address on
* 110    }
             bra.s     L113          *## and repeat test on next RAM byte
L123:        movem.l   (sp)+,d5/d4/a5
             rts
* 111  }
* 112
* 113  void ROM_test(void)
* 114  {
             .even
_ROM_test:   movem.l   d5/a5,-(sp)
* 115    register unsigned short * address;   /* Address of EPROM word                   */
* 116    register unsigned short sum=0;
             clr.w     d5            *## D5.W used for sum
* 117    *z = 0;                              /* Set digital output to all zeros to signal pass */
             move.l    _z,a1         *## Send out 00000000b to Z port to signal ok
             clr.b     (a1)
* 118    for(address=ROM_START; address<ROM_START+ROM_LENGTH; SUM+= *address++)   {;}
             suba.l    a5,a5         *## Funny clear of A5, ROM_START is 0 in this case
L143:        move.l    a5,d7         *## address moved to D7 for compare (why?)
             cmpi.l    #0x2000,d7    *## Gone over the top of ROM?
             bcc.s     L153          *## IF yes THEN exit for loop
L163:        add.w     (a5)+,d5      *## ELSE add word to sum
             bra.s     L143          *## and do again
* 119    if(sum) {*z = 0xFF;} /* IF a non-zero sum, signal error by digital o/p = 10101010b */
L114:        tst.w     d5            *## Is sum zero?
             beq.s     L104          *## IF yes THEN no problem
             move.l    _z,a1         *## ELSE send out 11111111b to Z port
             move.b    #0xff,(a1)    *## to signal an error has occurred
L104:        movem.l   (sp)+,d5/a5
             rts
* 120  }
             .globl    _update, _ROM_test, _RAM_test, _input_test, _output_test, _main
             .globl    _diagnostic, _diag_port, _a_d, _z, _y, _x
             .bss
             .even
_Oldest:     .         = .+1
             .globl    _Oldest
             .even
_Array:      .         = .+256
             .globl    _Array
```

(b) Coding for `diagnostic()` and supporting functions.

Table 15.7: An alternative RAM testing module for the 6809 system.

```
void RAM_test(void)
{
_asm("        .define RAM_START = 0, RAM_LENGTH = 800h\n");
_asm("        ldy     #0              ; i held in Y, =0\n");
_asm("        ldb     #10101010b      ; Test pattern in B\n");
_asm("        clr     _z              ; Send out all zeros to digital port to signal ok\n");
_asm("RLOOP:  lda     RAM_START,y     ; Get mem byte @ RAM_START+i (RAM_START from header)\n");
_asm("        stb     RAM_START,y     ; Put pattern back out to same location\n");
_asm("        cmpb    RAM_START,y     ; Did it get there?\n");
_asm("        bne     ERROR           ; IF not THEN break to error handler\n");
_asm("        sta     RAM_START,y     ; ELSE put byte back\n");
_asm("        leay    1,y             ; i++\n");
_asm("        cmpy    #RAM_LENGTH+1   ; Finished yet?\n");
_asm("        bne     RLOOP           ; IF not THEN test next byte\n");
_asm("        rts\n");
_asm("ERROR:  ldb     #11111111b      ; Put out the error code\n");
_asm("        stb     _z\n");
}                                     ; Exit via }'s RTS
```

15.3 In-Circuit Emulation

The simulation techniques explored in Section 15.1 offer a low-cost solution to software debugging. Similarly, the techniques covered in Section 15.2 give an inexpensive approach to verifying the target hardware. Testing the interaction of these two components is the subject of this section.

The low-cost approach to this problem is to take the hex file produced by the compiler and program an EPROM. With this firmware in situ, the operation of the system can be monitored using the normal hardware tools. A variation of this technique uses a **ROMulator**. This is a RAM pack with a flying lead and DIL plug masquerading as an EPROM. Machine code can be downloaded into the ROMulator which is plugged into the target EPROM socket. Such software is easier to change than firmware and some monitoring of target variables is possible.

Where a more extensive examination of both hardware and software is necessary, then an **in-circuit emulator (ICE)** is required [11]. An ICE is a microprocessor-based product which exercises the target hardware under the control of a microprocessor development system. A typical configuration is shown in Fig. 15.9. Here the ICE replaces the target microprocessor via an umbilical cord and plug. The ICE hosts the same processor as the target, often piggybacked onto the umbilical plug, to be as close to the target as possible. This slave (i.e. target) processor is controlled by the ICE master microprocessor, which also communicates with a computer via a serial link. Thus, typically the target MPU might be a 68008, with a Z80 master and 8086-based computer!

Many different configurations are possible. Historically the Intel corporation invented the ICE in 1975 as part of their development system for the 8080 MPU. The ICE-80 was a plug-in card to the Intel Microprocessor Development System (MDS) bus. An 8080 processor both emulated, controlled and communicated with

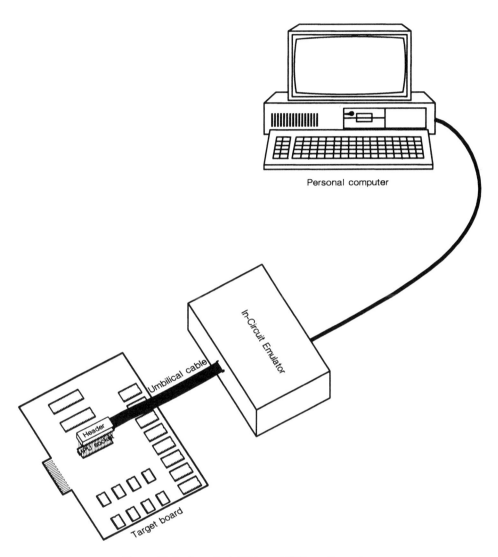

Figure 15.9: A typical PC-based ICE configuration.

the user. Most manufacturers followed with their own version, such as Motorola's EXORmacs MDS. Some of the large test equipment manufacturers, notably Tektronix and Hewlett Packard, developed a general purpose MDS, not tied to one specific manufacturer's product [12]. Here the ICE could be altered by changing the plug-in board, pod and associated software.

With the rise in popularity of the personal computer, the stand-alone configuration of Fig. 15.9 has become popular. The user interface can be anything from a dumb VDU terminal to a workstation or minicomputer. Firmware in the ICE itself communicates with this terminal and is used by the master processor. Generally, changing the target processor involves changing one or more of; the pod, firmware, ICE-board, terminal software.

Although most stand-alone ICEs will operate with a dumb terminal host, the internal ROM-based ICE commands are very basic and elementary. Using an intelligent terminal, such as a personal computer, allows a much more powerful and user-friendly software interface to insulate the user from the complexities of the ICE hardware. Aids such as menus and helpful prompts are useful to novice users. As with other software aids, the protocols and commands available are very product dependent, doubly so here as both hardware and software are involved. The following examples use the Noral SDT[1] product [13], but the facilities available are similar to most products [11].

All ICEs permit shadowing of the target's memory map. Thus memory is available to the slave emulator MPU on-board the ICE. As seen by this slave, its memory map can be set in chunks between local internal memory (known as overlay) or the target. As an example, consider a target with ROM between 3000 and 3FFFh, 6000 and 7FFFh, and E000 and FFFFh. The rest of its memory space is occupied by RAM or memory-mapped peripherals. Normally on power-up all memory is mapped to the target of type read/write. To 'move' the three ROM areas into the internal overlay ICE memory, use the MMO (Memory Map to Overlay) commands thus:

MMO 3000, 3FFF, P
MMO 6000, 7FFF, P
MMO E000, FFFF, P

where P stands for write Protected (an error will be printed if the software attempts to write to any of this overlay memory, that is simulating ROM). After this is done, the memory map is displayed as shown in Table 15.8. Sixteen blocks can be allocated in this manner, in minimum increments of 4 kbytes. Also shown in the listing is a memory test of the target RAM lying between C000 and CFFFh (MT is Memory Test). This writes $10101010b$ and $01010101b$ into all locations as specified. More sophisticated tests are available.

Using this overlay memory technique, resources may be gradually switched from the ICE to the prototype system. Thus a target A/D converter can be initially mapped to an overlay RAM location and used to test the software. The real peripheral can then be exercised by switching from overlay to target. Some ICEs even provide an optional local clock, which may be used instead of the target clock.

Facilities typically provided by an ICE are:

File handling
Downloading machine-code and symbol files into memory.

Register and memory examine/change
To examine and change any microprocessor register, overlay or target memory location.

---

[1]Noral Microelectronics, Logic House, Gate St., Blackburn, Lancs, BB1 3AQ, UK.

```
> mmo 03000h,03fffh,p                              <- Map Memory from
> mmo 06000h,07fffh,p                              <- target to
> mmo 0e000h,0ffffh                                <- Overlay
> mm                                               <- Memory Map disp
Ref  Type       Start Addr   End Addr   Mapping attribute
___  ____       _____   _____   _____

2    Target     000000       002FFF     Read / Write
3    Overlay    003000       003FFF     Overlay Read only (Write protected)
4    Target     004000       005FFF     Read / Write
5    Overlay    006000       007FFF     Overlay Read only (Write protected)
6    Target     008000       00DFFF     Read / Write
7    Overlay    00E000       00FFFF     Overlay Read / Write
1    Target     010000       0FFFFF     Read / Write
Free overlay RAM = 108K bytes    Mapping resolution = 4K bytes
>
> mt 0b000h,0bfffh,1                               <- Memory Test 1
Memory test complete (4096 tests) : 0 failure(s)
>
> m 0b000h,0b010h                                  <- display Memory
Address    0  1  2  3  4  5  6  7  8  9  A  B  C  D  E  F
_____    __ __ __ __ __ __ __ __ __ __ __ __ __ __ __ __

0B000      AA AA AA AA AA AA AA AA AA AA AA AA AA AA AA AA    ................
0B010      AA AA AA AA AA AA AA AA AA AA AA AA AA AA AA AA    ................

> ld a:list16_2.nor inthex                         <- LoaD program file
> $s                                               <- list Symbols

*** Module Symbols for : MAIN
00000400   MAIN
0000040C   CLOOP
00000414   FOREVER
0000041A   DISPLOOP
00000436   FLYBACK
0000044E   DELAY_LOOP
0000045A   UPDATE
00002000   ANALOG_X
00002001   ANALOG_Y
00006000   ANINPUT
0000A000   Z_BLANK
0000E000   Array
>
> rw pc 0400h                                      <- Reg Write to PC
>
> r                                                <- Reg display
D0=00000001  D1=00000000  D2=00000000  D3=00000000  D4=00000000  D5=00000000
D6=00000000  D7=0000FFFF  A0=00000000  A1=00000000  A2=00000000  A3=00006000
A4=00006001  A5=00000000  A6=00000000  PC=00000400  SSP=00010000  USP=00000000
SR= T:0  S:0  III:000  X:0  N:0  Z:0  V:1  C:1
```

Table 15.8: Memory Mapping and Testing.

Step execution

To execute the directed software in the target environment step by step, usually displaying registers and other information after each step.

Breakpoints

Insertion of conditions, which may be software and/or external hardware signals, to halt execution.

Execute
Full speed execution until a breakpoint is reached.

Trace analysis
This can be either a software or real-time trace. In the case of the latter the system runs to a breakpoint. At this point the contents of a display buffer can be read both before and after this event. The state of various external signals can be displayed as well as address, data and control bus signals. Unlike a software trace, this data is acquired in real time and only displayed when execution has terminated.

Most of the facilities described above are the same as those listed for software simulation in Section 15.1. However, in this case the software is being run in its real hardware environment using a real microprocessor, possibly in real time. Trace analysis is different, however, in that a 'snapshot' of bus cycles and bus activity can be obtained. This logic analysis feature is usually rather limited. Instead an external logic analyzer can be used, triggered by the ICE itself when a breakpoint is reached.

To illustrate some of these points, consider the example shown in Fig. 15.9. The source for the program is shown in the central window. The smaller window above this allows us to monitor selected memory locations or blocks as we step through the program. The **MONM** (MONitor Memory) commands setting this up is shown in the command area at the bottom of the page. The register window at the top right shows the state of the MPU's registers. The **Supervisor Stack Pointer (SSP)** has been set to $10000h$ by using the **RW** (Register Write) command. Below this is the state of the System stack. This is useful to examine parameters passed to the function or subroutine through the stack and monitoring frame data.

Clicking a mouse on the [S] or [GO] boxes causes the program to Step or GO and execute as appropriate. In the Step mode the line of code being executed is highlighted on the screen.

Although the data presented in Fig. 15.9 looks similar to that of Table 15.1, remember that the latter is a pure simulation whilst the former is running on an actual 68008 microprocessor.

High-level ICE driven packages are now becoming available, which have the same relationship as the low and high-level simulators discussed in Section 15.1. Some of these are extensions of existing simulation products which makes moving between a simulation and emulation environment easier.

Although an in-circuit emulator is versatile, it is expensive (typically $7000+), relatively bulky and fragile. They can also be cantankerous! Thus it makes sense to use a simulator at the outset to check out the purely software aspects of the project. If testability has been incorporated into the hardware, as described in Section 15.2, then the ICE can be left for the final phases of the testing and 'tough nut' servicing situations.

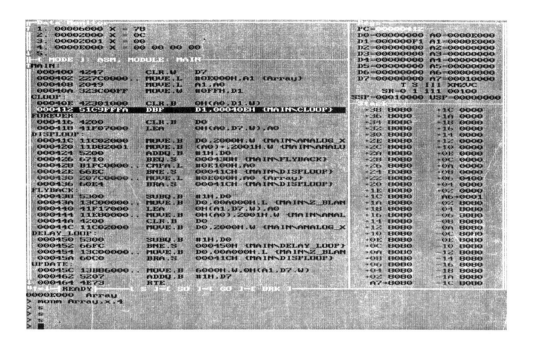

Table 15.9: A window into the hardware using an ICE.

References

[1] Wakerly, J.F.; *Microcomputer Architecture and Programming*, Wiley, 1989, Section 13.1.

[2] Atherton, W.A.; Pioneers: Grace M. Hopper, *Electronics World + Wireless World* (UK), **95**, no. 1646, Dec. 1989, pp. 1192 and 1194.

[3] MacClean, A.; The Great C Debugger Review, *.EXE* (UK), **1**, no. 9, March 1988,

pp. 12–25.

[4] Adams, M.; *Development without Development Systems*, from *Microprocessor Development and Development Systems*, ed. Tseng, V., Granada (UK), 1982, Chapter 8.

[5] Adams, M.; C, 68000 assembler and the IBM PC, *.EXE* (UK), 1, no. 9, March 1988, pp. 26–30.

[6] Ferguson, J.; *Microprocessor Systems Engineering*, Addison-Wesley, 1985, Section 8.3.

[7] Wilcox, A.D.; Bringing up the 68000 – A First step, *Dr Dobb's Journal*, 11, Jan. 1986, pp. 33–40.

[8] Stockton, J and Scherer, V.; Learn the Timing and Interfacing of MC68000 Peripheral Circuits, *Electronic Design*, 27, no. 26, Nov. 8, 1979, pp. 58–64.

[9] Ferguson, J.; *Microprocessor Systems Engineering*, Addison-Wesley, 1985, Section 4.2.

[10] Gilmour, P.S.; Caveat Tester, *Embedded Systems Programming*, 4, no. 7, July 1991, pp. 58–65.

[11] Ferguson, J.; In-Circuit Emulation, *Wireless World* (UK), 84, no. 1580, June 1984, pp. 53–55.

[12] Lejeuine, B.; *In-Circuit Emulation*, in *Microprocessor and Microprocessor Development Systems*, ed. Tseng, V., Granada (UK), 1982, Chapter 7.

[13] Ferguson, J.; *Microprocessor Systems Engineering*, Addison-Wesley, 1985, Section 5.1.

CHAPTER 16

C'est la Fin

Having designed and tested our project it only remains to wrap up by doing a comparative analysis of the various implementations and giving some suggestions on how the basic specification can be extended.

16.1 Results

One of the first questions asked is how will a C-coded program compare with its assembly-level equivalent? To try and answer this question I have coded both our systems at assembly level, so that we can contrast the two approaches. In defence of the expected outcome, it should be pointed out that small routines, especially those that intimately interact with hardware, are the forte of assembly-level code and the antithesis of high-level languages. Thus our results will be at the far end of the spectrum; however, this will at least give us a worst-case yardstick to balance the pros and cons of the two approaches.

Our first demonstration is the 6809-based coding of Table 16.1. This is structured after the C-level coding of Tables 14.3 and 14.4. Like the C program, the variables Oldest and Array[] are stored in absolute memory locations and so are globally known to both the routines MAIN and UPDATE.

At the beginning of the scan (lines 18–29) the address of the leftmost Y co-ordinate, Array[Oldest], is calculated and placed in Index register_X. This calculation is done in lines 18–21 by expanding out the 8-bit Oldest index, pointing X to Array[0] and using the instruction LEAX D,X to put the effective address Oldest + Array back in X.

The main scan routine simply uses the Post-Increment Index address mode automatically to advance this pointer once each time Array[i] is fetched, preparatory to sending it out to the Y plates. The stratagem of keeping the X count in Accumulator_A and fetching Array[i] down to Accumulator_B, means that both X and Y co-ordinates can be output together using a single Store Double instruction (line 35).

In incrementing the Array[] pointer, a check is made to detect the situation when its value reaches Array[256] (Array+256) and to reset back to the beginning Array[0]. This gives a pseudo circular structure. Thus if Oldest were 40h, then

Table 16.1: A 6809-based assembly-level coding.

```
 1                          .processor m6809
 2                     ;**********************************************************
 3                     ;* This is the background routine which spends its time sequentially *
 4                     ;* going through the 256 array bytes, sending out the value to the   *
 5                     ;* Y plates whilst ramping up the X plates                           *
 6                     ;**********************************************************
 7                          .define  ANALOG = 2000h, ANINPUT = 6000h, Z_BLANK = 0A000h
 8                          .psect   _text
 9  E000  10CE0400 MAIN:    lds      #400h          ; Set up stack top
10  E004  7F0000            clr      Oldest         ; Start new index at start of array
11                     ; for (i=0; i<256; i++) Clear the array
12  E007  8E0003            ldx      #Array         ; Point to bottom of array
13  E00A  6F80     CLOOP:   clr      ,x+            ; Clear Array[i], i++
14  E00C  8C0103            cmpx     #Array+256     ; Over the top yet?
15  E00F  26F9              bne      CLOOP          ; IF not THEN again
16                     ; while(1) Forever display data to oscilloscope
17                     ; First calculate the leftmost array element address as Array+Oldest
18  E011  4F       FOREVER: clra                    ; Keep X count in Acc.A (=00h)
19  E012  F60000            ldb      Oldest         ; The leftmost array index
20  E015  8E0003            ldx      #Array         ; Point IX to bottom of array
21  E018  308B              leax     d,x            ; Oldest+Array=leftmost address
22                     ; Now begin the scan, using X as a pointer to Array[i]
23  E01A  E680     DISPLOOP: ldb     0,x+           ; Get Array[i], i++
24  E01C  FD2000            std      ANALOG         ; Send out X and Y points together
25  E01F  8C0101            cmpx     #Array+256     ; Over the top?
26  E022  2603              bne      CONTINUE       ; IF not THEN continue
27  E024  8E0001            ldx      #Array         ; ELSE point back to the beginning
28  E027  4C       CONTINUE: inca                   ; Increment X count
29  E028  26F0              bne      DISPLOOP       ; IF not back to zero THEN again
30                     ; Flyback
31  E02A  4A                deca                    ; Make Acc.A = FFh
32  E02B  B7A000            sta      Z_BLANK        ; Send out to Z port to blank beam
33  E02E  4F                clra                    ; Prepare to return beam to leftside
34  E02F  E684              ldb      0,x            ; X back round to the start again
35  E031  FD2000            std      ANALOG
36  E034  4A       DELAY_LOOP: deca                 ; 1 ms minimal delay
37  E035  26FD              bne      DELAY_LOOP
38  E037  B7A000            sta      Z_BLANK        ; Send out 00000000 to enable beam
39  E03A  20D5              bra      FOREVER        ; and repeat scan
40
41                     ;**********************************************************
42                     ;* This is the NMI interrupt service routine which puts the          *
43                     ;* analog sample in the array and updates the Oldest index           *
44                     ;* ENTRY : Via NMI. Array[] and Oldest are global                    *
45                     ;* EXIT  : Value at ANALOG in Array[Oldest], Oldest inc'ed mod-256   *
46                     ;**********************************************************
47                     ; Array[Oldest++]=*a_d
48  E03C  4F       UPDATE:  clra                    ; Prepare to promote Oldest
49  E03D  F60000            ldb      Oldest         ; to a 16-bit quantity
50  E040  8E0003            ldx      #Array         ; Point IX to bottom of array
51  E043  308B              leax     d,x            ; Address of Array[Oldest] in IX
52  E045  B66000            lda      ANINPUT        ; Read analog sample
53  E048  A784              sta      0,x            ; Put it out to Array[Oldest]
54  E04A  7C0000            inc      Oldest         ; Oldest++, mod-256
55  E04D  3B                rti                     ; and exit
56
57                     ;**********************************************************
58                     ;* The Reset and NMI vectors, assuming a 2716 EPROM from E000-E7ffh  *
59                     ;**********************************************************
60                          .org     MAIN+7fch      ; Takes us up to E7fch, NMI vector
61  E7FC  E03C              .word    UPDATE         ; UPDATE is the start address of ISR
62  E7FE  E000              .word    MAIN           ; Reset address
63
64                     ;**********************************************************
65                     ;* RAM-based variables                                               *
66                     ;**********************************************************
67                          .psect   _data
68  0000     Oldest:        .byte    [1]            ; Reserve one byte for Oldest
69  0001     Array:         .byte    [256]          ; and 256 bytes for the array
70                          .public  MAIN, CLOOP, FOREVER, DISPLOOP, CONTINUE
71                          .public  DELAY_LOOP, UPDATE, Oldest, Leftmost, Array
72                          .public  ANALOG, ANINPUT, Z_BLANK
73                          .end
```

the leftmost value (X = 00h) would be `Array[40]`, and when X reached BFh the Y point would be `Array[255]`. The next point at X = C0h should be `Array[0]`.

The NMI ISR UPDATE simply computes the address of `Array[Oldest]` in the same way as the leftmost point was calculated, fetches the analog sample down into this element and increments the index Oldest with wrap around from FFh to 00h. As the NMI interrupt saves and restores all internal registers, there is no restriction on register usage.

The total length of the routine is 77 bytes plus vectors. The scan time for one screen of data, ignoring any interrupt service time, is 7.5 ms, giving a sweep rate of 133 Hz.

The 68008-based equivalent is shown in Table 16.2. Like the C program of Table 14.9, variables are preferentially located in registers, with only the global object `Array[256]` being located in memory. The X count is held in **D0.B** and the index to the oldest updated array element in **D7.B**. Address register A0 is used in the background program to point to the array element currently being fetched, whilst **A1** is a convenient way of holding the constant address of `Array[0]`.

On entry to the scan loop (lines 33–40), **A0** points to the leftmost array element to be displayed. After each point is displayed (lines 33–34) both the X count (in **D0.B**) and array pointer (**A0.L**) are incremented. In the case of the latter, wraparound occurs whenever the address reaches 256 above the array base (lines 37 and 39). This gives the necessary circular data structure.

During the flyback delay, the array pointer is reset to the new leftmost point in line 44 by using **D7.W** as an index (the oldest array element) with **A1.L** (pointing to the base of the array), that is $Y_{leftmost} = $ Array[Oldest]. As this index address mode uses a (sign-extended) word-sized index register (byte sized not allowed), **D7** was originally word-sized cleared in line 22 to ensure no non-zero bits in **D7[15:8]** will upset this calculation.

The level-7 ISR is called UPDATE, and simply uses **D7.W** (the oldest index) as an offset to **A1.L** (which is permanently pointing to the array base), to move the value from the A/D converter to `Array[Oldest]`. Adding one to **D7.B** ensures that this points to the array element furthest back in time on exit. The byte-sized increment automatically gives wraparound. As none of the registers are saved or retrieved by a 68000 MPU interrupt, using **D7** and **A1** as global register variables is legitimate.

The program totals 102 bytes excluding vectors and takes 5.9 ms for one screen's worth of data, ignoring any interrupt service time. This gives a sweep rate of 169 Hz.

The final figures then show a size factor of 2.4 for the 6809-based circuit and a speed factor of 2.7. The 68008 has a closer size factor of 1.7 together with a speed factor of 2.9. If we treat these figures as a worst-case scenario, then for realistic situations these factors are likely to be of the order of 1.5 at best; that is a C coding will have around 50% more code and be 50% slower than an equivalent assembly-level implementation. Against this must be ranged the high-level code advantages of cost, portability and reliability.

Figure 16.1 shows a typical set of X and Y traces captured on a Hewlett Packard 54501A digitizing oscilloscope. The upper trace shows the contents of the 256-byte array covering approximately two seconds. The bottom trace shows the X sweep.

Table 16.2: A 68008-based assembly-level coding.

```
  1                             .processor m68000
  2                   ;***********************************************************************
  3                   ;* This is the background routine which spends its time sequentially  *
  4                   ;* going through the 256 array bytes, sending out the value to the    *
  5                   ;* Y plates whilst ramping up the X plates                            *
  6                   ;* D0 holds X count, D7 holds Oldest                                  *
  7                   ;* A0 points to the leftmost element, A1 to the base of the array    *
  8                   ;***********************************************************************
  9                             .define ANALOG_X = 2000h, ANALOG_Y = 2001h, ANINPUT = 6000h,
 10                                     Z_BLANK = 0A000h
 11                             .psect _text
 12
 13                   ; The vector table
 14 000000 00010000   VECTOR:   .double 10000h           ; Initial value of Supervisor SP
 15 000004 00000400             .double MAIN             ; and of the PC
 16 000008 00000000             .double [29]             ; skip until
 17 00007C 0000045C             .double UPDATE           ; Level-7 vector
 18 000080 00000000             .double [224]            ; Skip until start of program @0400h
 19
 20                   ; Program proper starts here
 21 000400 4247       MAIN:     clr.w   d7               ; Start Oldest index as zero
 22 000402 227C0000E000         movea.l #Array,a1        ; The constant address of array start
 23 000408 2049                 movea.l a1,a0            ; is the leftmost point first time in
 24                   ; for (i=255; i>-1; i--) Clear the array
 25 00040A 323C00FF             move.w  #255,d1          ; Use D1 as a loop count i
 26 00040E 42301000   CLOOP:    clr.b   0(a0,d1.w)       ; Clear Array[i]
 27 000412 51C9FFFA             dbf     d1,CLOOP         ; IF not THEN again
 28                   ; while(1) Forever display data to oscilloscope
 29                   ; First calculate the leftmost array element address as Array+Oldest
 30 000416 4200       FOREVER:  clr.b   d0               ; X-count = 00h
 31 000418 41F07000             lea     0(a0,d7.w),a0    ; A0 holds leftmost element address
 32                   ; Now begin the scan with A0 pointing to Array[i]
 33 00041C 11C02000   DISPLOOP: move.b  d0,ANALOG_X      ; Send out X-co-ordinate to screen
 34 000420 11D82001             move.b  (a0)+,ANALOG_Y   ; & Y-co-ord; inc'ed array pointer
 35 000424 5200                 addq.b  #1,d0            ; Increment X-co-ordinate
 36 000426 6710                 beq     FLYBACK          ; IF back to zero THEN scan finished
 37 000428 B1FC0000E100         cmpa.l  #Array+256,a0    ; A0 over the array top?
 38 00042E 66EC                 bne     DISPLOOP         ; IF not THEN next point
 39 000430 207C0000E000         movea.l #Array,a0        ; ELSE go round to the first element
 40 000436 60E4                 bra     DISPLOOP         ; and go again
 41                   ; Flyback
 42 000438 5300       FLYBACK:  subq.b  #1,d0            ; Make D0.b = FFh
 43 00043A 13C00000A000         move.b  d0,Z_BLANK       ; Send out to Z port to blank screen
 44 000440 41F17000             lea     0(a1,d7.w),a0    ; A1+D7 is leftmost address, -> A0
 45 000444 11E800002001         move.b  0(a0),ANALOG_Y   ; Send it out to the Y-plates
 46 00044A 4200                 clr.b   d0               ; X-co-ordinate is zero
 47 00044C 11C02000             move.b  d0,ANALOG_X      ; at left side of screen
 48 000450 5300       DELAY_LOOP: subq.b #1,d0           ; 1 ms nominal delay
 49 000452 66FC                 bne     DELAY_LOOP
 50 000454 13C00000A000         move.b  d0,Z_BLANK       ; Send out 00000000 to enable beam
 51 00045A 60C0                 bra     DISPLOOP         ; and repeat scan
 52
 53                   ;***********************************************************************
 54                   ;* This is the level-7 interrupt service routine which puts the       *
 55                   ;* analog sample in the array and updates the Oldest index            *
 56                   ;* ENTRY : Via INT-7. Array[] is global and D7.B holds Oldest pointer*
 57                   ;* ENTRY : A1 points to the array bottom Array[0]                    *
 58                   ;* EXIT  : Value at ANALOG in Array[Oldest], Oldest inced mod-256    *
 59                   ;***********************************************************************
 60                   ; Array[Oldest++]=*a_d
 61                   ; Overwrite Array[Oldest] with latest data from A/D converter
 62 00045C 13B860007000 UPDATE: move.b  ANINPUT,0(a1,d7.w); Send sample to Array[Oldest]
 63 000462 5207                 addq.b  #1,d7            ; Increment Oldest index modulo-256
 64 000464 4E73                 rte
 65
 66                   ;***********************************************************************
 67                   ;* RAM-based variables                                                 *
 68                   ;***********************************************************************
 69                             .psect _data
 70 00E000       Array:          .byte   [256]           ; 256 bytes for the array
 71                             .public MAIN, CLOOP, FOREVER, DISPLOOP, DELAY_LOOP, UPDATE
 72                             .public FLYBACK, Array, ANALOG_X, ANALOG_Y, ANINPUT, Z_BLANK
 73                             .end
```

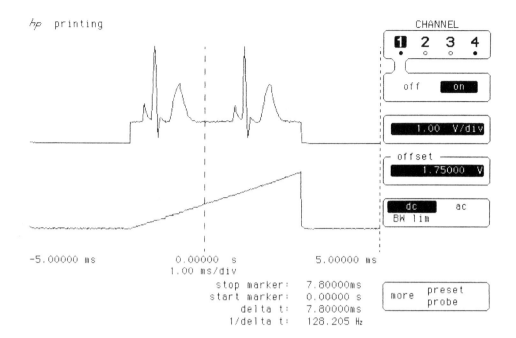

Figure 16.1: Typical X and Y waveforms, showing two ECG traces covering 2 s.

The flyback blank has been deliberately increased to give a refresh rate of 50 Hz.

16.2 More Ideas

The time-compressed memory project used to illustrate the use of a high-level language for an embedded target originated as part of a commercial project, but has been suitably watered down to avoid the perennial problem of 'not being able to see the wood for the trees'. That being so, there is plenty of scope for a more ambitious project based on this basic core. In this section a few ideas are presented which should help fertilize the reader's mind in planning any further work.

One relatively simple extension involving only a software change is to increase the amount of data presented on the screen. Four minutes worth can be displayed by using two traces, which are scanned in succession.

The overall double scan of the complete four minutes, stored in a 512-byte array, will still have to be accomplished in 20 ms (10 ms per trace) to give a flicker-free display.

The apparently two separate traces can be simulated by reducing the vertical resolution to seven bits. The top trace is displayed with a MSB of 1, whilst the bottom 256 data bytes have a MSB of 0. The Y-output D/A converter will then bias the first 256 data bytes by $\frac{1}{2}$ scale. Of course the data bytes must first be logic shifted once right (divided by two) before the MSB is tampered with.

As the 512 data points need to be displayed in the time previously required for 256 points, the processor will have to work twice as hard. If the software is coded in C then it is doubtful if there is sufficient reserve. Renovating the circuit to use a 2 MHz 68B09 (with faster EPROM) or a 12.5 MHz 68000 MPU will provide the additional horsepower if this is the case. However, this is probably a good case for using an assembly-level coding. It does work!

Using a bidirectional X-sweep would slightly reduce the scan time. By displaying, say, the bottom trace from left to right and then returning along the top trace right to left, the flyback and Z-blank delays are eliminated. Using a triangular instead of sawtooth timebase is a standard technique implemented by printers. It is feasible to use this bidirectional scan for the single trace of the basic project, but as the traces will be superimposed, the oscilloscope must not exhibit appreciable hysteresis, in my experience a problem with low-cost oscilloscopes.

Continuing with this theme, the freeze facility can be applied to only one of these traces, say, the upper, whilst the lower continues on as normal.

Another approach to displaying additional data is to use hidden pages with only one or two on-screen traces. Thus, for example, we could store all data for the last 32 minutes in a 32 kbyte RAM, but only display the last two minutes worth. However any of the 2-minute pages from time past could be displayed as commanded using the setting of the switch port. Furthermore a hard-copy routine could be written to dump the entire 32-minutes worth sequentially to a chart recorder or graphics printer, or even uploaded to a PC for further analysis. The option of invisibly acquiring data as this process is in progress, by using a shadow RAM, is useful. Indeed this option is also a possibility for the freeze process in the main project. Thus the freeze command makes the display static, but data continues to be acquired 'behind the scenes'.

Depending on the quality of the analog data, it may be desirable to apply a simple digital smoothing routine at the output (e.g. see the 3-point filter of page 250). Regardless of such processes, an 8-bit quantized system will look rather granular in hard copy. With a 12-bit A/D converter, the number of quantization levels increases from 256 to 4096. Unfortunately this will require a similar enhancement of RAM capacity from 256 byte-sized elements to 4096 word-sized elements for each 2-minute slot, a 32-fold increase!

Displaying a 4096 element data array in 20 ms is well above the capabilities of any of the processors used in this text. Rather, every other 16th element could be used for a conventional 8-bit oscilloscope display and the full resolution reserved for the hard-copy or uploaded version, where time is not an issue. Even so, the extra overhead of a ×16 interrupt rate, reading and writing 12-bit quantities over an 8-bit bus (e.g. see Fig. 6.1) and extracting one in sixteen bytes is onerous.

With the processor pretty well spending its entire time displaying the waveform, there is no spare capacity available to analyze the data. A second processor running in parallel with the display processor would enable both functions to be carried out in real time. Data acquired by the master processor could be sent to the slave by writing the latest sample to an output port, then interrupting the slave which reads this as an input port. Of course the option of uploading, say, via the serial link, is a viable alternative if the data rate is not too high.

Analysis tasks include detecting waveform peaks and calculating beat rates and beat-to-beat variations. A separate display of the appropriate data could be maintained by the slave or multiplexed on to the primary display. This display device need not be CRO-based, a liquid crystal panel is a viable alternative, and will probably contain its own microprocessor.

Appendix A

Acronyms and Abbreviations

A	Accumulator_A
A/D	Analog to Digital converter
B	Accumulator_B
BIOS	Basic Input/Output System
CCR	Code Condition Register
D	Accumulator_D
D/A	Digital to Analog converter
DMA	Direct Memory Access
ea	Effective Address
ECG	Electro-CardioGram trace
EKG	See ECG
EPROM	Erasable Programmable Read-Only Memory
ICE	In-Circuit Emulator
I/O	Input/Output
ISR	Interrupt Service Routine
K	$1024 = 2^{10}$
LSB	Least Significant Bit or Byte
M	$1,048,576 = 2^{20}$
MDS	Microprocessor Development System
MSB	Most Significant Bit or Byte
MSDOS	MicroSoft Disk Operating System
Op-code	Operation code
OS	Operating System
PC	Personal Computer
PC	Program Counter
PIA	Peripheral Interface adapter
PI/T	Parallel Interface Timer
RAM	Random Access Memory
ROM	Read-Only Memory
S	System Stack Pointer register
SBC	Single-Board Computer
S/N	Signal to Noise ratio
S/H	Sample and Hold
SP	Stack Pointer register
SSP	System Stack Pointer register
TOF	Top Of Frame
TOS	Top Of Stack
TTL	Transistor Transistor Logic
U	User Stack Pointer register
USP	User Stack Pointer register
VDU	Visual Display Unit
X	Index register X
Y	Index register Y

Index

Emphasized page numbers indicate major entries.

2's complement, 209, 220, *337*, 347

Access time, *16*
Accumulator
    A, *3*
    B, *3*, 22
    D, *5*, 22
Address, *see also* Pointer, *65*
    Comparison, 135
    Even, 65, 67, 270, 380, 406
    Odd, 65, 355
    Segmented, 178
    Sign extension of, *89*
Address bus, *6*, 11, *13*, 57, 62, *65*, *66*, 405
Address decoder, *13*, 16, 70, 77, *351*, 355, 358, 359
Address mode-6809
    Absolute, 5, *34*, 36
    Absolute indirect, *36*, *38*, 373
    Direct (short), *5*, *36*, 281, 301
    Extended (long), *5*, *36*
    Immediate, *36*
    Indexed, *37*
        Accumulator offset, *39*, 156
        Constant offset, *5*, *38*
        Post-increment, *38*, 41, 48, 56, 423
        Pre-decrement, *38*, 41, 140
        Program Counter offset, *40*, 51
    Inherent, *34*, *35*
    Register direct, *35*
    Relative, *37*, *40*
Address mode-68000
    Absolute, 108, *110*
        Long, *110*
        Short, *110*, 389
    Address register
        Direct, *110*
        Indirect, *111*, 246
        Indirect with displacement, *111*
        Indirect with Index, 63, *112*
        Post-increment, 91, 94, 101, 109, *112*, 114, 119, 126, 254
        Pre-decrement, 91, *112*, 114, 246
    Alterable, *115*
    Data alterable, *115*
    Data register
        Direct, *110*
    Immediate, *109*
    Inherent, *109*
    Program Counter
        Indirect with Displacement, *113*
        Indirect with Index, *114*
    Quick Immediate, *109*
Address register, *5*, *62*, 94, 110
    A7, *61*, 63, 91, 126
    Clearing, *95*
    Operations on, *89*, *94*
    Sign extension of, *94*, 100, 110
Analog to digital converter (A/D), 138, 428
Arithmetic logic unit (ALU), *3*, 58, 385
Array, 5, 35, 37, 65, *108*, 154
    Addition of, 112
    Filtering, *112*
    in C, 236, *249-62*
        2-dimensional, 224
        Definition of, *250*
        Initialization of, *259*
        Multi-dimensional, *251*, 254, *257*
    of functions, 286
    of structures, *265*, 275
    Overrun, *255*
    Pointer to, *255*
    `sizeof`, 223, *250*
    Multiplication of, *39*
    Passing to subroutine, 135
    Pointer to, *37*

sizeof, *256*
Walking through, 196, 254, 376, 389, 412
ASCII code, 48, 208, 228
   BEL, 292
   BS, 292
   CR, 292
   FF, 292
   HT, 292
   LF, *291*
   NULL, 48, 119
   VT, 292
Assembler, 20, 30, 38, 42, 116, *171–92*
   2-pass, 178
   Absolute, 45, *173–80*
   Comments, *43*, 209
   Cross, *190*
   Directive, *42*, 173
     .bss, *380*
     .byte, *40*, *45*, 212, 259
     .data, *210*
     dc, *182*, 212
     .define, *44*, 276
     .double, *45*, 56
     ds, *182*
     .end, *44*
     .endm, *179*
     equ, *182*, 276
     .even, *270*, 380
     .external, *155*, 248
     .globl, 248, 262, 380
     globl, 181
     ident, *182*
     .include, *44*, 276
     .macro, *179*
     opt, *182*
     .org, *45*, *173*, 178, 181
     .processor, *43*
     .psect, *44*, 155, 173, *181*, 301
     .public, *155*, *181*, 248
     sect, *182*
     .text, *45*, *210*
     .word, *45*, 56
     xdef, 181, *182*
     xref, *182*, 248
   Label, *see also* Symbols, *42*, 125, 178, 179, 181
   Relocatable, 45, 173, *180–92*
   Resident, *190*
Assembly-level language
   Incorporating with C, *287–92*

BCD code
   to ASCII, *221*
Binary, *see* Number base, Binary
Boolean expressions, *223*
Bootstrap program, *170*
Bus
   Address, *see* Address bus
   Control, *see* Control bus
   Data, *see* Data bus
   Double fault, *see* Double-bus fault
Bus Error, 69

C instruction, *see* Instruction, C
C constants
   Character ' ', 216
   Floating-point (F), *216*
   Hexadecimal 0x, 216
   Initialization of, *259*
   Integer, *215*, 291
   Long double floating-point (L), 216
   Long integer (L), 216
   Octal 0, 216
   Pointer-to type, *257*
   Promotion, *216*
   static const, *259*
   Unsigned integer (U), 216, 288
   Unsigned long integer (UL), 216
C directives
   #define, *276*, *278*
   @dir, *301*, 373
   #else, *279*
   #elseif, *279*
   #endif, *279*
   #error, *279*
   #if, *279*
   #ifndef, *279*
   #include, *279*, 280, 312
   $INTERRUPT, *294*
   @port, *295*, *369*, 381
   #pragma, *280*, 302
   #undef, *278*
C function, *see* Function
C objects, *see* Types
C operator, *see also* Instruction, C
   ! (bitwise NOT), 221
   ! (logic NOT), *222*
   ! (unary NOT), 215
   != (NOT equivalent), 224
   { (Begin statement), 204
   } (End statement), 204
   * (contents-of), 255, 258, *261*, 270
   * (multiplication of), 218

Index   433

* (pointer), 257
*/ (End comment), *204*
*= (compound multiplication assignment), 225
+, *193*
+ (addition), 218, 220
+ (unary positive), *220*
++, *193*
++ (increment), *221*, 229, 261, 278, 365
+= (compound addition assignment, 225
+= (compound addition assignment), *205*
, (concatenate), *225*, 230, 237
- (subtraction), 220
- (unary negative), 216, 220
-- (decrement), 205, *221*, 261, 278
. (structure member), *265*, *270*, *272*
/ (division), *224*
/* (Begin comment), *204*
/= (compound division assignment), *224*
; (statement terminator), 204, 215, *219*
= (assignment), 204, *223*
== (equivalent), *223*
?: (conditional assignment), *224*, 245
[] (array), *249*, 257
% (modulus), *224*
& (address-of), *255*, 270
& (bitwise AND), 214, *220*, 221
-> (structure pointer), *270-2*
<< (shift left), *221, 222*
>> (shift right), *221*, *222*, *225*, 342
&& (logic AND), *223*
> (greater than), *222*
^ (bitwise EOR), 221
~ (logic NOT), 219
|| (logic OR), *223*
| (bitwise OR), 221
Assignment
   l_value, *227*
   r_value, *227*, 363
Association of, 219, 220
Binary, *219*
Braces, *205*, 242
cast, *222*, 227, 256, 257, 272, 276, 291, 363
Compound assignment, *224*
Parentheses, *216*
Precedence of, *218*, 220
sizeof, *223*, 250, *256*, 270
typedef, *278*, *306*
Unary, *219*, 221, 261
C preprocessor, *276*
C qualifier

auto, *209*
const, *214-5*, 250, 251, *259*, 272, 298, 412
extern, 215, *248*, 279, 280, 288
register, *210*, 250, 255, 363, 364, 412
signed, *209*
static, 210
unsigned, *209*, 342
   char, 215
void, 256
volatile, *214-5*, 267, 363
C statement, *219*
   Compound, *205*, 210, *219*, 228
   Expression, *219*
   Null, 215, *225*, 230, 237
   Simple, *205*
C token
   %d, 305
   %f, 305
   %ld, 305
   %u, *305*
   %x, 305
   \a, 292
   \b, 292
   \f, 292
   \n, *291*, 305
   \r, 292
   \t, 291
   \v, 292
Checksum, 177, 411
Clock
   Wait states, 75, 77, 79, *81*
Co-processor, 207, 280, 295
   68020, 167
Code Condition Register (CCR)
   6809, *3*, 26
   68000, *58*, 63, 89, 94, 129
Comparison, *28*
   Signed, *28*
Compile time, *see* Load time
Compiler, 133, 135, 172, *192-200*, 209
   House rules, *288*, 292, 293, 380
   ROMable, *283*, *391*
Compiler dependency, *see* Implementation dependency
Constants in C, *see* C constants
Context switching, *149-51*, 292
   6800, 149
Control
   6800
      E, 71
      VMA, 71

6809
   BREQ, 351
   DMA/BREQ, *6, 8*
   E, *9*, 11, 17, 71, 405
   Halt, *6, 7*, 351
   MRDY, *10*, 72, 351
   Q, *9*, 11, 13, 16, 351, 355, 405
   R/$\overline{\text{W}}$, *7*, 11, 351, 405
   Reset, *8*, 351
68000
   AS, *66*, 75, 77, 407
   BERR, *69*, 161
   BG, 71
   BGACK, 71
   BR, 70
   DS, 407
   CLK, 71
   DTACK, *69, 70*, 72, 75, 77, 84, 160, 358, 407
   E, 71, 84
   Halt, *67, 69, 70, 162*, 355
   IPL, *69*
   LDS, *67*, 72, 79
   R/$\overline{\text{W}}$, *69*, 407
   Reset, *67, 162*, 355
   UDS, *67*, 72, 79, 81
   VMA, 71, *84*
   VPA, 71, *84*, 160, 355
68008
   AS, 67
   DS, 80, 84, 355, 359
   IPL, 355
   R/$\overline{\text{W}}$, 359
Control bus, 7
Control register
   PIA, 99, 103

Data bus, *6, 10*, 57, *66*, 405
   Synchronous, 13
Data register, *3, 62*
Debug
   Breakpoint, *392, 394*
   High-level, *395–404*
   Low-level, *391*
   Single-step, 69, *392*
Dedicated microprocessor, 190, 214
Device handler, *166*
Digital to analog converter (D/A), 154, *334–41*
Direct memory access (DMA), 6, 66, 69–71, 164, 170
Direct Page register, *see* Register, Direct Page

Directive, *see* Assembler, Directive
Disassembly, 392
Division, 119
Double-bus fault, 69, 70, *162*, 355, 406
Driver, *see* Function, Driver

Effective address (ea), 22, 36, 41, 62, 63, 108, 111, 114, 225
Embedded microprocessor, 58, 61, 65, 166, 181, 251, 259, 279, 298, 304, *306–12*
   I/O routines, *310*
Environment
   Cross, *190*, 390
   Hosted, 203, 240, 278, 279, 283
   Naked, 61, 203, 240, 279, *283, 304*, 390
   Resident, *190*, 390
EPROM programmer, 190
Error file, *see* File, Error
Error status, 51, 120, 135, 231, 258, 261
Exception, *see* Interrupt, Trap
Execution time, 67

False, *see* Logic value, False
File
   Error, *175*
   Header, *276, 312*, 362, 411
      <math.h>, *280*
      <stdio.h>, 305, 312
      <stdlib.h>, 305
   Hex, *see* File, Machine code
   Listing, *46, 175*
   Machine code, 46, *177–87*, 283, 299, 381
      Intel, *177*, 368
      Intel extended, 178
      Motorola S1/S9, *177*
      Motorola S2/S8, *178*, 188
      Motorola S3/S7, *178*
   Object, *181*
   Object code, 46
   Source code, *46, 173*
   Symbol, *176*
Firmware, *40*
Flag, 143
   C, *3*, 25, 26, *28*, 58, 61, 94
   E, *3*, 8, *149*, 164, 370
   H, *3*
   Monitoring in C, 214
   N, *3*, 27, *28*, 58, 101
   S, *see* Status register, 68000, S
   T, *see* Status register, 68000, T
   V, *3*, 25, 26, *28*, 58, 96, 98, 165

Index 435

X, *58*, *94*, 98, 102
Z, *3*, 27, *28*, 48, 58, *95*, 99–101, 200
Frame, *137-42*, 209, *245*, 373, 385
Frame Pointer, *137-42*, 209, 246, 395
Free-run, *405-7*
   6809, 354, *405*
   68008, *405*
Function, *see also* Subroutine, 203, 210, *240-9*
   Array of, 286
   _asm(), *288*, 370, 383, 416
   Body, *204*, 205, *242*
   Called at an absolute location, *290*
   Declaration of, *245*, 292
   Definition of, *242*, *245*
   Driver, *258*, *310*
   exit(), *365*, *380*
   External, 280, 288, 292
   Interrupt, *294*, 361, *366*, 379–81
   main(), *206*, *240*, 245, *285*, 365, 379
   malloc(), 365
   Mixing assembly and C, 246
   Nested, 206
   Parameter passing to
      by reference, 251
   Passing a pointer to, 142, *259*
   Passing a structure to, *267*, *270*
   Passing an array to, *251*, 269
   Passing parameters to, *245-6*, 287, 292
      by copy, *245*, 251, 269
      by reference, 270
   Prototype, 204, 205, *242*, 245, 280
   Recursive, 242
   Return from, 240
   Returning a structure, 272
   Returning pointer, 261
   static, 248, 249
   Value of, *142*, *205*, *240*, *242*, 245
   void, 125, 240, 242, 245

Header, 44, *276*, 291
Heap, 365
Hexadecimal, *see* Number base, Hex
High-level language, 88, 108, 124, *192-200*
   Address in, 65
   Algol, 134
   BASIC, 200
   C, 134, 142, 171
      ANSII, 203, 304, 377
      Auto variables, *see* Variables, Auto
      Comments, 204
      Flow control, *227-37*

      Function, *see* Function
      Old, 203, 245, 246, 250, 257, 279, 280, 304
      Origins of, *202-3*
      Register variables, *see* Variables, Reg
      Static, *see* Variables, Static
      Whitespace, 204
   Fortran, 125, 200
      Interrupts in, 146
   Pascal, 125, 134, 142, 196, 200, 204, 242, 265
   Portability, 133, 166, 203, 223, 273, *278*, 290, *303-64*, 376, 385
Hold time, 11

Implementation dependency
   asm, *289*
   Extension of char and short, 225
   of division, *224*, 306
   of floating-point types, 226
   of header directories, 279
   of int types, 223, 278, *305*, 385
   of interrupt functions, *294*
   of modulus, *224*, 306
   of parameter passing, *246*
   of signed shifts, *222*
   Passing parameters to a function, *287*
   #pragma, 280
   Side effects, 221
In-circuit emulator (ICE), 177, 192, 351, 368, 391, 405, *416-20*
Index register
   6809, *5*
   68000, *113*
Initialization, *see also* Variables, Initialization of
Instruction
   6800
      CLI, 27
      LDAA, 5
      PSHA, 19
      PSHB, 19
      SEI, 27
   6809
      ABX, *22*, *46*
      ADDA, *3*
      ADDD, *5*
      AND, *28*
      ANDCC, *26*, *146*
      ASR, *25*
      BCC, *29*
      BCS, *29*

BEQ, 27
BITA, *28*
BRN, *34*, 102
BSR, *126*
CLR, 19
CLRB, *405*
CWAI, *164*
DEC, 25
EXG, 20
INC, *3*, *25*, 295, 301
INCA, 5, 34, 35
INCB, 5
INX, *25*
JSR, *126*
LBRA, *29*
LBRN, *34*
LDA, 6, *10*
LDS, *21*
LDU, *21*
LDX, 20
LEA, *22*, *41*, 46, 48, *125*, 423
LSL, *25*
MUL, *25*
NOP, *34*, 35, 405
Null, *405*
ORCC, *26*, 146
PSHS, 19, 20, 366
PULS, 48, 131
RTI, *8*, *149*, *164*, 366, 370
RTS, 34, 41, 48, *126*
SEX, *22*
STA, *11*
STD, 423
STX, *20*
SWI, 3, *165*
SWI2, 35, *165*
SWI3, *165*
SYNC, 9, *164*, 288
TFR, 20, 35
TST, 27
8086
  CALL, 126
  HLT, 164
  INT, 166
  RET, 126, 134
68000
  ADD, 62, *93*, 116
  ADDA, 63, *94*, *110*
  ADDI, *94*, *109*
  ADDQ, *94*, 108, *109*, 110
  ADDX, *58*, *94*, 116
  AND, *98*

ANDI, 148
ANDI to CCR, *98*
ASL, *98*
ASR, *98*, 225
BCHG, 99
BCLR, 99
BEQ, *101*
BPL, *103*
BRA, *101*
BSET, 62, 99
BSR, *126*
BTST, *99*, 408
CHK, *166*
CLR, 62, 89, 95, 115, *303*
CMP, *100*
CMPA, 63, *100*
CMPI, *101*
CMPM, *101*
DBcc, *103*, *105*
DBF, *105*, 109, 118, 119, 132
DBNE, 135
DIVS, 91, *96*, 145, 164, 167
DIVU, 91, *96*, *119*, 145, 164, 167
EOR, *98*
EORI to CCR, 98
EXG, *91*
EXT, *96*, 225
ILLEGAL, *166*
Illegal, 60
JSR, *126*, 246
LEA, *63*, *95*, 114, *125*, 134
LINK, *141*, 210
LSL, *98*
LSR, *96*
MFSR, *89*
MOVE, 62, *67*, *89*, 94, 116
MOVE from SR, *89*, *129*
MOVE to CCR, *89*, 95, *129*
MOVE to SR, *89*
MOVEA, 62, *89*, *110*
MOVEM, *63*, *91*, 117, 129, *140*
MOVEP, 84
MOVEQ, *109*
MTCCR, *89*
MTSR, *89*
MULS, *96*
MULU, *96*
NEG, 95
NEGX, 95
NOP, 406
NOT, *98*
OR, *98*

ORI, *406*
ORI to CCR, 60, 98
PEA, 95
Privileged, 89, 164, 167
Privileged ANDI to SR, *60*, 98
Privileged EORI to SR, 60, 98
Privileged MOVE to SR, *60*
Privileged ORI to SR, 60, 98
Privileged RESET, *69*, 355
ROL, 91
ROXL, *98*
ROXR, *98*
RTE, *151*, 161, *164*, 295
RTR, *129*
RTS, 89, 108, 117, *126*
STOP, *164*
SUB, *93*
SUBA, 63, *94*, *110*
SUBI, *94*, *109*
SUBQ, *94*, *109*, 110
SUBX, *94*
SWAP, *91*, *98*, 119, 121
TST, *100*, *101*
UNLK, *141*
68008
   MOVE, 90
   Null, *405*
68010
   MOVE from CCR, 89, 129
   MOVEQ, *89*
   RTD, 134
68020
   CLR, *304*
   DIVU, 119
   LBcc, 102
   LINK, 141
   MULU, 122
   SWAP, 91, 98

C
   break, *232*, *236*, 237
   continue, *236*, 237
   do-while, *236*
   else, *227*, 230
   else-if, *230*, *231*
   for, *237*, 363
   goto, *234*
   if, *227*, *236*
   if-else, *227*
   if-else-nested, *229*
   if-nested, *228-30*
   return, *205*, 240, *242*
   switch-case, *231*

   while, 200, *205*, 215, 225, *234*, 237, 342
Fetch and execute, 3
Instruction decoder, *6*
Integrated circuit
   16L8 PAL, *355*
   20L10 PAL, 359
   2516 EPROM, *16*
   27128 EPROM, 77, 82
   2764 EPROM, *16*, 84
   555 timer, *351*, *355*
   6116 RAM, *16*
   6264 RAM, 16, 80
   6821 PIA, 71, 84, 99, 154, 158, *265-75*
   68230 PI/T, 77, *84*, 158, 161
   74LS05 buffer, 79
   74LS08 AND gate, 77
   74HCT125 3-state buffer, 351
   74LS138 decoder, 13, 351, 358
   74LS148 priority encoder, 148, 160
   74LS154 decoder, 77
   74LS164 shift register, *79*
   74LS377 register, 13, 79, 351
   AD7528 D/A, *338-41*, 359
   AD7576 A/D, *344-47*, 359
Interpreter, *200*
Interrupt, *143-67*, *292-7*
   6809
      FIRQ, 3, 146, 149, *154*, 164, 293, 370
      IRQ, 3, 146, 149, 154, 292
      NMI, *8*, 146, 351, 366
   68000, 60, 61, 293
      Acknowledge, *160*, 355
      Autovectoring, 72, *160*
      Non-maskable, 60, 69, *148*, 355, 380
      Priority level IPLn, *60*, *146*, *148*
      Spurious, *161*
      Trace, 151
      Uninitialized, *161*
   68008
      IPLn, *149*
   Disjoint globals, *295*
   Edge-triggered, 154
   Handler, *292*, *368*, 380, 381
   in C, 228, *292-7*, 361, 365, 366, 368, 373, 379-81
   Level-triggered, *154*
   Multiple sources, *158*
   Polling, *143*
   Service routine (ISR), 143, 149, 297
      in C, *294*, 361, 365, 366, 368, 373, 379-81

Simulation of, *395*
Software, *see* Software interrupt and Trap, 61, 145, *165*
Interrupt flag, *154*, 158, 161
Interrupt mask
   6809
      F, 3, *146*, 151, 164
      I, 3, *8, 9*, 26, *146*, 151, 164, 292
   68000, 60, 68, 69, *148*, 151, 160, 293, 380
Interrupt service routine, *see* Interrupt, Service routine

Label, *see* Symbols
   in C, *234*
Library, 124, 180, *188, 279-80*, 287, *304*
   getchar(), *310*
   gets(), 312
   Machine-level, 270, *279*
   printf(), *305, 310, 312*, 391
   putchar(), *310*
   puts(), 312
   scanf(), *305, 310*
   Support, 280
   User-callable, *280*
Linker, 46, 51, 155, 171, 173, *180-92*, 215, 248, 279, *286, 298*, 365, 380, 392
   Command file, *187*, 299
Load time, 41, 155, 214, 249, 259, 272, 297, 373
Loader, 177, *190*
   Absolute, *192*
   Hexadecimal, 170
   Relocatable, *192*
Loading, 283
   Initial values, 45, *212*, 297, 380
Logic value
   False, *205*, 214, *222, 223*, 227
   True, *205, 222, 223*, 227, 408
Look-up table, *39*, 48, *56*, 113, *123*
   in C, 214, *251*, 259, 298
Loop, 35, 48, *103, 108, 233*
   DO-WHILE, *105*, 132, 205, 373
   Endless, 154, 237, 262, 363
   REPEAT-UNTIL, *236*
   WHILE-DO, *105*

Machine code, *see also* File, Machine code, *170*, *188*
Machine state, *8*, 143, 164
Macro
   Assembler, *179*

C, *276*
Mask, *see* Interrupt mask
Maximal munch, *193*
Memory
   EPROM, 79, 407
      2716, 351, 403
      2764, *16*, 84, 286, 351
      27128, 77, 82
      Simulation of, *416*, 418
   RAM, 94, 298
      48Z02, 351
      6116, *13, 16*, 287, 351
      6264, *80*
      6809, 2, *13*
      68000, *80*, 118
      68008, 84
      Initial value of, 213, *297-303*
      Testing, *408*, 418
   ROM, 40, 48, 212
      6809, *16*
      68000, 81, 118
      68008, 84
      Testing, *411*
   Time-compressed, *see* Time-compressed memory
Microcomputer unit (MCU), *326*
Microprocessor development system (MDS), 155, 192, 416
Microprocessor unit (MPU)
   6309E, 9
   6809E, 9
   680x family
      Data ordering, *20*
   808x family
      Data ordering, *20*
   Z80, 2, 19, 391
   4004, 2
   6309, 7, 9
   6502, 5, 133
   6800, 2, 5, 9, 19, 37, 149, 177
   6801, 2
   6802, 2
   6805, 2
   6809, 63, 116, 350-5
      Hardware, *2-18*
      Software, *19-56*
      Hardware, *57-84*
      Software, *88-123*
   8008, 2
   8051, 2
   8080, 2, 19, 57, 177, 416
   808x family, 6, 13, 304

Index 439

8085, 2
8086, 57, 133, 303
80x86 family, 65, 100, 178, 242, 255, 310
80386, 57, 303
68000, 303
68HC000, 65, 72
68008, *65*, 67, 69, 71, 77, 80, 84, 88, 115, 355–9
68020, 57, 303
68030, 57, 391
68040, 57, 71
Embedded, *see* Embedded micro
Modular programming, 124
Monitor, *see* Operating system, Monitor
Multiple-precision operations, 58
    Addition, *94*, 145
    Decrement, *25*
    Increment, 25
    Negate, *95*
    Shift, *26*, *98*
    Subtraction, *94*
    Zero, *95*
Multitasking, 61, *149*

Naked environment, *see* Environment, Naked
Naked system, *192*
Number base
    Binary
        Suffix b, *3*
    Hexadecimal
        Prefix 0x, *176*, 187, 216
        Suffix h, *3*
    Octal
        Prefix 0, 216

Object code, *see* File, Object, 177
Operating system (OS), 61, 70, 280, 283, *304*, 310
    BIOS, *166*, 170
    DOS, 304
    Loader, *190*
    Monitor, *61*, 310
        INCH, *310*
        OUTCH, 290, *310*
    MSDOS, 166, 278, 310, 391
    UNIX, 202, 304
Operation
    AND, 26
    Arithmetic Shift, *25*, 98, 222, 225
    Block-Compare, *101*
    Branch, 5, *40*, 62, *101*
    Call, *126*

Circular Shift, *25*, *98*
Compare, 58, 205
Conditional Branch, *29*, 100, *101*
Decrement, 94
Division, 91
    by two, 118
EOR, 26
Exchange, 35
Increment, 48, 94
Jump, 5, 30, *40*, 126
Load, 20
Logic Shift, *25*, *98*, 222
Long Branch, *29*
Multiplication, *25*, 46, 51, 120, 279
NOT, 26
OR, 26
Pull, 35, 48, 63, 90, 119, 126, 129, 140
Push, 35, 48, 63, 90, *117*, 119, 129, 140, 246
Read-modify-write, *25*, 303
Return, *126*, 242, 366
Shift, 47, *96*, 221
Store, 20
Test, *27*, 205
Transfer, 35
Operation code (op-code), 6, 19, *29*, *34*, 35, 62, 88, 94, 102, 108, 116, 167
Post-byte, 20, 22
Post-word, 91
Optimization
    of volatile variables, *215*

Parameter passing, *see* Subroutine, Parameter passing to
Peripheral Interface Adapter (PIA), *see* Integrated circuit, 6821
Personal computer (PC), 166
    APPLE, 57, 58
    IBM, 57, 84, 391
Pointer, *see also* Address, 48, 110, 135, 215, 255–62
    Arithmetic, *256*, *261*, *275*
    Casting a constant to, *257*
    const, 257, 258, 363, 385
    Definition of, *256*
    External, 373
    Null, *261*, *286*, *365*
    sizeof, *256*
    static, 257, 373, 385
    to an array, *251*, 255
    to a function, 261, 286, *291*
    to a register object, 210

to a structure, *270*, *272*
Type, *255*
void, *256*, 261
Polling, see also Interrupt, Polling, 99, *103*, *158*, 215
  in C, 198
Port, 209
  6809
    Output, *13*
  68000
    Output, *79*
  68008
    Output, 80
  Defining in C, *258*
  Fixed address of, 111
  Input
    Simulation of, 404
    volatile, *214*
  Output
    Simulation of, 394, 404
    Read-only, 363
    Simulation of, *395*, *404*
    Write-only, 412
Portability, see High-level language, Portability
Position independent code (PIC), 29, 40, *51*, 96, 108, *114*, 126, 133
Program
  Background, *145*, *149*
  C
    3-point filter, 250
    7-segment decoder, *259*
    Analog to digital conversion, *345*
    Array display, *262*
    ASCII hex to decimal, 228
    BCD to ASCII, 221
    Binary to BCD, *224*
    Block copy, *251*
    Delay, 363
    Factorial, *231*, *251*
    Integral power, *242*, 400
    Modulus of, 228
    Real-Time Clock, 228, *230*, *295*
    Squaring, 242
    Sum of integers, 200, *203*, 209, *307*, 398
  Foreground, *149*
  Linker, see Linker
  Loader, see Loader
Program Counter (PC), 5, 43, 48, 125
  68000, *61*, 160
Program section, *44*, 176

0, 182, 187
9, 182, 187, 298
13, 298
14, 182, 187, 298
_bss, *181*, 298, 365, 380
bss, *298*
.data, 262
_data, 45, 46, 51, 54, 118, 155, *181*, 210, 298
dseg, 298
_text, 44, 46, 118, *181*, 214, 251, 298, 365, 373, 380
tseg, 298
_zpage, *181*, *281*, *301*
Program-Assembly
  6809
    7-segment decoder, 39
    Analog data averaging, *139*
    Array display, *154*
    Binary to BCD, 48
    Delay, 128, 129
    Factorial, 51
    Multiple-precision add, *145*
    Sum of integers, 45, 392
  68000
    3-point filter, 112
    7-segment decoder, 113, 114
    Analog data averaging, *141*
    Array clear, 108
    Array display, *160*
    Binary to BCD, 119
    Block copy, *135*
    Delay, 105, 128, 129
    Factorial, 120
    Modulus, *179*
    Square root, *288*
    Sum of integers, 117, 175, 392
Promotion, see also Sign extension, 156, 216
  char/int, 225, 365, 377
  int/pointer, *385*
  short/int, 377
  Signed, *22*, 96, 225
  Unsigned, 22, 46, 96, 118, 121, 142, 377

Race hazard, 13
Real-time operation, 173
Register
  Control, see Control register
  Data, see Data register
  Direct Page, 6, *36*, 303
  Qualifier in C, see also C qualifier, **register**
  Read-only

in C, 214, 215, 412
Status, 214
Register transfer language (RTL), 3
Reset
   6809, 6, 9, 351
      Interrupt mask, *146*, 293
   68000, 60, 61, *160*, *355*
      Bus Error during, *70*
      Interrupt mask, *148*, 293
      Program Counter, *68*
      Stack Pointer, *68*, 69, 380
      Status register, *68*
   in C, *283*
ROM, *see* Memory, ROM
ROMable code, 45, 115, 250, 251, 257, 259, *283-303*, 368, 373, 381, 390
   Initialization of static variables, *212*
   Initialization of variables, 135
   Look-up table, 214
ROMulator, 407, *416*
Run time, 41, 114, 155, 181, 214, 250, 272
   Initialization, 297, 380, 385

Sample and hold, *348*
Sampling theorem, *332*
Side effect, 131, *220*
Sign extension, 29, 39, 63, *see also* Promotion, 91, 96, 225
   of Accumulators, *22*
Simulation
   Low-level, *391*
   of interrupts, *395*, *401*
   of ports, *395*, *404*
   of ROM, 394
Simulator, 45, 177, *391-404*
   High-level, *395-404*
Single-board computer (SBC), 310
Software
   Portable, *see* High-level language, Portability
Software interrupt, *see* Interrupt, Software
Source code, 42, 155, *173*
Stack, 125
   Cleaning, 139, 142, 246
   Setting up, *20*
   System, *22*, 48, 54
   User, 22
Stack Pointer, 125
   6809
      Hardware, 5
      NMI, *146*
      System, 5, 8, 54, 139

User, *5*, 20, 138
68000, 63
   Evenness of, *69*, 91, 127, 129, 140, 162, 355, 406
   Supervisor, *61*, 151, 160, 420
   System, 126
   User, *61*
Startup routine, 240, *284-301*, 363, *365*, *380*, 401
   using `asm()`, *289*
   Initialization of variables, 365
State
   Machine, 145, *149*
   Supervisor, *60*, *61*, 67, 68, 70, 89, 91, 98, 151, 166
   User, *60*, 70, 91, 151, 166, 167, 287
Status register
   68000, *60*, *89*, 129, 148, 164
   S, *60*, 68, 151
   T, *61*, 68, 151, 167
Status signal
   6809
      BA BS, *9*, 158
   68000
      FC, *70*, 75, 160
   FC, 355
String, 45, 48, 119
   Compare, 101
   in C, 298, 312
Structure, *265-75*
   `C-sizeof`, 223
   `const`, *267*
   Declaration of, *265*
   Definition of, 265, 266
   FILE, 278
   Initialization of, *267*
   of pointers, *266*
   Passing to a function, *267*, *270*
   Pointer to, *270*, *272*
   `static`, *267*
   Template, *265*
Subroutine, 37, 41, 117, *124-42*, 203, 240
   Nested, *125*
   Parameter passing to, 22, 39, 46, 117, 119, *131-42*
      an array, 135
      by copy, *132*
      by reference, *135*
      via a stack, *134*
   Passing parameters to
      by reference, 260
   Re-entrant, 133

Recursive, *124*, *131*
Transparent, *48*, *128*, 131, 136
Void, *131*
Symbol table, *see* Table, Symbol
Symbols, *180-8*
   Absolute, *180*, 182, 188
   Defined, 188
   External, *180*, 181, 182, 368
   Global, *180*
   Local, *180*
   Public, *see* Symbols, Global, 188
   Relocatable, *180*, 181

Table
   Look-up table, *see* Look-up table
   Symbol, *178*, 182, 188, 392
Time-compressed memory, *316-20*
Time-out, *103*
Trace, 61, 68, 167
Trap, 61, *164-7*
   Address error, 67, *166*
   Bus error, 162
   Divide by zero, *96*, 145, *167*
   Illegal instruction, *166*
   Line A, *167*
   Line B, 167
   Privilege violation, *167*
   Trace, *167*
   V, *165*
True, *see* Logic value, True
Type
   `const`, *see* C qualifier, `const`
   Conversion
      char/int, 225, 242, 365, 377, 385
      int/short, 194
      short/int, 365, 377
   Floating-point, 295
      `double`, *207*, 280
      `float`, *207*
      `long`, *209*
      `long double`, *207*, 226
   Integer
      char, *208*, 225, 362, 363, *364*, 365, 373
      int, 194, 204, *209*, 220, 225, 278, 295, *306*, 385
      short, 194, *209*
      Signed, 227
      Unsigned, 204, *227*
   Mixed arithmetic, *226*
   `register`, *see* C qualifier, `register`
   `static`, *see* C qualifier, `static`

`void`, *209*

Union, *272*

Variables
   Automatic, 131, *134*, 142, *209*, 245, 248, 395
      Initialization of, 297
      Lifetime, *210*
   Booleans, *222*
   Declaration, 249
   Definition of, *204*, *212*, *245*, 249
   External, 215, *see* Variables, Global
   Global, *133*, 142, 146, 262, 292, 297, 361, 365, 379, 408
      Initialization of, *249*, 294, *298*, *363*
   Identifier, *215*
   Initialization of, 45, 135, 181, 204, *210-14*, *297-303*
      Automatic, *213*, 250
      Register, *213*
      Static, *212*, 250, 380
   Local, *134*, 136, 142, 204, 242
   Public, *248*, 365, 368, 408
   Register, 110, 131, 140, 142, *210*, 376, 380, 385
   Scope, *246-9*
      Local, *210*
   Static, 131, *133*, 248, 262, 365
      `const`, *214*, 251, 408
      External, *210*, *249*, 408
      Initialization of, *298*
      Lifetime, *210*
   Volatile, 198, 412
Vector
   6809
      FIRQ, *8*, 368
      IRQ, *9*, *154*, 292
      NMI, *8*, 366, 403
      Reset, *8*, 45, *154*, 160, 286
   68000
      Non-maskable interrupt (31), 355, 380
      Reset, *160*, 380
      TRAP, 166
   Dynamic, 154
Vector table
   6809, *154*, 158, 285
   68000, 69, 182, 287, 380
Volatile object, *see* Variables, Volatile

Watch-dog timer, 70, 162
Word size
   Byte (8), 58, 62, 67

in C, *209*, *215*, 258, 262, 362
Long (32), 51, 58, 62, 67, 94
Word (16), 58, 62, 65, 67, 94

Zero page addressing, 6, *301*, 373